Mathematics in Industry

Mathematics in Industry

Edited by

Angela Slavova

CAMBRIDGE SCHOLARS

PUBLISHING

Mathematics in Industry,
Edited by Angela Slavova

This book first published 2015

Cambridge Scholars Publishing

12 Back Chapman Street, Newcastle upon Tyne, NE6 2XX, UK

British Library Cataloguing in Publication Data
A catalogue record for this book is available from the British Library

ISBN (10): 1-4438-6401-3, ISBN (13): 978-1-4438-6401-5

TABLE OF CONTENTS

Chapter Three: Chaotic Systems and their Application in Industry

Chapter Four: Industrial Applications in Mechanics

Chapter Five: Algorithms in Industrial Mathematics

Chapter Six: Network Applications in Industry

PREFACE

This book describes a wide range of problems concerning recent achievements in the field of industrial and applied mathematics. The main goal is to provide new ideas and research both for scientists, who develop and study mathematical methods and algorithms, and for researchers, who apply them to solving real-life problems. The book promotes basic research in mathematics leading to new methods and techniques useful to industry and science. This volume will be a medium for the exchange of information and ideas between mathematicians and other technical and scientific personnel.

The main topics are: Numerical Methods and Algorithms; Control Systems and Applications; Partial Differential Equations and Applications; Neurosciences (Neural Networks); Equations of Mathematical Physics, etc.

Many important real-life applications of partial differential equations and equations of mathematical physics are presented in the book (Chapters 1,7 and 8). More precisely, a non-local version of a nonlinear Schrödinger equation is studied, which is a theoretical description of wave propagation in PT-symmetric coupled wave guides and photonic crystals. Continuity of the solution map for the cubic 1D periodic nonlinear wave (NLW) equation is investigated. The Cauchy problem to the generalized sixth-order Boussinesq equation is studied. This problem arises in a number of mathematical models of physical processes, for example in the modeling of surface waves in shallow waters and in the dynamics of nonlinear lattices. A family of Modified Korteweg-de Vrez (MKDV) equations is delivered and is related to simple Lie algebra. G-strand equations are studied and peakon-antipeakon collisions are solved analytically and can be applied in the theory of image registration. The three-soliton interactions for the Manakov system are modeled by a perturbed complex Toda chain.

A survey is presented concerning chaotic systems and their application in industry (Chapter 3). A receptor-based Cellular Nonlinear Network model with hysteresis is studied. The dynamics and stability of this model are studied from the point of view of local activity theory and an edge-of-chaos domain is obtained. Continuous feedback control is applied in order to stabilize the system. The coupled FitzHugh-Nagumo neural system is studied in this survey, and stabilization of the discretized models is proposed which is simple to implement.

Industrial Applications in mechanics are presented (Chapter 4). Li-ion batteries are widely used currently in the automotive industry, in electronic devices, etc. Pole scale simulations are provided on 3D CT images of the porous electrodes.

Algorithms in industrial mathematics are investigated (Chapter 5). An improved algorithm for generating primitive Pythagorean triples is proposed, an algorithm based on a well-known construction by Barning and Hall. A similar construction is considered in the four-dimensional case of Pythagorean quadruples and the generalized case of relatively prime quadruples.

Another topic which is considered in the volume is network applications in industry (Chapter 6). A neural network for classification of plastic and non-plastic materials with blasting action after blow-up with coherent signals in the optical range is proposed. Another network application is a graphical user interface created to study the static equation of linear Cellular Neural Networks (CNN). An interactive web tool is developed to explore associations in networks built with Affymetrix transcriptional profiling data and other sources of genomics data.

Linear algebra applications are considered (Chapter 9). A general parametric AE-solution set is obtained, which appears in the domain of various industrial applications. This chapter also presents a review of the main results of the component-wise stability of Wang's parallel partition method for banded and tri-diagonal linear systems.

High performance and scientific topics are included in this volume (Chapter 2). The importance of the computing infrastructure is unquestionable for the development of modern science. This chapter presents an approach to the installation and configuration of a high performance computing infrastructure with grid access. The cluster comprises a large pool of computational blades and two powerful GPGPU-enabled servers. The Danish Eulerian Model is a powerful and sophisticated air pollution model. Novel developments in the up-to-date parallel implementation of the model are presented in this chapter. A field fire model is proposed which is based on game modeling using hexagonal cells and rules. A parallel version of the algorithm is run on the Blue Gene supercomputer.

This book has a very important role in the promotion of interdisciplinary collaboration between applied mathematics and science, engineering and technology.

I would like to thank very much Dr. Maya Markova for her help in preparing this volume.

Sofia, May 2014 Angela Slavova

CONTRIBUTORS

Gregory Agranovich
Dept. of Electrical and Electronic Engineering
Ariel University Center of Samaria,
44837 Ariel, Israel
e-mail: agr@ariel.ac.il

Emanouil Atanassov
Institute of Information and Communication Technologies
BAS Acad. G. Bonchev St., Bl.25A
1113 Sofia, Bulgaria
e-mail: emanouil@parallel.bas.bg

Georgi Boyadzhiev
Institute of Mathematics and Informatics
Bulgarian Academy of Sciences
Sofia, Bulgaria
e-mail: georgi_boyadzhiev@yahoo.com

Danail S. Brezov
Department of Mathematics
University of Architecture Civil Engineering and Geodesy
1 Hristo Smirnenski Blvd.
1046 Sofia, Bulgaria
e-mail: danail.brezov@gmail.com

Dimitar Dimitrov
Institute of Information and Communication Technologies
BAS Acad. G. Bonchev St., Bl.25A
1113 Sofia, Bulgaria
e-mail: d.slavov@bas.bg

Milena Dimova
Institute of Mathematics and Informatics,
Bulgarian Academy of Sciences
Acad. Bonchev str., bl.8,
1113, Sofia, Bulgaria
e-mail: mkoleva@math.bas.bg

Mariya Durchova
Institute of Information and Communication Technologies
BAS Acad. G. Bonchev St., Bl.25A,
1113 Sofia, Bulgaria
e-mail: mabs@parallel.bas.bg

Kaloyan Dzhurov
University of Rousse
Rousse 7017, Bulgaria
e-mail:kdzhurov@uni-ruse.bg

Valerij Dzhurov
University of Rousse
Rousse 7017, Bulgaria
e-mail: vdzhurov@yahoo.com

Alexander Fabricant
Institute of Mathematics and Informatics
Bulgarian Academy of Sciences
Sofia 1113, Bulgaria
e-mail: fabrican@math.bas.bg

Stefka Fidanova
Institute of Information and Communication Technologies,
BAS Acad. G. Bonchev St.,
Bl.25A,
1113 Sofia, Bulgaria
e-mail: stefka@parallel.bas.bg

Dobromir Georgiev
Institute of Information and Communication Technologies
BAS Acad. G. Bonchev St.,
Bl.25A
1113 Sofia, Bulgaria
e-mail:
dobromir@parallel.bas.bg

Vladimir Georgiev
Department of Mathematics
University of Pisa
Largo Bruno Pontecorvo 5
56100 Pisa, Italy
e-mail: georgiev@dm.unipi.it

Vladimir S. Gerdjikov
Institute of Nuclear Research and Nuclear Energy, BAS
72 Tsarigradsko chausee
Sofia 1784, Bulgaria
e-mail: gerjikov@inrne.bas.bg

Todor Gurov
Institute of Information and Communication Technologies
BAS Acad. G. Bonchev St.,
Bl.25A
1113 Sofia, Bulgaria
e-mail: gurov@bas.bg

Darryl Holm
Department of Mathematics
Imperial College
London SW7 2AZ
UK
e-mail: d.holm@imperial.ac.uk

Oleg Iliev
Fraunhofer Institute for Industrial Mathematics
ITWM,
Kaiserslautern, Germany
and KAUST, Saudi Arabia
e-mail:
oleg.iliev@itwm.fraunhofer.de

Rossen Ivanov
School of Mathematical Sciences
Dublin Institute of Technology,
Kevin Street
Dublin 8
Ireland
e-mail: rossen.ivanov@dit.ie

Sofiya Ivanovska
Institute of Information and Communication Technologies
BAS Acad. G. Bonchev St.,
Bl.25A
1113 Sofia, Bulgaria
e-mail: sofia@parallel.bas.bg

Aneta Karaivanova
Institute of Information and Communication Technologies
BAS Acad. G. Bonchev St., Bl.25A
1113 Sofia, Bulgaria
e-mail: anet@parallel.bas.bg

Natalia Kolkovska
Institute of Mathematics and Informatics
Bulgarian Academy of Sciences
Acad. Bonchev str., bl.8
1113, Sofia, Bulgaria
e-mail: natali@math.bas.bg

Jordanka Paneva - Konovska
Faculty of Applied Mathematics and Informatics
Technical University of Sofia
1000 Sofia, Bulgaria
e-mail: yorry77@mail.bg
Associated at:
Institute of Mathematics and Informatics
Bulgarian Academy of Sciences
Acad.G.Bonchev str., bl.8
Sofia 1113, Bulgaria

Ivan G. Koprinkov
Department of Applied Physics
Technical University of Sofia
1000 Sofia, Bulgaria
e-mail: igk@tu-sofia.bg

Milena Kostova
University of Rousse
Rousse 7017, Bulgaria
e-mail: mpk@mail.bg

Nikolai Kutev
Institute of Mathematics and Informatics
Bulgarian Academy of Sciences
Sofia 1113, Bulgaria
e-mail: kutev@math.bas.bg

Assen Kyuldjiev
Institute for Nuclear Research and Nuclear Energy
BAS, 1784
Sofia, Bulgaria
e-mail:A.Kyuldjiev@gmail.com

Arnulf Latz
DLR Institute of Technical Thermodynamics,
Meyerhoferstrasse N25,
89081 Ulm, Germany
e-mail:arnulf.latz@dlr.tt.de

Elena Litsyn
Department of Mathematics
Ben-Gurion University of the Negev
Beer-Sheva, Israel
e-mail: elitsyn@gmail.com

Svetozar Margenov
Institute of Information and Communication Technologies
Bulgarian Academy of Sciences
Acad. G. Bonchev St., Bl. 25A
1113 Sofia, Bulgaria
e-mail:
margenov@parallel.bas.bg

Pencho Marinov
Institute of Information and
Communication Technologies
BAS, Acad. G. Bonchev St.,
Bl.25A
1113 Sofia, Bulgaria
e-mail: pencho@parallel.bas.bg

Dimitar M. Mladenov
Theoretical Physics Department,
Faculty of Physics,
Sofia University "St. Kliment
Ohridski"
5 James Bourchier Blvd
1164 Sofia, Bulgaria
e-mail:
dimitar.mladenov@phys.uni-
sofia.bg

Ivailo M. Mladenov
Institute of Biophysics
Bulgarian Academy of Sciences
Acad. G.Bonchev Str., Bl. 21
Sofia 1113, Bulgaria
e-mail:mladenov@bio21.bas.bg

Clementina D. Mladenova
Institute of Mechanics,
Bulgarian Academy of Sciences
Acad. G. Bonchev Str., Bl. 4
Sofia 1113, Bulgaria
e-mail: clem@imbm.bas.bg

Tzvetan Ostromsky
Institute of Information and
Communication Technologies
Bulgarian Academy of Sciences
Acad. G. Bonchev St., Bl. 25A
1113 Sofia, Bulgaria
e-mail: ceco@parallel.bas.bg

Velizar Pavlov
Department of Applied
Mathematics and Statistics
University of Ruse
7017 Ruse, Bulgaria
e-mail: vpavlov@uni-ruse.bg

Peter Popov
Institute of Information and
Communication Technologies
Bulgarian Academy of Sciences
Acad. G. Bonchev St., Bl. 25A
1113 Sofia, Bulgaria
e-mail: ppopov@parallel.bas.bg

Tsviatko Rangelov
Institute of Mathematics and
Informatics
Bulgarian Academy of Sciences
Sofia 1113, Bulgaria
e-mail: rangelov@math.bas.bg

Victoria Rashkova
University of Ruse
Ruse 7017, Bulgaria
e-mail: vkr@ami.uni-ruse.bg

Angela Slavova
Institute of Mathematics and
Informatics
Bulgarian Academy of Sciences
Sofia 1113, Bulgaria
e-mail: slavova@math.bas.bg

Aleksandar A. Stefanov
Theoretical Physics Department
Faculty of Physics,
Sofia University
"St. Kliment Ohridski"
5 James Bourchier Blvd
1164 Sofia, Bulgaria
e-mail: astefanov@phys.uni-sofia.bg

Rosangela Sviercoski
Institute of Information and Communication Technologies
Bulgarian Academy of Sciences
Acad. G. Bonchev St., Bl. 25A
1113 Sofia, Bulgaria
e-mail:
rosviercoski@parallel.bas.bg

Maxim Taralov
Fraunhofer ITWM,
Fraunhofer-Platz-1,
67663 Kaiserslautern, Germany
e-mail:
maxim.taralov@itwm.fraunhofer.de

Vasilena Taralova
Fraunhofer ITWM,
Fraunhofer-Platz-1,
67663 Kaiserslautern, Germany
e-mail:
vasilena.taralova@itwm.fraunhofer.de

Michail D. Todorov
Department of Applied Mathematics and Computer Science
Technical University of Sofia
1000 Sofia, Bulgaria
e-mail: mtod@tu-sofia.bg

Todor P. Todorov
Department of Applied Physics
Technical University of Sofia
1000 Sofia,Bulgaria
e-mail: totodorov@tu-sofia.bg

Maria E. Todorova
College of Energetics and Electronics
Technical University of Sofia
1000 Sofia, Bulgaria
e-mail: metodorova@tu-sofia.bg

Tihomir Valchev
Dublin Institute of Technology
Kevin Street Lower,
Dublin 8, Ireland
e-mail: Tihomir.Valchev@dit.ie; tihov@yahoo.com

Stanislav K. Varbev
Theoretical Physics Department,
Faculty of Physics
Sofia University "St. Kliment Ohridski"
5 James Bourchier Blvd
1164 Sofia, Bulgaria
e-mail:
stanislavvarbev@phys.uni-sofia.bg

Daniela Vasileva
Institute of Mathematics and Informatics
Bulgarian Academy of Sciences
Acad. Bonchev str., bl.8
1113, Sofia, Bulgaria
e-mail: vasileva@math.bas.bg

Alexander B Yanovski
Department of Mathematics and Applied Mathematics
University of Cape Town
7700 Rondebosch
Cape Town, South Africa
e-mail:
Alexandar.Ianovsky@uct.ac.za

Jochen Zausch
Fraunhofer ITWM,
Fraunhofer-Platz-1,
67663 Kaiserslautern, Germany
e-mail:
jochen.zausc@itwm.fraunhofer.de

CHAPTER ONE

REAL-LIFE APPLICATIONS OF PDF

Local Well-posedness for the Cubic 1D Periodic Square-root Klein Gordon Equation

Vladimir Georgiev and Mirko Tarulli

Introduction

We consider the Cauchy problems associated with the following square-root KG equation

$$(i\,\partial_t - \sqrt{-\Delta + m^2})u = \sigma|u|^2 u \text{ for } t \geq 0, \tag{1}$$

where $m > 0$, $\sigma = \pm 1$, and $u(t,x)$ is 2π-periodic in x. If we have solutions $u(t,x) \in C([0,T]; H^s(0,2\pi))$, with $s > 1/2$, then the equation enjoys two conservation laws

$$\| u(t) \|_{L^2(0,2\pi)} = const$$

and

$$\frac{1}{2} \| (-\Delta + m^2)^{1/4} u(t) \|_{L^2}^2 + \frac{\sigma}{4} \| u(t) \|_{L^4}^4 = const. \tag{2}$$

Moreover we state:

Definition 1.1 *The problem (1) is well-posed in $H^s(0,2\pi)$ with $s \in (0,1)$ if for any $R > 0$ one can find $T = T(R) > 0$ such that for any initial data $u(0) = f \in H^s$ with $\| f \|_{H^s} \leq R$ one can define unique solution $u(t,x) \in C([0,T]; H^s)$ so that the solution map*

$$f \in B(R) = \{g \in H^s; \| g \|_{H^s} \leq R\} \to u(t,x) \in C([0,T]; H^s),$$

is continuous.

A stronger property is the uniform continuity of the solution map. In this direction we have our main result, that is

Theorem 1.2 *If one selects* $s \in (1/3,1/2)$, *then the Cauchy problem associated with*

$$(i\partial_t - \sqrt{-\Delta + m^2})u = |u|^2 u \; for \; t \geq 0, \tag{3}$$

cannot have a uniformly continuous solution map in H^s.

From now let us select $m = 1$ and indicate by $\sqrt{-\Delta + 1} = \langle D_x \rangle$. Then the above result is valid also for

$$(i\,\partial_t - \langle D_x \rangle)u = -|u|^2 u,$$

but in this case one expects some blow-up effect similar to the one obtained in [3]. To explain the idea of the proof, let us look at a solution of the form

$$u(t,x) = u(t,x;s,\varepsilon) = v_{\geq 0}(t,x;s,\varepsilon) + w_\varepsilon(t,x), \tag{4}$$

where $w(t,x) = w_\varepsilon(t,x)$ satisfies

$$(i\,\partial_t - |D_x|)w = (|v+w|^2(v+w) - P_{\geq 0}(|v|^2)v) + S(D_x)(v+w), \tag{5}$$

with the smoothing operator

$$S(D_x) = \langle D_x \rangle - |D_x|,$$

and with zero initial data. In addition we choose $v(t,x) = v_{\geq 0}(t,x)$, as the solution of the modified equation for the NLW (see [2] for more details)

$$i(\partial_t - \partial_x)v = P_{\geq 0}(v|v|^2),$$

where $P_{\geq 0}$ is the operator

$$P_{\geq 0}\left(\sum_{k\in\mathbb{Z}} \hat{f}(k)e^{ikx}\right) = \hat{f}(0) + \sum_{k=1}^{\infty} \hat{f}(k)e^{ikx}.$$

We shall construct a family of $v_{\geq 0}(t,x)$ defined for any positive ε and for $j = 0,1$ as follows

$$v_\varepsilon^{(j)}(t,x) = v_{\geq 0}(t,x;j,s,\varepsilon) = \frac{a_j e^{-i\alpha_j t}}{1-c_0 e^{i(t(1-\gamma_j)+x)}}, \tag{6}$$

where

$$c_0 = c_0(\varepsilon) = \sqrt{1-\varepsilon}, a_j = a_j(\varepsilon) = m_j \varepsilon^{s+1/2}, \quad 0 < s < \frac{1}{2}, \tag{7}$$

$$m_j = m_j(\varepsilon) = 1 + \frac{j}{|\log\varepsilon|}, \ j = 0,1, \tag{8}$$

and with

$$\gamma_j = \frac{c_0^2 a_j^2}{1-c_0^2}, \ \alpha_j - \gamma_j = \frac{a_j^2}{(1-c_0^2)^2}. \tag{9}$$

We choose $\varepsilon > 0$ small and use the fact that $v_\varepsilon^{(j)}(t,x) = v_{\geq 0}^{(j)}(t,x;s,\varepsilon)$ introduced in (6) could be used in the proof of the fact that the solution map is not uniformly continuous; that is the statement of Theorem 1.2. Note that the relations (7) and (9) imply

$$2\gamma = \gamma(\varepsilon) = m^2\varepsilon^{2s}(1+O(\varepsilon)),$$

$$\alpha = \alpha(\varepsilon) = \ m^2\varepsilon^{2s-1}(1+O(\varepsilon)). \tag{10}$$

Furthermore we can apply the argument of Section 5 in [1] (see also [5]) so we shall obtain the following estimates

$$\| v_\varepsilon^{(1)}(0,\cdot) - v_\varepsilon^{(2)}(0,\cdot) \|_{H^s(0,2\pi)} \leq C\frac{1}{|\log\varepsilon|}, \tag{11}$$

with $C > 0$ and consequently, for suitable interval of type

$$\{t \sim \varepsilon^{1-2s} |\log\varepsilon|\},$$

we have:

$$\| v_\varepsilon^{(1)}(t,\cdot) - v_\varepsilon^{(2)}(t,\cdot) \|_{H^s(0,2\pi)} \geq D > 0, \quad (12)$$

with some $D > 0$ independent of $\varepsilon > 0$. Our main goal is to construct functions $u_\varepsilon^{(1)}(t,x), u_\varepsilon^{(2)}(t,x)$ so that

- $u_\varepsilon^{(1)}(t,x), u_\varepsilon^{(2)}(t,x)$ are solutions of the square - root KG equation (3) of the form (4),

- they have slightly smoother than $H^{1/2}$ regularity, i.e.

$$u_\varepsilon^{(1)}, u_\varepsilon^{(2)} \in C([0,T(\varepsilon)]; H^{s_1}(0,2\pi)),$$

for suitable choices of $s_1 > 1/2,\ T(\varepsilon) > 0$,

- the functions $u_\varepsilon^{(1)}, u_\varepsilon^{(2)}$ satisfy the estimates (11) and (12) with some $s \in (1/3,1/2)$.

The inequalities (11) and (12) for these solutions will imply the conclusion of Theorem 1.2.

Solutions of 1D square - root NLKG system as perturbations of Szegö type solutions

For any function $a: \mathbb{Z} \to \mathbb{C}$ we set

$$a(D)f(x) = \sum_{k\in\mathbb{Z}} a(k)\hat{f}(k)e^{ikx},$$

where $\hat{f}(k)$ is the Fourier coefficient of f. In particular we have

$$|D|f(x) = \sum_{k\in\mathbb{Z}} |k|\hat{f}(k)e^{ikx}. \quad (13)$$

We have also the partition of unity

$$I = P_+ + P_0 + P_-, \tag{14}$$

where

$$P_+(k) = \begin{cases} 1, & if\ k > 0; \\ 0, & otherwise, \end{cases} \quad \mathrm{P}_-(k) = \begin{cases} 1, & if\ k < 0; \\ 0, & otherwise, \end{cases}$$

by this we obtain

$$|D| = DP_+ - DP_- = (P_+ - P_-)D. \tag{15}$$

We shall use also the operators

$$P_{\geq 0} = P_+ + P_0, \quad P_{\leq 0} = P_- + P_0.$$

Let use recall from [2] that if one just picks up $m = 0$, then the square root Klein Gordon equation (1) can be transformed into a system

$$2i(\partial_t - \partial_x)u_{\geq 0} = Q_{\geq 0}(u_{\geq 0}, u_-), \tag{16}$$

$$i(\partial_t + \partial_x)u_- = Q_-(u_{\geq 0}, u_-).$$

The system (16) is simplified essentially when $u_- = 0$ and becomes a simple scalar equation of type

$$i(\partial_t - \partial_x)v_{\geq 0} = P_{\geq 0}(|v_{\geq 0}|^2 v_{\geq 0})\mathrm{for} t \geq 0. \tag{17}$$

Lemma 2.1 *For any* $s \in (1/3,1/2)$ *one can find solutions*

$$v_\varepsilon(t,x) \in C(\mathbb{R}; H^s(0,2\pi)),$$

of (17) having the form (6), i.e.

$$v_\varepsilon(t,x) = v_{\geq 0}(t,x; s,\varepsilon),$$

such that $P_-(v_\varepsilon) = 0$.

Making the substitution

$$u = v + w, \quad v(t,x) = v_\varepsilon(t,x), \tag{18}$$

in (16) we arrive at the following equation

$$(i\,\partial_t - |D_x|)w = (w+v)^2\overline{(w+v)} - v^2\overline{v} + P_-(v^2\overline{v}) + S(D)(v+w), \tag{19}$$

where

$$(w+v)^2\overline{(w+v)} - v^2\overline{v} = 2wv\overline{v} + \overline{w}v^2 + w^2\overline{v} + 2w\overline{w}v + w^2\overline{w}. \tag{20}$$

It is important to classify all terms on the right hand side of (19). First of all we notice that the last term, because of the smoothing nature of the operator $S(D)$ (see for instance [4] and reference therein) fulfills

Lemma 2.2 *Assume* $f_\varepsilon \in C([0,1], H^s)$, *for some* ε, $s = \frac{1}{2} - \varepsilon$, $\varepsilon > 0$, *then one has*

$$\| S(D)f_\varepsilon \|_{H^{a+s}} = \| (\langle D_x \rangle - |D_x|)f \|_{H^{a+s}} \leq C \| f_\varepsilon \|_{H^s},$$

for all $a \in [0,1/2]$ and with $C > 0$ independent from ε.

The above lemma suggests that we can concentrate on the remaining terms. Then we have linear combinations of the following:

1. Term $w^2\overline{w}$ cubic in w.
2. Terms $w^2\overline{v}$ and $w\overline{w}v$ quadratic in w.
3. Terms $wv\overline{v}$ and $\overline{w}v^2$ linear in w.
4. Term of type $P_-(vv\overline{v}) = P_-(v|v|^2)$.

In this way we have

$$(i\,\partial_t - |D_x|)w = \sum_{j=1}^{3} Q_j(v,w) + P_-(vv\overline{v}), \tag{21}$$

where

$$3Q_1(v,w) = 2wv\overline{v} + \overline{w}v^2, \tag{22}$$

$$Q_2(v,w) = w^2\overline{v} + 2w\overline{w}v,$$

$$Q_3(w) = w^2\overline{w}.$$

We can rewrite (21) as

$$(i\,\partial_t - |D_x|)(w + w_0) = \textstyle\sum_{j=1}^{3} Q_j(v,w), \tag{23}$$

where w_0 is a solution to the linear equation

$$(i\,\partial_t - |D_x|)w_0 = P_-(vv\overline{v}). \tag{24}$$

Since $v(t,x) = v_\varepsilon(t,x)$ is a family of smooth solutions, such that

$$\| v_\varepsilon \|_{C([0,1];\dot{H}^\sigma)} \leq C\varepsilon^{s-\sigma}, \quad \| v_\varepsilon \|_{C([0,1];L^2)} \leq C\varepsilon^{s}, \tag{25}$$

and

$$\| v_\varepsilon \|_{C([0,1];L^\infty)} \leq C\varepsilon^{s-1/2}$$

(see Proposition 5.1), we see that is seems difficult to derive an estimate of type

$$\| w_0 \|_{C([0,1];H^\sigma)} \leq C\varepsilon^{\theta} < \infty, \tag{26}$$

with some $\sigma > 1/2, \theta > 0$. However, we shall overcome this obstacle and establish (26) by using suitable modified Bourgain spaces and estimates for v leading to the smoothing property (26). Once the smoothing property (26) is verified, we can follow the approach based on semi-classical estimates and show that (23) has solution w satisfying better error estimates.

Semi-classical estimates for the solution

We refer for this Section to the paper [2]. Given any $\sigma > 1/2$ and any smooth function $w(t,x)$ consider the semiclassical energy

$$E_{\sigma,\varepsilon}(w(t)) = \varepsilon^{-1} \parallel w(t,\cdot)) \parallel_{L^2}^2 + \varepsilon^{2\sigma-1} \parallel w(t,\cdot) \parallel_{\dot{H}^\sigma}^2, \qquad (27)$$

by this fact along the section we shall assume

$$E_{\sigma,\varepsilon}(w_0(t)) = O(\varepsilon^\theta),$$

for some $\varepsilon > 0, \theta > 0$. It is easy to compare this semiclassical norm with the standard Sobolev norm. This is given by the

Lemma 3.1 *For any* $s, 0 < s \leq \sigma$

$$\parallel f(x) \parallel_{H^s} \leq C\varepsilon^{1/2-s}\sqrt{E_{\sigma,\varepsilon}(f)}. \qquad (28)$$

The construction of Szegö type solutions $v(t,x) = v_\varepsilon(t,x)$, *described in the previous section together with (25), guarantees that*

$$E_{\sigma,\varepsilon}(v(t)) \leq C\varepsilon^{2s-1}. \qquad (29)$$

For the equation (23) we have the following semi-classical estimate

Lemma 3.2 *For any* $\sigma \geq 0$ *there exists a constant* $C > 0$ *so that for any* $T \in (0,1)$ *we have*

$$2 \sup_{0\leq t\leq T} \sqrt{E_{\sigma,\varepsilon}(w(t))} \leq C \sup_{0\leq t\leq T} \sqrt{E_{\sigma,\varepsilon}(w_0(t))} + \qquad (30)$$

$$+C\left(\sum_{j=1}^3 \int_0^T \sqrt{E_{\sigma,\varepsilon}(Q_j(t))}\ dt\right).$$

Proof: We have the following estimates

$$\parallel w(t,\cdot) \parallel_{L^2} \leq C \parallel w_0(t,\cdot) \parallel_{L^2} + \sum_{j=1}^3 \int_0^t \parallel Q_j(\tau,\cdot) \parallel_{L_x^2} d\tau,$$

$$\| w(t,\cdot) \|_{\dot{H}^\sigma} \leq C \| w_0(t,\cdot) \|_{\dot{H}^\sigma} + \sum_{j=1}^{3} \int_0^t \| Q_j(\tau,\cdot) \|_{\dot{H}^\sigma} d\tau.$$

From these estimates we get out (30).

Thus we need the following

Lemma 3.3 *For any* $\sigma > 1/2$ *one can find a constant* $C = C(\sigma) > 0$ *so that for any* $t, 0 \leq t \leq 1$ *we have*

$$\sqrt{E_{\sigma,\varepsilon}(Q_3(t))} \leq C\big(E_{\sigma,\varepsilon}(w(t))\big)^{3/2}. \tag{31}$$

We have also

Lemma 3.4 *For any* $\sigma > 1/2$ *one can find a constant* $C > 0$ *so that for any* $t, 0 \leq t \leq 1$ *we have*

$$\sqrt{E_{\sigma,\varepsilon}(Q_2(t))} \leq CE_{\sigma,\varepsilon}(w(t)) \;\varepsilon^{s-1/2}. \tag{32}$$

We quote Proposition 5.1 and we can write the following semi-classical estimate for the Szegö type solutions $v(t,x) = v_\varepsilon(t,x)$

$$\| v(t,\cdot) \|_{L^\infty} \leq C\varepsilon^{s-1/2}, \;\; \sqrt{E_{\sigma,\varepsilon}(v(t))} \leq C\varepsilon^{s-1/2}. \tag{33}$$

In a similar way we find

Lemma 3.5 *For any* $\sigma > 1/2$ *one can find a constant* $C > 0$ *so that for any* $t, 0 \leq t \leq 1$ *we have*

$$\sqrt{E_{\sigma,\varepsilon}(Q_1(t))} \leq C\sqrt{E_{\sigma,\varepsilon}(w(t))} \;\varepsilon^{2s-1}. \tag{34}$$

The semi-classical estimate (30) shows that we can set

$$g(T) = \sup_{0\leq t\leq T} \sqrt{E_{\sigma,\varepsilon}(w(t))}, \;\; \varepsilon_1 = \varepsilon^{1-2s}, \; \varepsilon^\theta = \varepsilon_1^{\theta_1}$$

and derive the following estimate

$$g(T) \leq C\left(\varepsilon_1^{\theta_1} + \int_0^T g(t)^3 dt + \int_0^T g(t)^2 \frac{dt}{\sqrt{\varepsilon_1}} + \int_0^T g(t)\frac{dt}{\varepsilon_1}\right). \quad (35)$$

Lemma 3.6 *If $g(t)$ is a continuous non - negative function satisfying (35) with $\theta_1 \in (0,1/2)$, then one can find $\varepsilon_0 > 0$ so that for $0 < \varepsilon_1 < \varepsilon_0$ we have the inequality*

$$g(T) \leq 2C\varepsilon_1^{\theta_1/2} \quad (36)$$

for

$$0 \leq T \leq T^*(\varepsilon_1) = \frac{\theta_1 \varepsilon_1}{2} \ |\log\varepsilon_1|.$$

Proof of Theorem 1.2

In this section, following the spirit of the paper [2], we shall complete the proof of Theorem 1.2 provided that the smoothing estimate (26), with some $\sigma > 1/2, \theta > 0$, is satisfied. As already shown in Lemma 3.6 we have the estimate

$$g(t) = \sup_{0 \leq s \leq t} \sqrt{E_{\sigma,\varepsilon}(w_\varepsilon(s))} \leq 2C\varepsilon_1^{\theta_1/2} \quad (37)$$

for

$$\varepsilon_1 = \varepsilon^{1-2s}, \ \theta_1 = \frac{\theta}{1-2s} > 0$$

and for

$$0 \leq t \leq T^*(\varepsilon_1) = \frac{\theta_1 \varepsilon_1}{2} \ |\log\varepsilon_1|.$$

Note that the estimate (28) implies

$$\| w_\varepsilon(t) \|_{H^s} \leq C\varepsilon^{1/2-s}. \quad (38)$$

The substitution (18) yields that for any $\varepsilon \in (0,1/2)$ and for any $j = 0,1$

$$u_{\varepsilon}^{(j)}(t,x) = v_{\geq 0}^{(j)}(t,x;s,\varepsilon) + w_{\varepsilon}^{(j)}(t,x),$$

is a solution to the equation (3) (with $m = 1$), i.e. for $u = u_{\varepsilon}^{(j)}(t,x)$ it solves

$$(i\,\partial_t - < D_x >)u_{\varepsilon}^{(j)} = |u_{\varepsilon}^{(j)}|^2 u_{\varepsilon}^{(j)} \text{for} t \geq 0,$$

with initial data

$$u_{\varepsilon}^{(j)}(0,x) = \frac{a_j}{1 - c_0 e^{ix}}, \tag{39}$$

such that

$$c_0 = c_0(\varepsilon) = \sqrt{1-\varepsilon}, a_j = a_j(\varepsilon) = \left(1 + \frac{j}{|\log\varepsilon|}\right)\varepsilon^{s+1/2}. \tag{40}$$

Using Lemma 5.2, it is easy to verify that

$$\| u_{\varepsilon}^{(1)}(0,\cdot) - u_{\varepsilon}^{(0)}(0,\cdot) \|_{H^s}^2 = \sum_{k\geq 0} \langle k \rangle^{2s} |a_1 - a_0|^2 (1-\varepsilon)^k \leq \frac{C}{|\log\varepsilon|^2}. \tag{41}$$

The estimates (33) show that

$$\| v(t) \|_{H^s} \leq C\varepsilon^{1/2-s}\sqrt{E_{\sigma,\varepsilon}(v(t))} \leq C_1. \tag{42}$$

Our principal target shall be to establish that for some $D > 0$ we have

$$\| v_{\geq 0}^{(1)}(t,x;s,\varepsilon) - v_{\geq 0}^{(0)}(t,x;s,\varepsilon)) \|_{H^s} \geq D, \tag{43}$$

for t in a suitable interval; the estimates (37), (38) guarantee that the term

$$v_{\varepsilon}^{(j)}(t,x) = v_{\geq 0}^{(j)}(t,x;s,\varepsilon),$$

is dominant in the representation $u_{\varepsilon}^{(j)} = v_{\varepsilon}^{(j)} + w_{\varepsilon}^{(j)}$ so the estimates (41) and (43) will complete the proof of Theorem 1.2. For this we have to verify (41) only. To do this we shall concentrate the proof on the estimate of a suitable Fourier coefficient in the Fourier expansion of

$$v_\varepsilon^{(j)}(t,x) = \frac{a_j(\varepsilon)e^{-i\alpha_j(\varepsilon)t}}{1-c_0(\varepsilon)e^{i(x+t-\gamma_j(\varepsilon)t)}},$$

where the parameters $\alpha_j(\varepsilon), \gamma_j(\varepsilon)$ satisfy the asymptotical expansions

$$2\gamma_j(\varepsilon) = \varepsilon^{2s}\left(1+\frac{j}{|\log\varepsilon|}+o\left(\frac{1}{|\log\varepsilon|}\right)\right),$$

$$\alpha_j(\varepsilon) = \varepsilon^{2s-1}\left(1+\frac{j}{|\log\varepsilon|}+o\left(\frac{1}{|\log\varepsilon|}\right)\right), \tag{44}$$

as $\varepsilon \searrow 0$. One has the following

Lemma 4.1 *For any $d_0 > 0$ one can find constants $D, d_1, d_2 > 0$ with $d_1 < d_2 < d_0$ so that for any $\varepsilon \in (0,1/2)$, any $j = 0,1$ and any integer $k, 0 \le k \le N(\varepsilon) = (log2)/\varepsilon$, the Fourier coefficient*

$$C_k^{(j)}(t,\varepsilon) = \frac{1}{2\pi}\int_{-\pi}^{\pi}\frac{a_0(\varepsilon)e^{-i\alpha_j(\varepsilon)t}}{1-c_0(\varepsilon)e^{i(x+t-\gamma_j(\varepsilon)t)}}\ e^{-ikx}dx, \tag{45}$$

satisfies the estimate

$$\left|C_k^{(1)}(t,\varepsilon) - C_k^{(0)}(t,\varepsilon)\right| \ge D(1-\varepsilon)^{k/2}\varepsilon^{s+1/2}, \tag{46}$$

for

$$d_1\varepsilon^{1-2s}|log\varepsilon| \le t \le d_2\varepsilon^{1-2s}|log\varepsilon|. \tag{47}$$

It is easy to compare the Fourier coefficients in (45) with the standard Fourier coefficients

$$\widehat{v_\varepsilon^{(j)}}(t,k) = \frac{1}{2\pi}\int_{-\pi}^{\pi} v_\varepsilon^{(j)}(t,x)\ e^{-ikx}dx = \frac{a_j}{a_0}C_k^{(j)}(t,\varepsilon).$$

From the fact that

$$a_j(\varepsilon) - a_0(\varepsilon) = \varepsilon^{s-1/2}\left(\frac{1}{|\log\varepsilon|} + o\left(\frac{1}{|\log\varepsilon|}\right)\right),$$

thus Lemma 4.1 gives the following

Lemma 4.2 *For any* $d_0 > 0$ *one can find constants* $D, d_1, d_2 > 0$ *with* $d_1 < d_2 < d_0$ *so that for any* $\varepsilon \in (0,1/2)$, *any* $j = 0,1$ *and any integer* $k, 0 \leq k \leq N(\varepsilon) = (log2)/\varepsilon$ *the Fourier coefficients*

$$\widehat{v_\varepsilon^{(j)}}(t,k) = \frac{1}{2\pi}\int_{-\pi}^{\pi} v_\varepsilon^{(j)}(t,x)\ e^{-ikx}dx, \tag{48}$$

satisfy the estimate

$$\left|\widehat{v_\varepsilon^{(1)}}(t,k) - \widehat{v_\varepsilon^{(0)}}(t,k)\right| \geq D(1-\varepsilon)^{k/2}\varepsilon^{s+1/2}, \tag{49}$$

for

$$d_1\varepsilon^{1-2s}|log\varepsilon| \leq t \leq d_2\varepsilon^{1-2s}|log\varepsilon|. \tag{50}$$

To complete the proof of (41) we use Lemma 4.2 and the relations

$$\| v_{\geq 0}^{(1)}(t,x;s,\varepsilon) - v_{\geq 0}^{(0)}(t,x;s,\varepsilon)) \|_{H^s}^2 \geq$$

$$\geq C\sum_{k=0}^{N(\varepsilon)} \left|\widehat{v_\varepsilon^{(1)}}(t,k) - \widehat{v_\varepsilon^{(0)}}(t,k)\right|^2 \langle k\rangle^{2s} \geq D\sum_{k=0}^{N(\varepsilon)} \varepsilon^{2s+1}(1-\varepsilon)^k\langle k\rangle^{2s},$$

combined with the following estimate, which implies by the way the optimality of the one in Lemma 5.2:

Lemma 4.3 *There is a positive constant* $C > 0$ *so that for any* $d > 0$, *and* $\theta \in [0,1]$ *and for any* $\varepsilon \in (0,1/2)$ *we have*

$$\sum_{k=0}^{d/\varepsilon} (1+k)^\theta(1-\varepsilon)^k \geq \frac{C}{\varepsilon^{1+\theta}}. \tag{51}$$

Invariant space for Szegö type equation

In this section we present some known facts about the space where the functions (6) are defined. More precisely we see that typical elements in

$$L^2_{\geq 0}(0,2\pi) = \{f \in L_2(0,2\pi); P_-(f) = f_- = 0\},$$

are functions of type

$$f(z;c) = \frac{1}{1-cz}, \ c \in \mathbb{C}, |c| < 1, \ z = e^{ix}.$$

Arguing as [1] one can look for solutions to the equation (17) having the form

$$\varphi_{a,c}(x) = af(e^{ix};c), \ a \in \mathbb{C}.$$

The solution $v_+(t,x)$ shall be defined by the formula

$$v_{\geq 0}(t,x) = \frac{a_0 e^{-i\alpha t}}{1-c_0 e^{i(t(1-\gamma)+x)}}. \tag{52}$$

We choose the parameters c_0, a_0 according to (7). Moreover the parameters γ, α satisfy the asymptotic expansions given by (10). Our first observation about the smooth functions

$$v_{\geq 0}(t,x) = v_\varepsilon(t,x) = \sum_{k=0}^{\infty} a(t)^k c(t)^k e^{ik(t+x)},$$

is the following:

Proposition 5.1 *For any* $p \in [2,\infty]$ *we have the estimate*

$$\| v_\varepsilon(t,\cdot) \|_{L^p(0,2\pi)} \leq C\varepsilon^{s-1/2+1/p}. \tag{53}$$

For any $\sigma \geq 0$

$$\| v_\varepsilon(t,\cdot) \|_{\dot{H}^\sigma(0,2\pi)} \leq C\varepsilon^{s-\sigma}. \tag{54}$$

The main ingredient to prove the above Proposition (5.1) is the estimate

$$|v(t,x)| \leq \sum_{k=0}^{\infty} \varepsilon^{s+1/2}(1-\varepsilon)^{k/2}, \tag{55}$$

in connection with the following:

Lemma 5.2 *For any* $\theta > 0$ *there is a constant* $C_0 = C(\theta) > 0$ *so that for any* $\varepsilon \in (0,1/2)$ *we have*

$$\sum_{k=0}^{\infty} (1+k)^{\theta}(1-\varepsilon)^{k} \leq \frac{C_0}{\varepsilon^{1+\theta}}. \tag{56}$$

Furthermore we have

Lemma 5.3 *If* $a \in \mathbb{C}$ *and* $c \in \mathbb{C}$ *with* $|c| < 1$ *then*

$$v_{\geq 0}(x) = \frac{a}{1-cz}, \ z = e^{ix}, \tag{57}$$

which satisfies the relations

$$P_{\geq 0}(v_{\geq 0}|v_{\geq 0}|^2) = \frac{a|a|^2(1-|c|^2cz)}{(1-cz)^2(1-|c|^2)^2}, \tag{58}$$

$$P_{-}(v_{\geq 0}|v_{\geq 0}|^2) = \sum_{k=1}^{\infty} \frac{a|a|^2}{(1-|c|^2)^2}\overline{c^k}e^{-ikx}. \tag{59}$$

Smoothing estimates by means of modified Bourgain spaces

The modified Bourgain spaces $X^{s,\delta}_{\pm,\alpha,\gamma}$ associated with the system (16) and then with NLKG equations can be defined as the completion of the space

$$S(\mathbb{R}, C^{\infty}[0,2\pi]),$$

with respect to the norms

$$\| f \|_{X^{s,\delta}_{\pm,\alpha,\gamma}} = \left(\int_{\mathbb{R}} \sum_{k\in\mathbb{Z}} \langle \tau + \alpha \mp k(1-\gamma)\rangle^{2\delta} \langle k \rangle^{2s} |\tilde{f}(\tau,k)|^2\right)^{1/2}. \tag{60}$$

Here s, δ are fixed, α, γ are parameters depending on $\varepsilon > 0$, and the modified symbols are of the form

$$\langle \tau + \alpha \mp k(1-\gamma) \rangle,$$

where α and γ are appropriate parameters associated with the sequences defined according to (6) with $\frac{1}{3} < s < \frac{1}{2}$ (then the above spaces and their norms might depend on ε). The main observation concerning the solutions $v_{\varepsilon,s}(t,x)$ is that

$$\varphi(t) v_{\varepsilon,s}(t,x) \in X^{s,N}_{+,\alpha,\gamma},$$

for any integer $N \geq 1$ and for any $\varphi(t) \in C_0^\infty(\mathbb{R})$. In addition we have

$$\| \varphi(t) v_{\varepsilon,s}(t,x) \|_{X^{s,N}_{+,\alpha(\varepsilon),\gamma(\varepsilon)}} \leq C_N,$$

with some constant C_N independent of ε. We have the following useful ancillary tools:

Lemma 6.1 *For any $s, 1/4 \leq s < 1/2$, and any real number $D \in (1/2,1)$ there exists a constant $C = C(s,D) > 0$ independent of $\varepsilon > 0$ so that*

$$\| \varphi_T(t) v_{\varepsilon,s}(t,x) \|_{X^{s,D}_{+,\alpha,\gamma}} \leq C(s,D) T^{1-D}, \tag{61}$$

i.e.

$$\int_{\mathbb{R}} d\tau \sum_{k=0}^{\infty} \langle \tau + \alpha - k(1-\gamma) \rangle^{2D} (1+k)^{2s} |\widetilde{(\varphi_T v_{\varepsilon,s})}(\tau,k)|^2 \leq C^2 T^{2-2D}. \tag{62}$$

Our next step is to evaluate the $X_-^{s_1,\delta_1-1}$ norm of the term

$$g(t,x) = P_-(v_{\varepsilon,s} |v_{\varepsilon,s}|^2).$$

Lemma 6.2 *If $s \in (1/3,1/2)$ and $s_1 > 1/2, 1 > \delta_1 > 1/2$, $0 < \theta < 1/2$, satisfy*

$$s_1 + \delta_1 + \theta \leq 3s,$$

then for any $T \in (0,1)$ *and any* $\varphi(t) \in C_0^\infty(\mathbb{R})$, *we have*

$$\varphi_T(t) g_{\varepsilon,s}(t,x) = \varphi\left(\frac{t}{T}\right)) P_-(v_{\varepsilon,s}|v_{\varepsilon,s}|^2) \in X_-^{s_1,\delta_1-1},$$

and

$$\| \varphi_T(t) g^{\varepsilon,m}(t,x) \|_{X_-^{s_1,\delta_1-1}} \leq C\varepsilon^\theta,$$

with some constant $C > 0$ *independent of* ε, δ_1, T.

Finally, we turn to the solution w_0 to the linear equation (24) with zero initial data. We have already established that

$$F = \varphi\left(\frac{t}{T}\right)) P_-(v_\varepsilon|v_\varepsilon|^2) \in X_-^{\sigma,\delta_1-1},$$

for some $\sigma > 1/2, \delta_1 > 1/2$. To derive the smoothing property (26), we need only the classical estimate

Lemma 6.3 *For any* $T > 0$ *and for any* $\sigma \geq 0$, $1/2 < \delta_1 < 1$, *one can find a constant* $C = C(T, \delta_1) > 0$ *so that for any*

$$F \in X_-^{\sigma,\delta_1-1},$$

the solution w_0 *to*

$$(i\partial_t - |D_x|) w_0 = F, \quad w_0(0,x) = 0,$$

is

$$w_0 \in C([-T,T]; H^\sigma(0,2\pi)),$$

and satisfies the estimate

$$\| w_0 \|_{C[-T,T];H^\sigma(0,2\pi))} \leq C \| F \|_{X_-^{\sigma,\delta_1-1}}. \tag{63}$$

Bibliography

[1] Gérard, P. and Grellier, S. The cubic Szegö equation. *Annales scientifiques de l'ENS*, 43, No. 5:761-810, 2010.

[2] Georgiev, V. and Tzvetkov, N and Visciglia, N. On continuity of the solution map for the cubic 1d periodic NLW equation. *Preprint 2014*, 2014.

[3] Krieger, J. and Lenzmann, E. and Raphaël, P. Nondispersive solutions of the L^2 critical half-wave equation. *Arch. Ration. Mech. Anal.*, 209:61 - 129, 2013.

[4] Ruzhansky, M. and Turunen, V. Pseudo-Differential Operators and Symmetries. *Birkhüser Basel-Boston-Berlin*, 2:1-697, 2010.

[5] Tzvetkov, N. A la frontière entre EDP semi et quasi lineaires. *Memoire d' habilitation á diriger les recherches, Université Paris-Sud, Orsay*, 2003.

An Estimate from Below for the First Eigenvalue of P-Laplacian via Hardy Inequalities

Alexander Fabricant, Nikolai Kutev and Tsviatko Rangelov

1 Introduction

The aim of this paper is to estimate from below the first eigenvalue λ of p--Laplacian $\Delta_p u = div\,(|\nabla u|^{p-2}\nabla u)$, $p > 1$ in a bounded simply connected domain $\Omega \subset R^n$, $n \geq 2$ with smooth boundary $\partial\Omega$

$$\begin{cases} -\Delta_p u = \lambda |u|^{p-2}u & in\ \ \Omega, \\ u = 0 & on\ \ \partial\Omega \end{cases}. \tag{1}$$

Problem (1) is the Euler--Lagrange equation minimizing the Reyleigh quotient

$$\frac{\int_\Omega |\nabla u|^p dx}{\int_\Omega |u|^p dx}. \tag{2}$$

among functions $u \in W_0^{1,p}$, and is introduced in [22] and independently in [11].

The first eigenvalue λ is positive and simple; the corresponding eigenfunction is positive and unique (up to multiplication with a constant), and belongs to the class $W_0^{1,p}(\Omega)$, see [3, 4, 23, 6, 20]. Below, for the first eigenvalue λ of (1) we will use also the notation $\lambda_{p,n}(\Omega)$.

For $p > 1$ and $n = 1$ for $\Omega = (a, b)$, the value of $\lambda_{p,1}(\Omega)$ is

$$\lambda_{p,1}(a,b) = (p-1)\left(\frac{\pi_p}{b-a}\right)^p, \quad \pi_p = 2\int_0^1 \frac{ds}{\sqrt[p]{1-s^p}}, \tag{3}$$

see [28, 9].

For $n \geq 2$ only in the case $p = 2$, i.e. for the Laplace operator, the value of $\lambda_{2,n}(\Omega)$ is known with analytical formulae for domains Ω with simple geometry like a ball, a shell, a parallelepiped etc. and with a numerical approximation for more general domains; see the review [14], chapter 3. For example if Ω is a ball centered at zero $B_R \subset R^n$ then

$$\lambda_{2,n}(B_R) = (p-1)\left(\frac{\mu_1^{(\alpha)}}{R}\right)^2, \quad \alpha = \frac{n}{2} - 1, \tag{4}$$

where $\mu_1^{(\alpha)}$ is the first positive zero of the Bessel function J_α, see [14] and [19].

If $p \neq 2$ the explicit value of $\lambda_{p,n}(\Omega)$ is not known even for domains Ω like a ball or a parallelepiped. That is why setting an explicit lower bound for $\lambda_{p,n}(\Omega)$ is an important task. Different numerical methods are applied in order to approximate $\lambda_{p,n}(\Omega)$ from below, see [21, 9, 8] and references herein. Note that, in this case, $p \neq 2$; an analytical expression for the lower bound of $\lambda_{p,n}(\Omega)$ by the Cheeger's constant is given in [18]. As for the estimates from above, every function $\phi \in W_0^{1,p}(\Omega)$ replaced in (2) is an upper bound for $\lambda_{p,n}(\Omega)$.

Different methods for lower bounds of $\lambda_{p,n}(\Omega)$ are developed in the literature. Examples are: isoperimetric estimates and Cheeger's constant, [10, 21, 20], or inverse power method by means of iterative technique and corresponding numerical calculations in [9] (see section 2 for more details).

In section 1.3 we estimate $\lambda_{p,n}(\Omega)$ from below via different Hardy--type inequalities with kernels singular at an interior point or on the boundary and with double singular kernels. We compare in section 4 the results obtained by the well known methods described in sections 2 and the new estimates by Hardy--type inequalities obtained in section 3.

2 State of the art

2.1 Faber--Krahn type inequality

Let us recall the Faber--Krahn type inequality which gives an estimate from below of $\lambda_{p,n}(\Omega)$ for an arbitrary bounded domain $\Omega \subset R^n$ with $\lambda_{p,n}(\Omega^*)$, where Ω^* is the n--dimensional ball of the same volume as Ω, see [24, 7, 16, 18].

Theorem 2.1 *([18]) Among all domains of given n--dimensional volume the ball Ω^* with the same volume as Ω minimizes every $\lambda_{p,n}(\Omega)$; in other words*

$$\lambda_{p,n}(\Omega) \geq \lambda_{p,n}(\Omega^*). \tag{5}$$

2.2 Cheeger's constant

Another lower bound for $\lambda_{p,n}(\Omega)$ is based on the Cheeger's constant

$$h(\Omega) = \inf \frac{|\partial D|}{|D|}.$$

Here D varies over all smooth subdomains of Ω whose boundary ∂D does not touch $\partial\Omega$, and with $|\partial D|$ and $|D|$ denoting the $(n-1)$- and n-dimensional Lebegue measure of ∂D and D respectively. The following theorem is proved in [10] for $p = 2$ and in [20] for $p > 1$.

Theorem 2.2 *([10,21]) For every $p \in (1, \infty)$ the first eigenvalue of (1) can be estimated from below via*

$$\lambda_{p,n}(\Omega) \geq \left(\frac{h(\Omega)}{p}\right)^p = \Lambda^{(1)}_{p,n}(\Omega). \tag{6}$$

Inequality (6) is sharp for $p \to 1$, because $\lambda_{p,n}(\Omega)$ converges to the Cheeger's constant $h(\Omega)$, see [18], Corollary 6.

The Cheeger's constant $h(\Omega)$ is known only for special domains. For example, if Ω is a ball $B_R \subset R^n$, then $h(\Omega) = \frac{n}{R}$ and Theorem 2.2 gives the following lower bound for $\lambda_{p,n}(B_R)$

$$\lambda_{p,n}(B_R) \geq \left(\frac{n}{pR}\right)^p = \Lambda_{p,n}^{(1)}(B_R), \ for \ \ p > 1, n \geq 2. \tag{7}$$

Thus combining the above results the following inequality holds for $p \to 1$, see [18], Corollary 15

$$\lambda_{1,n}(\Omega) \geq n\left(\frac{\omega_n}{|\Omega|}\right)^{1/n} = \Lambda_{1,n}^{(1)}(\Omega), n \geq 2. \tag{8}$$

where ω_n is the volume of the unit ball in R^n. If Ω is a ball B_R then (8) becomes equality

$$\lim_{p\to 1}\lambda_{p,n}(B_R) = \frac{n}{R}. \tag{9}$$

In the other limit case $p \to \infty$ the result in [17] says that

$$\lambda_{\infty,n}(\Omega) = \lim_{p\to\infty}\left(\lambda_{p,n}(\Omega)\right)^{1/p} = (\max\{\, dist\ (x, \partial\Omega), x \in \Omega\})^{-1}.$$

In particular for $\Omega = B_R$

$$\lambda_{\infty,n}(B_R) = \lim_{p\to\infty}\left(\lambda_{p,n}(B_R)\right)^{1/p} = \frac{1}{R}. \tag{10}$$

2.3 Sobolev's constant

It is not difficult to estimate $\lambda_{p,n}(\Omega)$ from below in a bounded domain Ω for $1 < p < n$ by the well-known Sobolev and Hölder inequalities

$$\| \nabla u \|_p \geq C_{n,p} \| u \|_{\frac{np}{n-p}} \geq C_{n,p} \| u \|_p \, |\Omega|^{-1/n}. \tag{11}$$

The best Sobolev's constant is $C_{n,p}$ which is obtained in [5] and in [29]; see [26] for bounded domains

$$C_{n,p} = n^{1/p}\omega_n^{1/n}\left(\frac{n-p}{p-1}\right)^{\frac{p-1}{p}}\left[\frac{\Gamma\left(\frac{n}{2}\right)\Gamma\left(n+1-\frac{n}{p}\right)}{\Gamma(n)}\right]^{1/n},$$

where ω_n is the volume of the unit ball in R^n. From (11) the estimate from below of the first eigenvalue becomes

$$\begin{aligned}\lambda_{p,n}(\Omega) &\geq \frac{C_{n,p}^p}{|\Omega|^{p/n}} \\ &= n\left(\frac{\omega_n}{|\Omega|}\right)^{p/n}\left(\frac{n-p}{p-1}\right)^{p-1}\left[\frac{\Gamma\left(\frac{n}{2}\right)\Gamma\left(n+1-\frac{n}{p}\right)}{\Gamma(n)}\right]^{p/n},\end{aligned} \tag{12}$$

for $1 < p < n$. For the ball B_R the estimate (12) has the form

$$\begin{aligned}\lambda_{p,n}(B_R) &\geq \frac{n}{R^p}\left(\frac{n-p}{p-1}\right)^{p-1}\left[\frac{\Gamma\left(\frac{n}{2}\right)\Gamma\left(n+1-\frac{n}{p}\right)}{\Gamma(n)}\right]^{p/n} \\ &= \Lambda_{p,n}^{(S)}(B_R),\end{aligned} \tag{13}$$

for $1 < p < n$. As for the limit for $p \to 1$ and fixed n we obtain from (13)

$$\lim_{p\to 1}\Lambda_{p,n}^{(S)}(B_R) = \frac{n}{R^p}\left[\frac{\Gamma\left(\frac{n}{2}\right)}{\Gamma(n)}\right]^{1/n} < \frac{n}{R}.$$

For $p > n$ we have from [23] in the parallelepiped $P = \{0 < x_1 < a_1, \ldots, 0 < x_n < a_n\}$ with $a_{min} = \min_{1\leq i\leq n} a_i$ the estimate

$$\lambda_{p,n}(P) \geq \frac{p}{a_{min}^p}. \tag{14}$$

If Ω is an arbitrary bounded domain in R^n and R is the radius of the largest ball inscribed in the smallest parallelepiped (with minimal a_{min}) containing Ω then

$$\lambda_{p,n}(\Omega) \geq \lambda_{p,n}(B_R) \geq \frac{p}{R^p} = \Lambda_{p,n}^{(L)}(B_R), \quad for \quad p > n. \tag{15}$$

For the limit $p \to \infty$ in (15) we obtain

$$\lim_{p\to\infty}\sqrt[p]{\Lambda_{p,n}^{(L)}(B_R)} = \frac{1}{R}.$$

2.4 Numerical methods

A different method for computing $\lambda_{p,n}(\Omega)$, inspired by the inverse power method in finite dimensional algebra, is developed in [9, 8]. By means of an iterative technique the authors define two sequences of functions, one of them monotone decreasing, the other one monotone increasing. The first eigenvalue $\lambda_{p,n}(\Omega)$ is between the limits of these sequences. In the case of a ball, the two limits are equal and $\lambda_{p,n}(\Omega)$ coincides with them.

In [21] a finite element technique is used for numerical approximation of the first eigenfunction and the first eigenvalue for (1).

3 Estimates from below via different Hardy inequalities

Hardy-type inequalities are with kernels singular either at some interior point of Ω, usually at the origin, or with kernels singular on the boundary or combined double singularity at 0 and at $\partial\Omega$. We will apply these three Hardy-type inequalities in order to estimate from below the first eigenvalue $\lambda_{p,n}(\Omega)$. We are concentrating only on those Hardy inequalities (among the large number of results in the literature) which have explicitly given constants.

3.1 Kernels with singularity in an internal point

The classical Hardy inequality, see [15, 27], reads

$$\int_\Omega |\nabla u|^p dx \geq \left|\frac{n-p}{p}\right|^p \int_\Omega \frac{|u|^p}{|x|^p} dx \tag{16}$$

for every $p > 1$ and every $u \in W_0^{1,p}(\Omega)$, $0 \in \Omega$. Using the trivial estimate

$$\int_\Omega \frac{|u|^p}{|x|^p} dx \geq \frac{1}{D^p} \int_\Omega |u|^p dx,$$

where $D = \sup\{|x|, x \in \Omega\}$ we get

$$\frac{\int_\Omega |\nabla u|^p dx}{\int_\Omega |u|^p dx} \geq \left(\frac{|n-p|}{pD}\right)^p = \Lambda_{p,n}^{(2)}(\Omega), \quad for \quad n \neq p,$$

and the first trivial estimate from below is $\lambda_{p,n}(\Omega) \geq \Lambda^{(2)}_{p,n}(\Omega)$. If $\Omega = B_R$ then $D = R$ and

$$\lambda_{p,n}(B_R) \geq \left(\frac{|n-p|}{pR}\right)^p = \Lambda^{(2)}_{p,n}(B_R), \ \ for \ \ n \neq p. \tag{17}$$

For the limits $p \to 1$ and $p \to \infty$ for fixed n from (17) we obtain

$$\lim_{p\to 1}\Lambda^{(2)}_{p,n}(B_R) = \frac{n-1}{R}, \quad \lim_{p\to\infty}\sqrt[p]{\Lambda^{(2)}_{p,n}(B_R)} = \frac{1}{R}. \tag{18}$$

Different improved Hardy inequalities were obtained in [2, 13, 1] but since the constants in the additional terms are not given explicitly we will not use them in the comparison in section 4.

3.2 Kernels with singularity on the boundary

Let $\Omega \subset R^n$ be a bounded domain and $d(x) = dist\,(x, \partial\Omega)$ be the distance to the boundary. In [30] the following Hardy inequality in convex domain Ω was obtained

$$\int_\Omega |\nabla u|^p dx \geq \left|\frac{p-1}{p}\right|^p \int_\Omega \frac{|u|^p}{d^p(x)} dx + \frac{a(p,n)}{|\Omega|^{p/n}} \int_\Omega |u|^p dx, \quad u \in W_0^{1,p}(\Omega), \tag{19}$$

where

$$a(p,n) = \frac{(p-1)^{p+1}}{p^p}\left(\frac{\omega_n}{n}\right)^{p/n} \frac{\sqrt{\pi}\,\Gamma\left(\frac{n+p}{2}\right)}{\Gamma\left(\frac{p+1}{2}\right)\Gamma\left(\frac{n}{2}\right)}.$$

For the first eigenvalue for p-Laplacian in the ball B_R from (19) we obtain the inequality

$$\begin{aligned}\lambda_{p,n}(B_R) \ &\geq \ \left(\frac{p-1}{Rp}\right)^p \left[1 + \frac{p-1}{n^{p/n}} \frac{\sqrt{\pi}\,\Gamma\left(\frac{n+p}{2}\right)}{\Gamma\left(\frac{p+1}{2}\right)\Gamma\left(\frac{n}{2}\right)}\right] \\ &= \ \Lambda^{(3)}_{p,n}(B_R), \ \ for \ \ n \geq 2, p > 1.\end{aligned} \tag{20}$$

For the limits $p \to 1$ and $p \to \infty$ for fixed n from (20) we obtain

$$\lim_{p\to 1}\Lambda_{p,n}^{(3)}(B_R) = 0, \quad \lim_{p\to\infty}\sqrt[p]{\Lambda_{p,n}^{(3)}(B_R)} = \frac{1}{R}. \tag{21}$$

3.3 Kernels with double singularity

Now we will use a Hardy-type inequality with double singular kernel, a particular case of [12], Theorem 1.

Theorem 3.1 *For every $p \neq n$, $p > 1$, $n \geq 2$ and the ball B_R the estimate*

$$\lambda_{p,n}(B_R) \geq \left(\frac{1}{Rp}\right)^p \left[\frac{(n-1)^{n-1}}{(p-1)^{p-1}}\right]^{\frac{p}{n-p}} = \Lambda_{p,n}^{(4)}(B_R). \tag{22}$$

holds.

Proof. In the notations of Theorem 1 in [12] we choose $\alpha = 1$, $\beta = 1$, $\lambda(x) = R$, $\Omega = B_R$, $\gamma = \frac{p-1}{p}$, $m = -k = \frac{p-n}{p-1} \neq 0$, $s(x) = \frac{|x|}{R}$, $g(s(x)) = \frac{R^m - |x|^m}{m|x|^m}$. Then $v(x) = 1$, $w(x) = |x|^{-1}|g(s(x))|^{-1}$ and we get the inequality

$$\int_{B_R} |\nabla u|^p dx \geq \left|\frac{p-n}{p}\right|^p \int_{B_R} \frac{|u|^p}{|x|^{n-m}|R^m - |x|^m|^p} dx. \tag{23}$$

for every $u \in W_0^{1,p}(B_R)$.

Let us denote $|x| = \rho \in [0, R)$ for $x \in B_R$. From the estimate

$$\int_{B_R} \frac{|u|^p}{|x|^{n-m}|R^m - |x|^m|^p} dx \geq \inf_{\rho\in(0,R)} (|x|^{n-m}|R^m - \rho^m|^p)^{-1} \int_{B_R} |u|^p dx$$

and (23) we have

$$\begin{aligned}\lambda_{p,n}(B_R) &\geq \left|\frac{p-n}{p}\right|^p \inf_{\rho\in(0,R)} [\rho^{1-m}|R^m-\rho^m|]^{-p} \\ &= \left|\frac{p-n}{p}\right|^p \left[\sup_{\rho\in(0,R)} (\rho^{1-m}|R^m-\rho^m|)\right]^{-p}.\end{aligned} \tag{24}$$

This is because $m-n=(m-1)p$ and $1-m=\frac{n-1}{p-1}>0$.

For the function $z(\rho)=\rho^{1-m}|R^m-\rho^m|$ in the interval $(0,R)$ we have $z'(\rho)=[(1-m)R^m\rho^{-m}-1]\,sign\,(m)$ and $z'(\rho)=0$ only for $\rho_0=R(1-m)^{1/m}=R\left(\frac{n-1}{p-1}\right)^{1/m}$. Since $0<\left(\frac{n-1}{p-1}\right)^{1/m}<1$, then $0<\rho_0<R$ and from $z''(\rho_0)=-|m|(1-m)R^m\rho_0^{-m-1}<0$ it follows that function $z(\rho)$ has a maximum at the point ρ_0 and

$$z(\rho_0)=R\left(\frac{n-1}{p-1}\right)^{\frac{n-1}{p-1}}\left|\frac{p-n}{p-1}\right|. \tag{25}$$

Hence from (24) and (25) we get

$$\begin{aligned}\lambda_{p,n}(B_R) &\geq \left(\frac{p-1}{p}\right)^p\left(\frac{n-1}{p-1}\right)^{\frac{(n-1)p}{p-n}} R^{-p} \\ &= \left(\frac{1}{Rp}\right)^p\left[\frac{(n-1)^{n-1}}{(p-1)^{p-1}}\right]^{\frac{p}{n-p}} = \Lambda_{p,n}^{(4)}(B_R).\end{aligned} \tag{26}$$

Thus Theorem 3.1 is proved.

For the limits $p\to 1$ and $p\to\infty$ for fixed n from (26) we obtain

$$\lim_{p\to 1}\Lambda_{p,n}^{(4)}(B_R)=\frac{n-1}{R},\quad \lim_{p\to\infty}\sqrt[p]{\Lambda_{p,n}^{(4)}(B_R)}=\frac{1}{R}. \tag{27}$$

Corollary 3.2 *For every $p\neq n$, $p>1$, $n\geq 2$ and every bounded domain $\Omega\subset R^n$ the estimate*

$$\lambda_{p,n}(\Omega)\geq\left(\frac{\omega_n}{|\Omega|}\right)^{p/n}\frac{1}{p^p}\left[\frac{(n-1)^{n-1}}{(p-1)^{p-1}}\right]^{\frac{p}{n-p}}=\Lambda_{p,n}^{(4)}(\Omega) \tag{28}$$

holds.

Proof. From Theorem 2.1 with $\Omega^* = \{|x| < R_*\}$ if $|\Omega^*| = |\Omega|$ then $R_* = \left(\frac{|\Omega|}{\omega_n}\right)^{1/n}$, and (28) follows from (22) and (5).

In order to obtain an estimate from below for $\lambda_{n,n}(B_R)$, i.e. for $p = n$ we apply Theorem 1 in [11] for $n = p$.

Theorem 3.3 *For every $n \geq 2$ and the ball B_R the estimate*

$$\lambda_{n,n}(B_R) \geq \left(\frac{1}{Rn}\right)^n (n-1)^n e^n = \Lambda_{n,n}^{(4)}(B_R) \tag{29}$$

holds, where $e \approx 2.713$ is the Euler's constant.

Proof. In the notations of Theorem 1 in [12] we choose $\alpha = 1$, $\beta = 1$, $\lambda(x) = R$, $\Omega = B_R$, $\gamma = \frac{n-1}{n}$. Then $v(x) = 1$, $w(x) = \left(|x|\ln\frac{R}{|x|}\right)^{-1}$ and we get the inequality

$$\int_{B_R} |\nabla u|^n dx \geq \left(\frac{n-1}{n}\right)^n \int_{B_R} \frac{|u|^n}{\left(|x|\left|\ln\frac{R}{|x|}\right|\right)^n} dx, \tag{30}$$

for every $u \in W_0^{1,n}(B_R)$.

As in the proof of Theorem 3.1 we use that

$$\int_{B_R} \frac{|u|^n}{\left(|x|\left|\ln\frac{R}{|x|}\right|\right)^n} dx \geq \left[\sup_{\rho\in(0,R)} \rho\ln\frac{R}{\rho}\right]^{-n} \int_{B_R} |u|^n dx.$$

The function $z(\rho) = \rho\ln\frac{R}{\rho}$ has a maximum $z(\rho_0) = R/e$ at the point $\rho_0 = R/e$. From (29) we get

$$\lambda_{n,n}(B_R) \geq \left(\frac{n-1}{n}\right)^n \left(\frac{R}{e}\right)^{-n} = \left(\frac{1}{Rn}\right)^n (n-1)^n e^n,$$

and Theorem 3.3 is proved.

Remark 3.4 Simple computations give by means of L'Hospital rule the equality

$$\lim_{p\to n}\Lambda_{p,n}^{(4)}(B_R) = \Lambda_{n,n}^{(4)}(B_R).$$

Corollary 3.5 *For every $n \geq 2$ and every bounded domain $\Omega \subset R^n$ the estimate*

$$\lambda_{n,n}(\Omega) \geq \frac{\omega_n}{|\Omega|}\left(\frac{n-1}{n}\right)^n e^n = \Lambda_{n,n}^{(4)}(\Omega) \tag{31}$$

holds where ω_n is the volume of the unit ball in R^n and e is the Euler's constant.

4 Comparison of different estimates

We will compare the estimates from below of the first eigenvalue in the unit ball B_1 and denote $\Lambda_{p,n}^{(j)}(B_1) = \Lambda_{p,n}^{(j)}$ and $\Lambda_{p,n} = \max_j \Lambda_{p,n}^{(j)}$.

4.1 Using analytical formulae

Let us start with some general statements.

Proposition 4.1 *For every $n \geq 2$ there exists p_{0n}, $1 < p_{0n} < 2$ such that $\Lambda_{p,n}^{(1)} > \Lambda_{p,n}^{(4)}$ for $1 < p < p_{0n}$ and $\Lambda_{p,n}^{(1)} < \Lambda_{p,n}^{(4)}$ for $p_{0n} < p$.*

Proof. Let us define a function

$$f_n(p) = \frac{1}{n-p}[(n-1)\ln(n-1) - (p-1)\ln(p-1)] - \ln n.$$

The inequality $\Lambda_{p,n}^{(4)}(B_R) > \Lambda_{p,n}^{(1)}(B_R)$ holds *if and only if* $f_n(p) > 0$. We will show that for every fixed $n \geq 2$ the function $f_n(p)$ is a strictly increasing one for $p > 1$ and $f_n(1) < 0$, $f_n(2) > 0$. Thus, there exists $p_{0n} \in (1,2)$ such that $f_n(p) < 0$ for $1 < p < p_{0n}$, $f_n(p_{0n}) = 0$ and $f_n(p) > 0$ for $p_{0n} < p$.

For the first derivative of $f_n(p)$ we have

$$\begin{aligned} f'_n(p) &= \frac{1}{(n-p)^2}[(n-1)\ln(n-1)-(n-1)+(p-1)-(n-1)\ln(p-1)] \\ &= \frac{g_n(p)}{(n-p)^2}. \end{aligned}$$

Since $g'_n(p)=\frac{p-n}{p-1}$, $g''_n(p)=n-1(p-1)>0$ then $g_n(p)$ has a minimum at the point $p=n$ and $g_n(n)=0$. Using L'Hospital rule we obtain $\lim_{p\to n}f'_n(p)=\frac{1}{2(n-1)}>0$ and hence $f'_n(p)>0$ for every $p>1$. Moreover,

$$\lim_{p\to 1}f_n(p)=\ln(n-1)-\ln n<0, \quad and$$

$$f_n(2)=\frac{1}{n-2}[(n-1)\ln(n-1)-(n-2)\ln n]>0.$$

The second inequality holds because for the function $h(n)=(n-1)\ln(n-1)-(n-2)\ln n$ we have $h'=\frac{2}{n}+\ln(n-1)-\ln n$, $h''=\frac{n-2}{n^2(n-1)}\geq 0$, i.e., h' is an increasing function, $h'(n)\geq h'(2)=1-\ln 2>0$. Hence $h(n)$ is strictly an increasing function and $h(n)>h(2)=0$.

Proposition 4.2 *For every* $n\geq 3$, *we have* $\Lambda^{(2)}_{p,n}<\Lambda^{(1)}_{p,n}$.

Proof:

$$\Lambda^{(2)}_{p,n}=\left(\frac{|n-p|}{pR}\right)^p<\left(\frac{n}{pR}\right)^p=\Lambda^{(1)}_{p,n}.$$

For $p\in(1,4]$ and $n=2,3,4$ comparison of $\Lambda^{(1)}_{p,n}$ and $\Lambda^{(4)}_{p,n}$ is shown in Figure 1.

4.2 Using numerical estimates

The following comparisons are made by the Mathematica program for the complicated formulas (13), (20), (22) and the numerical approximations of the first eigenvalue in [9]. The analytical proof of the statements in the first two subsections below is an open problem.

4.2.1 Numerical comparison of $\Lambda_{p,n}^{(3)}$ and $\Lambda_{p,n}^{(4)}$

For every $n \geq 2$, $p > 1$ we have $\Lambda_{p,n}^{(3)} < \Lambda_{p,n}^{(4)}$. This is shown for $p \in (1,4]$ and $n = 2,3,4$ in Figure 1.

4.2.2 Numerical comparison of Sobolev constant (13) and $\Lambda_{p,n}^{(1)}$, $\Lambda_{p,n}^{(4)}$

Let $n \geq 2$ be fixed and $1 < p < n$. For the Sobolev constant $\Lambda_{p,n}^{(S)}$ in the ball B_R, defined in (13), we have

for $n = 2,3,4$, and $p > 1$ there exists p_{1n}, p_{2n}, $1 < p_{1n} < p_{0n} < p_{2n} < 2$ such that $\Lambda_{p,n}^{(1)} < \Lambda_{p,n}^{(4)} < \Lambda_{p,n}^{(S)}$ for $0 < p < p_{1n}$; $\Lambda_{p,n}^{(4)} < \Lambda_{p,n}^{(S)} < \Lambda_{p,n}^{(1)}$ for $p_{1n} < p < p_{0n}$; $\Lambda_{p,n}^{(1)} < \Lambda_{p,n}^{(4)} < \Lambda_{p,n}^{(S)}$ for $p_{0n} < p < p_{2n}$; $\Lambda_{p,n}^{(S)} < \Lambda_{p,n}^{(1)} < \Lambda_{p,n}^{(4)}$ for $p_{2n} < p$;

If $n = 5, \dots, 12$ then $\Lambda_{p,n}^{(S)} < \Lambda_{p,n}^{(4)}$;

If $n \geq 13$ then there exist p_{3n}, p_{4n}, $2 < p_{3n} < p_{4n}$ such that $\Lambda_{p,n}^{(S)} < \Lambda_{p,n}^{(4)}$ for $2 < p < p_{3n}$ and $p_{4n} < p$, and $\Lambda_{p,n}^{(4)} < \Lambda_{p,n}^{(S)}$ for $p_{3n} < p < p_{4n}$.

4.2.3 Comparison of the formulae in [9] and $\Lambda_{p,n}^{(3)}$, $\Lambda_{p,n}^{(4)}$

As was mention in section 1.2, a numerical method for evaluating the first eigenvalue of the problem (1) was developed in [9] and for $p \in (1,4]$ and $n = 2,3,4$. The approximate values of the first eigenvalue $\lambda_{p,n}$ in the ball B_1 together with the analytical values of $\Lambda_{p,n}^{(3)}$, $\Lambda_{p,n}^{(4)}$ is shown in Figure 2.

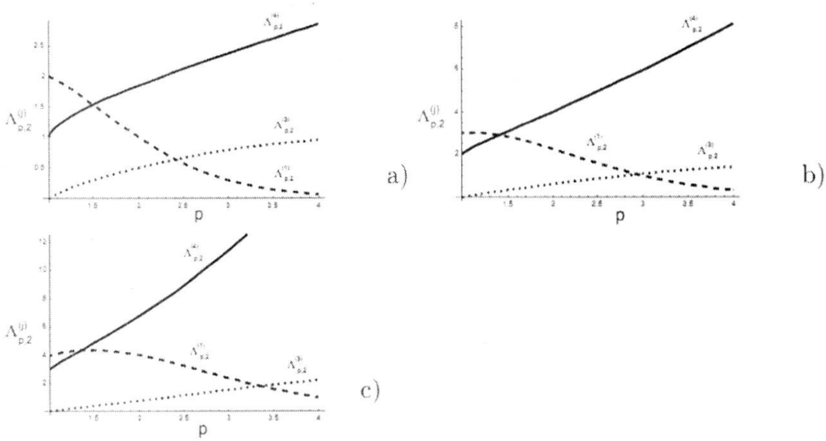

Figure 1: Comparison of $\Lambda_{p,n}^{(j)}$ for $p \in (1,4)$ and: a) $n = 2$; b) $n = 3$; c) $n = 4$.

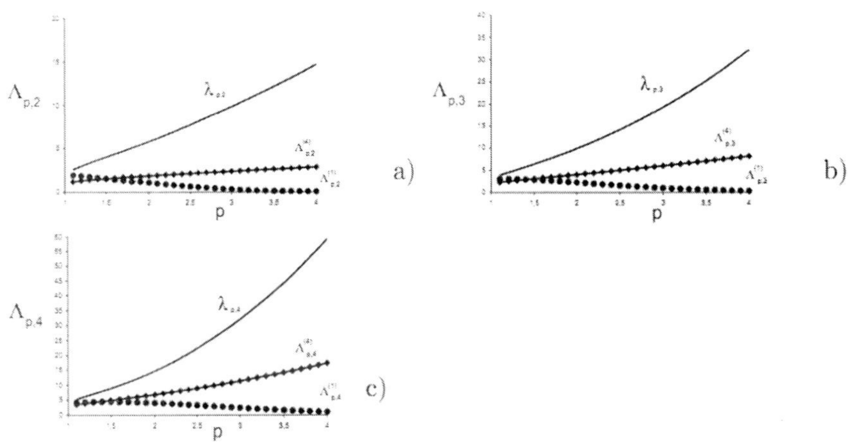

Figure 2: Comparison between the approximate value of $\lambda_{p,n}$, see [9] and the estimate from below $\Lambda_{p,n}^{(1)}$ in section 2, see [18] and the estimate from below $\Lambda_{p,n}^{(4)}$ by using Hardy inequality in section 3, see [12], for $p \in (1,4)$ and : a) $n = 2$; b) $n = 3$; c) $n = 4$.

Bibliography

[1] Adimurthi. Hardy--Sobolev inequality in $H^1(\Omega)$ and its applications. *Commum. Contemp. Math.*, 4:409–434, 2002.

[2] Adimurthi and S. Fillipas and A. Tertikas. On the best constant of Hardy--Sobolev inequalities. *Nonlinear Analysis*, 70:2826–2833, 2009.

[3] W. Allegretto and Y. Huang. A Picone's identity for the p--Laplacian and applications. *Nonlin. Anal.*, 32:819–830, 1998.

[4] A. Anane. Simplicité et isolation de la première valeur propre du p--Laplacien avec poids. *C.R. Acad. Sci. Paris, Sér. I Math.*, 305:725–728, 1987.

[5] T. Aubin. Problèmes isopérimétriques et espaces de Sobolev. *J. Diff. Geom.*, 11:573–598, 1976.

[6] M. Belloni and B. Kawohl. A direct uniqueness proof for equations involving the p-Laplace operator. *Manuscripta Math.*, 109:229-231, 2002.

[7] T. Bhattacharia. A proof of the Faber--Krahn inequality for the first eigenvalue of the p-Laplacian. *Ann. Mat. Pura Appl. Ser. 4*, 177:225--240, 1999.

[8] R. Biezuner and J. Brown and G. Ercole and E. Martins. Computing the first eigenpair of the p-Laplacian via the inverse iteration of sublinear supersolutions. *J. Sci. Comp.*, 51(1):180-201, 2012.

[9] R. Biezuner and G. Ercole and E. Martins. Computing the first eigenvalue of the p-Laplacian via the inverse power method. *J. Funct. Anal.*, 257:243-270, 2009.

[10] J. Cheeger. A lower bound for smallest eigenvalue of the Laplacian. In R. C. Gunning, editors, *Problems in Analysis, A Symposium in Honor of Salomon Bochner*, pages 195-199, Princeton Univ. Press, 1970.

[11] A. Fabricant and N. Kutev and T. Rangelov. Hardy-type inequality with weights. *Centr. Eur. J. Math.*, 11(9):1689--1697, 2013.

[12] S. Fillipas and A. Tertikas. Optimizing improved Hardy inequalities. *J. Funct. Anal.*, 192:186–233, 2002.

[13] D. S. Grebenkov and B.-T. Nguyen. Geometrical structure of Laplacian eigenfunctions. *SIAM Review*, 55(4):601-667, 2013.

[14] G. Hardy. Notes on some points in the integral calculus. *Messenger Math.*, 48:107-112, 1919.

[15] Y. X. Huang. On the eigenvalues of the p-Laplacian with varying p. *Proc. Amer. Math. Soc.*, 125:3347-3354, 1997.

[16] P. Juutinen and P. Lindqvist and J. Manfredi. The ∞-eigenvalue problem. *Arch. Rat. Mech. Anal.*, 148:89-105, 1999.

[17] B. Kawohl and V. Fridman. Isoperimetric estimates for the first eigenvalue of the p-Laplace operator and the Cheeger constant. *Comment. Math. Univ. Carolinae*, 44(4):659-667, 2003.

[18] B. Kawohl and S. Krömer and J. Kurtz. Radial eigenfunctions for the game-theoretic p-Laplacian. Technical report, No. 9,2013/2014, Fall, Institut Mittag--Leffler, 2013.

[19] B. Kawohl and P. Lindqvist. Positive eigenfunctions for the p--Laplace operator revisited. *Analysis (Münich)*, 26:545-550, 2006.

[20] L. Lefton and D. Wei. Numerical approximation of the first eigenpair of the p-Laplacian using finite elements and penalty method. *Numer. Funct. Anal. Optim.*, 18:389-399, 1997.

[21] E. Lieb. On the lowest eigenvalue of the Laplacian for the intersection of two domains. *Inv. Math.*, 74:441-448, 1983.

[22] P. Lindqvist. A note on the nonlinear Rayleigh quotient. In M. Gyllenberg and L. E. Persson, editors, *Analysis, Algebra and Computers in Mathematical Research*, pages 223-231, Marcel Dekker Lecture Notes in Pure and Appl. Math. 156, 1994.

[23] P. Lindqvist. On a nonlinear eigenvalue problem. *Fall School Anal. Lect., Ber. Univ. Jyväskylä Math. Inst.*, 68:33-54, 1995.

[24] P. Lindqvist. On the equation $div\,(|\nabla u|^{p-2}\nabla u)+\lambda|u|^{p-2}u=0$. *Proc. AMS*, 109:157--164, 1990. Addendum, ibid. 116:583--584, 1992.

[25] M. Ludwig and J. Xiao and G. Zhang. Sharp convex Lorentz--Sobolev inequalities. *Math. Anal.*, 350:169-197, 2011.

[26] B. Opic and A. Kufner. *Hardy-type Inequalities*. Pitman Research Notes in Math., 219, Longman, 1990.

[27] M. Del Pino and M. Elgueta and R. Manasevich. A homotopic deformation along p of a Leray--Shauder degree result and existence for $(|u'|^{p-2}u')'+f(t,u)=0,\ u(0)=u(T)=0,\ p>1$. *J. Diff. Eq.*, 80:213-222, 1995.

[28] G. Talenti. Best constant in Sobolev inequality. *Ann. Mat. Pura Appl.*, 110:353-372, 1976.

[29] F. de Thélin. Sur l'espace propre associé à la première valeur propre du pseudo-laplacien,. *C.R. Acad. Sci. Paris Sér. I Math.*, 303:355-358, 1986.

[30] J. Tidblom. A geometric version of Hardy's inequality for $W_0^{1,p}$. *Proc. Amer. Math. Soc.*, 132(8):2265-2271, 2004.

ON A NONLOCAL NONLINEAR SCHRÖDINGER EQUATION

TIHOMIR VALCHEV

Introduction

Nonlinear Schrödinger equation (NLS)

$$iq_t + q_{xx} \pm 2|q|^2 q = 0, \qquad q: R^2 \to C \tag{1}$$

is one of the classic integrable nonlinear equations. It appears in a variety of physical areas [8, 13] like nonlinear optics, plasma physics, and fluid mechanics as well as in a purely mathematical context like the differential geometry of curves [3]. Although having been extensively studied and a subject of numerous monographs like [4, 7, 8, 13], it is still stimulating further research activity; see [1, 10, 17]. Finding new integrable reductions of known nonlinear equations is one important trend in the theory of integrable systems. Nonlocal nonlinear Schrödinger equation (NNS "$\pm$")

$$iq_t + q_{xx} \pm 2q^2(x,t)q^*(-x,t) = 0, \tag{2}$$

recently introduced by Ablowitz and Musslimani [9], is a significant contribution to that area. Like the local NLS, equation (2) is *PT*-symmetric, i.e. it is invariant under the transform

$$x \to -x, \qquad t \to -t, \qquad q \to q^*. \tag{3}$$

This motivated the authors to propose NNS as a theoretical model to describe wave phenomena observed in *PT* symmetric nonlinear media [5, 11, 12].

The purpose of this report is to study some basic properties of the NNS "+" equation and its scattering operator. In doing this we shall make use of the covariant approach [4], being better suited for treating multi-component generalizations of NNS in a uniform way than that of Ablowitz and Musslimani. The report is organized as follows. Section 2 is dedicated to the direct scattering problem for NNS and the spectral properties of the corresponding scattering operator. In the next section we demonstrate how one can apply Zakharov-Shabat's dressing method to obtain special solutions to NNS. This way one easily reproduces the breathing solution obtained by Ablowitz and Musslimani [9]. In Section 4 we establish Hamiltonian formalism for NNS. For that purpose we derive a recursion operator which can generate the hierarchy of higher nonlinear equations, integrals of motion and symmetries associated with NNS. Then we apply the method of diagonalization of Lax pair [16] to describe the conserved densities of NNS through a recursion formula to generate all of them. We point a Hamiltonian to NNS and a Poisson structure assigned to it. Finally, Section 5 contains conclusions and additional remarks.

Direct Scattering Problem

NNS is an S-integrable equation, i.e. it is equivalent to compatibility condition $[L(\lambda), A(\lambda)] = 0$ for matrix differential operators $L(\lambda)$ and $A(\lambda)$ being chosen in the form:

$$L(\lambda) = \mathrm{i}\,\partial_x + Q - \lambda\sigma_3, \tag{4}$$

$$A(\lambda) = \mathrm{i}\,\partial_t + \frac{\mathrm{i}}{2}[\sigma_3, Q_x] + qp\sigma_3 + 2\lambda Q - 2\lambda^2\sigma_3, \tag{5}$$

where matrix coefficients are defined as follows:

$$Q(x,t) = \begin{pmatrix} 0 & q(x,t) \\ p(x,t) & 0 \end{pmatrix}, \qquad \sigma_3 = \begin{pmatrix} 1 & 0 \\ 0 & -1 \end{pmatrix}. \tag{6}$$

We shall restrict ourselves to the simplest case of zero boundary conditions

$$\lim_{|x|\to\infty} Q(x,t) = \mathbf{0}$$

for potential, i.e. we assume that q and p are Schwartz type functions.

To obtain a scalar NNS one has to impose an extra symmetry condition on Q so that p and q are no more independent. For instance, Ablowitz-Musslimani's NNS can be derived if one requires that $p(x,t) = q^*(-x,t)$. This idea can be made precise if one slightly extends the notion of Mikhailov's reduction group [2] by allowing action on x and t. Let us denote by $\{\psi(x,t,\lambda)\}$ the set of all fundamental matrices of the linear problem

$$L(\lambda)\psi(x,t,\lambda) = 0. \tag{7}$$

Let a finite group $\mathrm{G_R}$ act on $\{\psi(x,t,\lambda)\}$, i.e. it maps a solution ψ to linear problem (7) onto another solution

$$\tilde{\psi}(x,t,\lambda) = \mathcal{K}_{\mathrm{g}}\{\psi[k_{\mathrm{g}}(x,t,\lambda)]\}, \qquad \mathrm{g} \in \mathrm{G_R} \tag{8}$$

where $k_{\mathrm{g}}: R^2 \times C \to R^2 \times C$ is a smooth transform and $\mathcal{K}_{\mathrm{g}}$ is a group automorphism. As a result we see that the Lax operator $L(\lambda)$ must fulfill certain algebraic conditions, hence the potential Q acquires certain symmetries. Let us consider an example:

Example 3.1 Ablowitz-Musslimani's reduction

In this case the reduction group is Z_2. It maps an arbitrary fundamental solution ψ onto

$$\tilde{\psi}(x,t,\lambda) = \sigma_1\psi^*(-x,t,-\lambda^*)\sigma_1, \qquad \sigma_1 = \begin{pmatrix} 0 & 1 \\ 1 & 0 \end{pmatrix}. \tag{9}$$

Therefore the potential Q satisfies the symmetry condition

$$Q(x,t) = \sigma_1 Q^*(-x,t)\sigma_1. \tag{10}$$

Hence relation $p(x,t) = q^*(-x,t)$ holds true.

Most of our considerations in this section are general, i.e. we shall not fix a particular reduction.

Let $\psi(x,t,\lambda)$ be any fundamental solution to Zakharov-Shabat's system. Since $[L(\lambda),A(\lambda)]=0$ it satisfies

$$A(\lambda)\psi(x,t,\lambda)=\psi(x,t,\lambda)C(\lambda) \tag{11}$$

for some arbitrary matrix $C(\lambda)$. An important class of solutions to (7) is given by Jost solutions ψ_+ and ψ_- defined through:

$$\lim_{x\to\pm\infty}\psi_\pm(x,t,\lambda)\mathrm{e}^{\mathrm{i}\lambda\sigma_3 x}=\mathbb{1}\,. \tag{12}$$

To make sure that (12) is correct we require that $C(\lambda)=-2\lambda^2\sigma_3$. Scattering matrix is introduced as usual being the transition matrix

$$\psi_-(x,t,\lambda)=\psi_+(x,t,\lambda)T(t,\lambda) \tag{13}$$

between the Jost solutions. One can present the scattering matrix in the following way

$$T(t,\lambda)=\begin{pmatrix} a^+(t,\lambda) & -b^-(t,\lambda) \\ b^+(t,\lambda) & a^-(t,\lambda)\end{pmatrix}. \tag{14}$$

Its time evolution is driven by linear equation:

$$\mathrm{i}\,\partial_t T-2\lambda^2[\sigma_3,T]=0. \tag{15}$$

It is easily integrated to give

$$T(t,\lambda)=\mathrm{e}^{-2\mathrm{i}\lambda^2\sigma_3 t}T(0,\lambda)\mathrm{e}^{2\mathrm{i}\lambda^2\sigma_3 t}. \tag{16}$$

Equation (16) represents a linearization of NNS. Due to (16) the functions $a^\pm$ do not depend on t hence they could serve as generating functions of integrals of motion for NNS.

Following well-known procedures developed for local equations [4, 13] one can prove the existence of fundamental solutions χ^+ and χ^- analytic in the upper half plane C_+ and the lower half plane C_- respectively. To do this one introduces another pair of functions $\xi_{\pm} = \psi_{\pm}\exp(\mathrm{i}\lambda\sigma_3 x)$ satisfying linear equation

$$\mathrm{i}\,\partial_x \xi_{\pm} + Q\xi_{\pm} - \lambda[\sigma_3, \xi_{\pm}] = 0. \tag{17}$$

Equivalently, $\xi_{\pm}$ can be viewed as solutions to the following Voltera type integral equations

$$\xi_{\pm}(x,t,\lambda) = \mathbb{1} + \mathrm{i}\int_{\pm\infty}^{x} \mathrm{e}^{-\mathrm{i}\lambda\sigma_3(x-y)} Q(y,t)\xi_{\pm}(y,t,\lambda)\mathrm{e}^{\mathrm{i}\lambda\sigma_3(x-y)}\mathrm{d}\,y. \tag{18}$$

By analyzing it one notices that the first column of ξ_+ and the second one of ξ_- allow for analytic continuation in C_-. Thus combining those in a matrix $\eta^-(x,t,\lambda)$ we obtain a new fundamental solution with analytical properties in that domain. Similarly, the second column of ξ_+ and the first one of ξ_- allow for such continuation in C_+. They are used to construct another solution η^+ analytic there.

The same result holds for the initial solutions ψ_+ and ψ_-. That way one constructs fundamental analytic solutions χ^+ and χ^-. Making use of LDU decomposition of the scattering matrix

$$T(t,\lambda) = T^{\mp}(t,\lambda)D^{\pm}(\lambda)(S^{\pm}(t,\lambda))^{-1}, \tag{19}$$

$$T^- = \begin{pmatrix} 1 & 0 \\ b^+/a^+ & 1 \end{pmatrix}, \quad S^+ = \begin{pmatrix} 1 & 0 \\ b^-/a^+ & 1 \end{pmatrix},$$

$$T^+ = \begin{pmatrix} 1 & -b^-/a^- \\ 0 & 1 \end{pmatrix}, \quad S^- = \begin{pmatrix} 1 & 0 \\ -b^+/a^- & 1 \end{pmatrix},$$

$$D^+ = diag\ (a^+, 1/a^+), \quad D^-(\lambda) = diag\ (1/a^-, a^-)$$

one can construct χ^+ and χ^- through the formulae:

$$\chi^{\pm} = \psi_- S^{\pm} = \psi_+ T^{\mp} D^{\pm}. \tag{20}$$

The ratios $b^{\pm}/a^{\pm}$ and $b^{\mp}/a^{\pm}$ that appeared above play the role of reflection coefficients.

Remark 3.2 We shall assume that a^{+} and a^{-} have a finite number of simple zeros in C_{+} and C_{-} respectively. The zeros of a^{+} and a^{-} correspond to pole singularities of χ^{+} and χ^{-} respectively.

Remark 3.3 As in the local NLS case [4, 7, 13] fundamental solutions η^{+} and η^{-} can be viewed as solutions to a local Riemann-Hilbert factorization problem

$$\eta^{-}(x,t,\lambda)=\eta^{+}(x,t,\lambda)G(x,t,\lambda), \qquad \lambda\in R, \tag{21}$$

$$G(x,t,\lambda)=\mathrm{e}^{-\mathrm{i}\lambda\sigma_3 x}[S^{+}(t,\lambda)]^{-1}S^{-}(t,\lambda)\mathrm{e}^{\mathrm{i}\lambda\sigma_3 x}.$$

with canonical normalization

$$\lim_{|\lambda|\to\infty}\eta^{\pm}(x,t,\lambda)=\mathbb{1}\,. \tag{22}$$

The fundamental analytic solutions allow one to study the spectral properties of the scattering operator. Following [4, 15] we define the resolvent operator of $L(\lambda)$ as an operator $R(\lambda)$ satisfying

$$L(\lambda)\circ R(\lambda)=\mathbb{1}\,,$$

where $\circ$ means composition of operators. One can write down $R(\lambda)$ in the form

$$(R(\lambda)\,F)(x,t)=\int_{-\infty}^{\infty}\mathcal{R}(x,y,t,\lambda)F(y)\,\mathrm{d}\,y \tag{23}$$

for $F\colon\mathbb{R}\to\mathbb{C}^2$ being a continuous vector-valued function. The integral kernel of $R(\lambda)$ can be expressed through the fundamental analytic solutions as follows [4]:

$$\mathcal{R}(x,y,t,\lambda)=\begin{cases}\mathrm{i}\chi^{+}(x,t,\lambda)\Theta^{+}(x-y)[\chi^{+}(y,t,\lambda)]^{-1}, & \lambda\in C_{+},\\ -\mathrm{i}\chi^{-}(x,t,\lambda)\Theta^{-}(x-y)[\chi^{-}(y,t,\lambda)]^{-1}, & \lambda\in C_{-}\end{cases} \tag{24}$$

where $\Theta^{\pm}$ are matrix-valued functions defined by

$$\Theta^{\pm}(x-y)=\theta(\pm(y-x))P-\theta(\pm(x-y))(\mathbb{1}-P), \quad P= diag\ (1,0).$$

The locus in λ-plane where $\mathcal{R}$ is unbounded constitutes the continuous part of a spectrum of $L(\lambda)$. It is determined by the requirement $\mathrm{Im}\,\lambda=0$, i.e. it coincides with the real axis in the λ-plane. According to Remark 3.2 the fundamental analytic solutions may have a finite number of pole singularities determined by the zeros of the diagonal elements of $T(t,\lambda)$. These in turn determine pole singularities of $\mathcal{R}$ to form the discrete spectrum of $L(\lambda)$, see [4, 15]. In the presence of reduction the discrete spectrum belongs to orbits of the reduction group, i.e. discrete eigenvalues go together in certain symmetric configurations. To illustrate this let us consider an example.

Example 3.4 Ablowitz-Musslimani's reduction

In this case the resolvent operator $R(\lambda)$ obeys symmetry condition

$$\sigma_1 R^*(-\lambda^*)\sigma_1 = R(\lambda) \tag{25}$$

while its kernel satisfies

$$\sigma_1 \mathcal{R}^*(-x,y,t,-\lambda^*)\sigma_1=\mathcal{R}(x,y,t,\lambda). \tag{26}$$

Relation (25) means that if μ is a discrete eigenvalue of $L(\lambda)$ so is $-\mu^*$, i.e. eigenvalues are located symmetrically to an imaginary axis (in particular, they can lie on the imaginary axis itself).

Special Solutions

In this section we aim at demonstrating how one can apply Zakharov-Shabat's dressing method to construct particular solutions to NNS. We shall start with a brief reminder of the concept underlying the dressing method [13, 18]. Then we shall illustrate all general ideas with an example.

The dressing method is an indirect way to generate a solution to a S-integrable equation, i.e. we construct solutions to an equation starting from a known one (seed solution)

$$Q_0(x,t) = \begin{pmatrix} 0 & q_0(x.t) \\ p_0(x,t) & 0 \end{pmatrix}.$$

It plays the role of a potential for the scattering operator

$$L_0(\lambda) = \mathrm{i}\,\partial_x + Q_0(x,t) - \lambda\sigma_3. \tag{27}$$

Let $\psi_0(x,t,\lambda)$ be an arbitrary fundamental solution to the linear problem

$$\mathrm{i}\,\partial_x\psi_0 + (Q_0 - \lambda\sigma_3)\psi_0 = 0. \tag{28}$$

Now let us construct function $\psi_1(x,t,\lambda) = g(x,t,\lambda)\psi_0(x,t,\lambda)$ and assume it satisfies a similar linear problem

$$\mathrm{i}\,\partial_x\psi_1 + (Q_1 - \lambda\sigma_3)\psi_1 = 0 \tag{29}$$

for some other potential Q_1 to be found. The multiplier g bears the name dressing factor and satisfies linear equation:

$$\mathrm{i}\,\partial_x g + Q_1 g - g Q_0 - \lambda[\sigma_3, g] = 0. \tag{30}$$

Due to Remark 3.3 g must be normalized as follows:

$$\lim_{|\lambda|\to\infty} g(x,t,\lambda) = \mathbb{1}\,. \tag{31}$$

Then the simplest nontrivial choice possible for the dressing factor is

$$g(x,t,\lambda) = \mathbb{1} + \frac{A(x,t)}{\lambda-\mu}, \qquad \mu \in C \tag{32}$$

while its inverse is sought in the form

$$[g(x,t,\lambda)]^{-1} = \mathbb{1} + \frac{B(x,t)}{\lambda-\nu}, \qquad \nu \in C. \tag{33}$$

After substituting the ansatz for g into (30) and setting $|\lambda| \to \infty$ we derive an interrelation between the seed potential and the dressed one, namely:

$$Q_1(x,t) = Q_0(x,t) + [\sigma_3, A(x,t)]. \tag{34}$$

The residues A and B are not independent. Indeed, from the identity $gg^{-1} = \mathbb{1}$ one can see that

$$A = -B = (\mu - \nu)P \tag{35}$$

for some projector P ($P^2 = P$). Since P is a projector of rank 1 it can be presented in the form:

$$P = \frac{XF^T}{F^T X},$$

where $X(x,t)$ and $F(x,t)$ are some 2-component column vectors. In order to find them one should analyse equation (31) and its counterpart satisfied by g^{-1}. As a result one can convince oneself that X and F are expressed in terms of fundamental solutions ψ_0 and $\tilde{\psi}_0$ to the bare linear problem defined in a vicinity of the poles μ and ν respectively:

$$F^T(x,t) = F_0^T[\psi_0(x,t,\mu)]^{-1}, \qquad X(x,t) = \tilde{\psi}_0(x,t,\nu)X_0. \tag{36}$$

2-vectors X_0 and F_0 are x-independent but evolve with time. It can be shown that their t-evolution is driven by the dispersion law of nonlinear equations through formulae:

$$X_0(t) = \mathrm{e}^{\mathrm{i}f(\nu)t}X_{0,0}, \qquad F_0^T(t) = F_{0,0}^T\mathrm{e}^{-\mathrm{i}f(\mu)t}. \tag{37}$$

We are particularly interested in the case when $Q_0(x,t) = 0$. In this case as a seed fundamental solution we can take

$$\psi_0(x,t,\lambda) = \mathrm{e}^{-\mathrm{i}\lambda\sigma_3 x}. \tag{38}$$

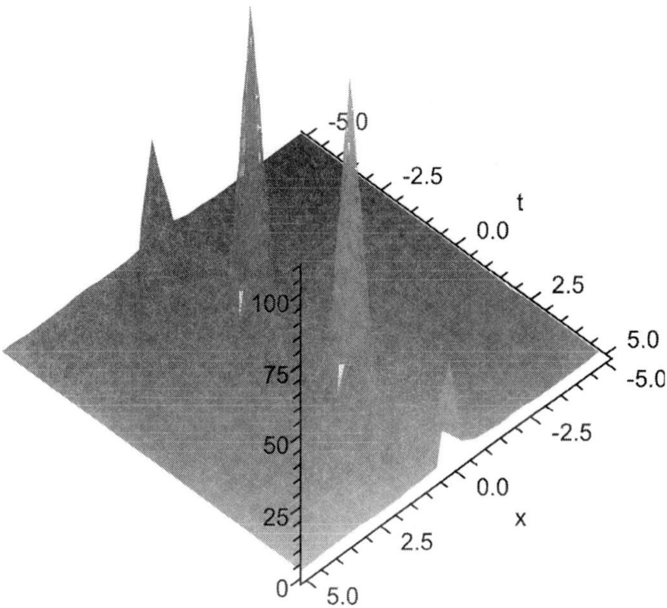

Figure 1: 3D plot of the module of the dressed solution (44) when $\gamma = 1$, $\kappa = 3$ and $\delta = 0$.

Our further considerations depend on the reduction imposed on a Lax pair. Let us assume we have Ablowitz-Musslimani's reduction (9). Then the dressing factor is subject to the symmetry condition:

$$\sigma_1 g^*(-x, t, -\lambda^*)\sigma_1 = g(x, t, \lambda). \tag{39}$$

This means that the poles of the dressing factor and its inverse are imaginary[1], i.e. $\mu = \mathrm{i}\gamma$ and $\nu = -\mathrm{i}\kappa$ and the projector P obeys the equality

$$\sigma_1 P^*(-x, t)\sigma_1 = P(x, t). \tag{40}$$

The vectors X and F are given by:

[1] In order to ensure proper asymptotic behaviour of the dressed solution we shall require that poles μ and ν are located at different half planes of λ-plane.

$$X(x,t) = \mathrm{e}^{-\kappa\sigma_3 x} X_0(t), \qquad F(x,t) = \mathrm{e}^{-\gamma\sigma_3 x} F_0(t). \tag{41}$$

Due to (40) X_0 and F_0 have one independent component only, namely we have

$$X_0 = \sigma_1 X_0^* \quad \Rightarrow \quad X_{0,2} = X_{0,1}^*, \tag{42}$$

$$F_0 = \sigma_1 F_0^* \quad \Rightarrow \quad F_{0,2} = F_{0,1}^*. \tag{43}$$

After substituting all information into (34) we get the following result:

$$q_1(x,t) = \frac{2\mathrm{i}(\gamma+\kappa)\mathrm{e}^{2\gamma x}\mathrm{e}^{4\mathrm{i}\kappa^2 t}}{\mathrm{e}^{2(\gamma+\kappa)x} + \mathrm{e}^{2\mathrm{i}[2(\kappa^2-\gamma^2)t+\delta]}}. \tag{44}$$

where $\delta \in R$ corresponds to the phases of $X_{0,1}$ and $F_{0,1}$. Thus we have reproduced Ablowitz-Musslimani's solution. (44) is not a travelling wave, it is a breather, see Fig. 1. The breather solution has singularities for $x = 0$ and

$$t_{sing} = \frac{(2m+1)\pi - 2\delta}{4(\kappa^2-\gamma^2)}, \qquad m \in Z.$$

One can apply the dressing procedure described above on and on thus generating a series of more complicated solutions.

Hamiltonian Formulation

Like the local NLS, equation (2) is an infinite-dimensional Hamiltonian system. In this section we shall consider the Hamiltonian properties of NNS. We shall start with an analytic description of the hierarchy of higher integrable equations in terms of recursion operators. The recursion operator plays a fundamental role in the theory of nonlinear equations since it paves the way to proving complete integrability [4].

Let us consider the generic Lax pair

$$L(\lambda) = \mathrm{i}\,\partial_x + Q(x,t) - \lambda\sigma_3, \tag{45}$$

$$A(\lambda) = \mathrm{i}\,\partial_t + \sum_{k=0}^{N} A_k(x,t)\lambda^k \tag{46}$$

for $Q(x,t)$ being in the form (6). Since the compatibility condition of (45) and (46) holds identically with respect to λ it yields to a series of recurrence relations for coefficients A_k, $k=0,\dots,N$. Starting to resolve them from the highest term downwards allows one to find all coefficients [4]. In doing this it is convenient to make use of splitting $A_k = A_k^{\perp} + a_k\sigma_3$ into a non-diagonal part $A_k^{\perp}$ and a diagonal one $a_k\sigma_3$. Proceeding that way we see that $A_N = c_N\sigma_3$ for some constant c_N while $A_{N-1}^{\perp} = -c_N Q$. Similarly, the non-diagonal part of the generic recurrence relation allows one to express $A_{k-1}^{\perp}$ through $A_k^{\perp}$ as follows:

$$A_{k-1}^{\perp} = \frac{\mathrm{i}}{4}[\sigma_3, \partial_x A_k^{\perp}] - a_k Q. \tag{47}$$

On the other hand from the diagonal part one is able to find a_k

$$a_k = c_k + \frac{\mathrm{i}}{2}\int_{\pm\infty}^{x} \mathrm{d}\, y \operatorname{tr}\left([Q(y), A_k^{\perp}(y)]\sigma_3\right). \tag{48}$$

After substituting (48) into (47) we obtain the recursion formula:

$$A_{k-1}^{\perp} = \Lambda_{\pm} A_k^{\perp} - c_k Q. \tag{49}$$

The integro-differential operators

$$\Lambda_{\pm} = \frac{\mathrm{i}}{4}[\sigma_3, \partial_x(.)] + \frac{\mathrm{i}Q}{2}\int_{\pm\infty}^{x} \mathrm{d}\, y \operatorname{tr}\left(Q[\sigma_3,(.)]\right) \tag{50}$$

are named recursion operators. By using them one can describe the integrable hierarchy of equations associated with the scattering operator (45) in the following way

$$\frac{\mathrm{i}}{4}[\sigma_3, Q_t] + f(\Lambda_{\pm})Q = 0, \tag{51}$$

where

$$f(\Lambda_{\pm}) = \sum_{k=0}^{N} c_k \Lambda_{\pm}^{k}, \qquad c_k \in C. \tag{52}$$

The polynomial $f(\lambda) = \sum_k c_k \lambda^k$ is the dispersion law of nonlinear equations. The NNS equation is obtained from (51) when $f(\lambda) = -2\lambda^2$.

Each member of the integrable hierarchy (51) is a Hamiltonian equation. So there is an infinite family of integrals of motion which can be regarded as Hamiltonians. We shall demonstrate how one can apply the method of diagonalization of a Lax pair [16] to derive the integrals of motion of NNS. For this to be done we apply a gauge transform

$$\mathcal{P}(x,t,\lambda) = \mathbb{1} + \sum_{k=1}^{\infty} \mathcal{P}_k(x,t)\lambda^{-k}, \tag{53}$$

where all coefficients $\mathcal{P}_k(x,t)$ are off-diagonal 2×2 matrices. The $L-A$ pair transforms into

$$\mathcal{L} = \mathcal{P}^{-1}L\mathcal{P} = \mathrm{i}\,\partial_x + \sum_{k=-1}^{\infty} \mathcal{L}_k(x,t)\lambda^{-k},$$

$$\mathcal{A} = \mathcal{P}^{-1}A\mathcal{P} = \mathrm{i}\,\partial_t + \sum_{k=-N}^{\infty} \mathcal{A}_k(x,t)\lambda^{-k}.$$

We require that the coefficients $\mathcal{L}_k(x,t)$ and $\mathcal{A}_k(x,t)$ are diagonal matrices. Then the zero curvature condition for the transformed Lax pair is written down as:

$$\partial_t\mathcal{L}_k - \partial_x\mathcal{A}_k = 0. \tag{54}$$

From those equations we deduce that $\mathcal{L}_k$ play the role of conserved quantities and $-\mathcal{A}_k$ are the corresponding currents. To find $\mathcal{L}_k$ one considers the relation

$$\mathcal{P}\mathcal{L} = L\mathcal{P}. \tag{55}$$

After comparing the coefficients before equal powers of λ in (55) we obtain an infinite series of recurrence relations. To resolve them one splits each of them into a diagonal and a non-diagonal part. Thus the generic relation leads to the following recursion formula:

$$\mathcal{L}_k = Q\mathcal{P}_k, \tag{56}$$

where

$$\mathcal{P}_k = \frac{1}{4}\mathrm{ad}_{\sigma_3}\left(\mathrm{i}\,\partial_x\mathcal{P}_{k-1} - \sum_{l=1}^{k-1} \mathcal{P}_l\mathcal{L}_{k-1-l}\right). \tag{57}$$

We list here the first three conserved quantities:

$$\mathcal{C}_1 = pq, \qquad \mathcal{C}_2 = \frac{\mathrm{i}}{2}(pq_x - qp_x), \qquad \mathcal{C}_3 = pq_{xx} + (pq)^2. \quad (58)$$

For us to obtain a true conserved quantity for NNS we need to impose an extra reduction. In the case of NNS of the Ablowitz-Musslimani type one has to require that $p(x,t) = q^*(-x,t)$ while for a local NLS we have interrelation $p(x,t) = q^*(x,t)$ holding true. The corresponding integrals of motion are found from the formula:

$$I_a(t) = \int_{-\infty}^{\infty} \mathrm{d}\, x \mathcal{C}_a(x,t), \qquad a = 1,2, \ldots. \quad (59)$$

$\mathcal{C}_3$ is a Hamiltonian density for NNS provided the Poisson bracket is defined as follows:

$$\{F, G\} = \mathrm{i} \int_{-\infty}^{\infty} \mathrm{d}\, y \left(\frac{\delta F}{\delta q(y)} \frac{\delta G}{\delta p(y)} - \frac{\delta F}{\delta p(y)} \frac{\delta G}{\delta q(y)} \right) \quad (60)$$

for F and G being functionals of q and p.

Conclusion

We have formulated and discussed the direct scattering problem for the scalar NNS. We have shown that in a quite similar manner to the local NLS case one can introduce Jost solutions, scattering matrix, fundamental analytic solutions etc. All that machinery allows one to study the spectral properties of the scattering operator by constructing its resolvent operator. As for the local NLS the operator L has a continuous spectrum which coincides with a real axis and a discrete spectrum of points symmetrically located with respect to an imaginary axis.

We have applied Zakharov-Shabat's dressing method to linear bundles with nonlocal reduction imposed. It has proved to be sufficient to use dressing factors with simple poles. As a special case we have considered in more detail the simplest case when the dressing factor has a single simple pole. This has allowed us to construct solutions in a way alternative to the approach used by Ablowitz and Musslimani [9]. In order to find more complicated solutions one can either dress several times using a single pole factor or can use a dressing factor with multiple poles.

We have derived a recursion operator for linear bundles with nonlocal reductions. This allows one to generate all higher equations belonging to the integrable hierarchy under consideration. We have shown that there exist an infinite number of integrals of motion for NNS. A recursion formula to generate all conserved quantities has been obtained by using the Lax pair diagonalization method. We have explicitly calculated the first three of them. A Poisson bracket to establish a Hamiltonian formulation of NNS has been given.

The results presented here can be extended in several directions. First, one may consider potentials obeying more complicated boundary conditions, say constant nonzero boundary conditions or time-dependent boundary conditions (nontrivial background). Such solutions could play an important role similar to that of Peregrine or Ma solutions for the local NLS.

Another promising direction of further developments is the study of multi-component NNS and the corresponding linear bundles associated with Hermitian symmetric spaces. An example of a multi-component NNS related to symmetric spaces of the type **A. III** is given by:

$$\mathrm{i}\mathbf{q}_t + \mathbf{q}_{xx} + 2\mathbf{q}(x,t)(\mathbf{q}^\dagger(-x,t)\mathbf{q}(x,t)) = 0,$$

where $\mathbf{q}$ is a matrix-valued smooth function. It has recently become known that certain nonlinear Schrödinger equations related to **A. III** and **BD. I** symmetric spaces find applications in Bose-Einstein condensation [6, 14], so their nonlocal counterparts could find similar applications too.

Acknowledgement

The author would like to thank Prof. V. Gerdjikov and Dr. R. Ivanov for fruitful discussions and moral support. This work is financially supported by the Government of Ireland Postdoctoral Fellowship in Science, Engineering and Technology.

Bibliography

[1] Hone A. Crum Transformation and Rational Solutions of the Non-focusing Nonlinear Schrödinger Equation. *J. Phys. A Math. Gen.*, 30:7473-7483, 1997.

[2] Mikhailov A. Reductions in Integrable Systems. The Reduction Groups. *Lett. in Jour. of Exper. and Theor. Phys.*, 32:187-192, 1980.

[3] Rogers C. and Schief W. *Bäcklund and Darboux Transformations. Geometry and Modern Applications in Soliton Theory*. Cambridge Univ. Press, Cambridge, 2002.

[4] Gerdjikov V. Vilasi G. and Yanovski A. *Integrable Hamiltonian Hierarchies. Spectral and Geometric Methods*. Lecture Notes in Physics **748** Springer Verlag, Berlin, Heidelberg, New York, 2008.

[5] Guo A. et al. Observation of *PT*-Symmetry Breaking in Complex Optical Potentials. *Phys. Rev. Lett.*, 103:093902, 2009.

[6] Uchiyama M. Ieda J. and Wadati M. Dark Solitons in $F = 1$ Spinor Bose-Einstein Condensate. *J. Phys. Soc. Jpn.*, 75:064002, 2006.

[7] Takhtadjan L. and Faddeev L. *The Hamiltonian Approach to Soliton Theory*. Springer Verlag, Berlin, 1987.

[8] Ablowitz M. and Segur H. *Solitons and the Inverse Scattering Transform*. SIAM, Philadelphia, 1981.

[9] Ablowitz M. and Musslimani Z. Integrable Nonlocal Nonlinear Schrödinger Equation. *Phys. Rev. Lett.*, 110:064105(5), 2013.

[10] Tajiri M. and Watanabe Y. Breather Solutions to the Focusing Nonlinear Schrödinger Equation. *Phys. Rev. E*, 57:3510-3519, 1998.

[11] Regensburger A. et al. Parity-Time Synthetic Photonic Lattices. *Nature*, 488:167-171, 2012.

[12] Rüter C. E. et al. Observation of Parity-Time Symmetry in Optics. *Nat. Phys.*, 6:192-195, 2010.

[13] Zakharov V. E. Manakov S. Novikov S. and Pitaevskii L. *Theory of Solitons: The Inverse Scattering Method*. Plenum, New York, 1984.

[14] Ieda J. Miyakawa T. and Wadati M. Exact Analysis of Soliton Dynamics in Spinor Bose-Einstein Condensates. *Phys. Rev Lett.*, 93:194102, 2004.

[15] Gerdjikov V. On Spectral Theory of Lax Operators on Symmetric Spaces: Vanishing versus Constant Boundary Conditions. *Jour. Geom. & Symm. in Phys.*, 15:1-41, 2009.

[16] Drinfel'd V. and Sokolov V. Lie Algebras and Equations of Korteweg-de Vries Type. *Sov. J. Math.*, 30:1975-2036, 1985.

[17] Zakharov V. and Gelash A. Soliton on Unstable Condensate. e-print arXiv:1109.0620, available at http://arxiv.org/pdf/1109.0620.pdf.

[18] Zakharov V. and Mikhailov A. On the Integrability of Classical Spinor Models in Two-dimensional Space-time. *Commun. Math. Phys.*, 74:21-40, 1980.

Bicharacteristic Curves in 3D Modeling of the Lithosphere

Georgi Boyadzhiev

Introduction

The main sources of seismic hazard are earthquakes generated in the Lithosphere - the crust and upper mantle of the Earth. Traditional analytical methods such as modal summation have been developed for modeling the Lithosphere as a structure of horizontal homogeneous layers. It is a 1D model that can be extended to a 2D one (see [5]). Further extension to 3D modeling by classical methods is impossible due to some natural restrictions of the models.

In this paper a new approach is given to 3D modeling of elastic piecewise homogeneous media, such as the Lithosphere. The method is based on classical tomography and the main sources of information are seismic waves, generated by a point source S and recorded by a set of seismic stations on the Earth's surface. The irregularity of earthquakes is counterpoised by the density of seismic stations, and plenty of data is available for geophysical surveys.

Let us suppose that the Earth is an elastic body that is a continuum, i.e. the matter is continuously distributed in space. Furthermore, since seismicity has a relatively local effect, from the geophysical point of view the planet can be represented with no loss of generality by the half-space $\Omega = \{(x, y, z) \in R^3 : z \geq 0\}$ with free surface boundary $\{z = 0\}$ and axis z positive downward. If the elastic parameters depend only on the vertical coordinate z then the wave propagating in solid media satisfies the following strongly coupled linear hyperbolic system (see [1] and [5])

$$\rho\frac{\partial^2 u_x}{\partial t^2} = X + (\lambda + 2\mu)\frac{\partial^2 u_x}{\partial x^2} + \mu\frac{\partial^2 u_x}{\partial y^2} + \mu\frac{\partial^2 u_x}{\partial z^2} + +(\lambda + \mu)\frac{\partial^2 u_y}{\partial x\,\partial y} + (\lambda + \mu)\frac{\partial^2 u_z}{\partial x\,\partial z} + \frac{\partial \mu}{\partial z}\frac{\partial u_x}{\partial z} + \frac{\partial \mu}{\partial z}\frac{\partial u_z}{\partial x}$$

$$\rho\frac{\partial^2 u_y}{\partial t^2} = Y + \mu\frac{\partial^2 u_y}{\partial x^2} + (\lambda + 2\mu)\frac{\partial^2 u_y}{\partial y^2} + \mu\frac{\partial^2 u_y}{\partial z^2} + +(\lambda + \mu)\frac{\partial^2 u_x}{\partial x\,\partial y} + (\lambda + \mu)\frac{\partial^2 u_z}{\partial y\,\partial z} + \frac{\partial \mu}{\partial z}\frac{\partial u_y}{\partial z} + \frac{\partial \mu}{\partial z}\frac{\partial u_z}{\partial y}$$

$$\rho\frac{\partial^2 u_z}{\partial t^2} = Z + \mu\frac{\partial^2 u_z}{\partial x^2} + \mu\frac{\partial^2 u_z}{\partial y^2} + (\lambda + 2\mu)\frac{\partial^2 u_z}{\partial z^2} + +(\lambda + \mu)\frac{\partial^2 u_x}{\partial x\,\partial z} + (\lambda + \mu)\frac{\partial^2 u_y}{\partial y\,\partial z} + \frac{\partial \lambda}{\partial z}\left(\frac{\partial u_x}{\partial x} + \frac{\partial u_y}{\partial y} + \frac{\partial u_z}{\partial z}\right) + 2\frac{\partial \mu}{\partial z}\frac{\partial u_z}{\partial z} \tag{1}$$

where $\lambda, \mu\ and\ \rho$ are piecewise continuous functions of $z\ and\ u_x, u_y, u_z, \sigma_{zz}, \sigma_{zx}\ and\ \sigma_{zy} \in C(\Omega)$. Function $u = (u_x, u_y, u_z)$ is called in physics the "displacement function". The boundary conditions of system (1) at the free surface $z = 0$ are as follows

$$\sigma_{zz} = (\lambda + 2\mu)\frac{\partial u_z}{\partial z} + \lambda\left(\frac{\partial u_x}{\partial x} + \frac{\partial u_y}{\partial y}\right) = 0$$

$$\sigma_{zx} = \mu\left(\frac{\partial u_x}{\partial z} + \frac{\partial u_z}{\partial x}\right) = 0$$

$$\sigma_{zy} = \mu\left(\frac{\partial u_y}{\partial z} + \frac{\partial u_z}{\partial y}\right) = 0. \tag{2}$$

Initial data are given by

$$u(S)|_{t=0} = \delta, \frac{du}{dt}(S)|_{t=0} = c.\zeta_0 \tag{3}$$

i.e. at the point source S ∈ Ω there is an impulse alongside the given vector $\zeta^0 = (\zeta_1^0, \zeta_2^0, \zeta_3^0)$.

Coefficients ρ, λ and μ depend on the geological properties of the rock. One reasonable approximation of the Lithosphere is a 3Dimensional structure of homogeneous blocks in welded contact $B_{i,j,k}$, where i and j are integers, and k is a natural number. Blocks $B_{i,j,k}$ may not be rectangular ones or their sides may not be parallel to the coordinate axis. Without loss of generality we assume the boundary $\partial B_{i,j,k}$ to be a piecewise smooth surface and that the source S of the seismic signal is in block $B_{0,0,0}$.

This way in $\Omega = \{B_{i,j,k}\}$ the system (1) with constant coefficients in every block $B_{i,j,k}$ is a realistic approximation to the propagation of waves in the Lithosphere.

Solving system (1), (2), (3) numerically is limited by such natural constraints as the size of the domain Ω. If it is not relatively small, as is generally the case, the grid is too large and computational time is too costly, or the approximation error is too high. On the other hand the fundamental solution of (1) can be explicitly written in integral form since so-called Rayleigh and Love modes give a good approximation to the solution when the distance from the source is large enough compared to the wavelength (see [1]). Numerical computing of the integral faces the same problems as pure numerical methods in solving (1), (2), (3) directly, namely the cost of computations and the error ratio. In another standard analytical approach widely used in geophysics, if the body forces are neglected the solutions of (1) are considered as plane harmonic waves propagating along the positive x axis

$$u(x,t) = F(z).e^{i(\omega t - kx)}$$

where ω is the angular frequency and k is the wave number corresponding to the phase velocity c, i.e. $k=\omega/c$. (see for instance [5]). The main disadvantage of this approach is that the plane wave is a two - dimensional one, living in the plane $y=0$ only, and all information on the y coordinate is lost. Therefore it is impossible to build a reasonable 3D model using a plane wave of such type, which is the reason a new approach to 3D modeling is suggested in this paper. Since an earthquake generates a singularity at point S, the method suggested is built on propagation of the singularities of system (1) itself.

There are alternative points of view on wave propagating in multi - layered solid media. For instance, [7] is a study of evolution systems for paraxial equations with non-smooth equations, applied in reflection seismic imaging. Solutions for the Cauchy problem of a system with low regularity of coefficients are given in integral form. In our paper we deal with the problem in a completely different way. We adopt so-called "train" solutions, i.e. the solution in one block determines the boundary conditions of neighboring blocks.

System (1) in Ω has step - wise coefficients and the classical results for the wave front set (Theorem 8.3.1, Hormander, v. I, p.271) are not applicable. Hence we use a "train solutions" construction in our model. The initial data at the point source S determine the solution of system (1) in block $B_{0,0,0}$, which induces the boundary conditions in the neighbouring block and so on. This way, instead of a system (1) with piecewise constant coefficients, we consider a series of related problems (1) with constant coefficients, which is a much easier task.

This method is based on the features of the bicharacteristic curves of system (1). As the principal part is real with constant coefficients, the wave front set is invariant under the bicharacteristic flow. Having in mind the source model described above, a point source with a seismic impulse in some direction, actually the singularities of the solution carry all the information about the wave. On the other hand, the singularities propagate over bicharacteristic curves within every homogeneous block. At the boundary between two blocks the bicharacteristics could be reflected or refracted. According to geometrical optics and microlocal analysis, if a bicharacteristic curve reflects off the sides of every block the angle of incidence to the surface is equal to the angle of reflection. As for refraction at the surface, it is computed in the usual way; more details and exact computations are given in Chapter 1 below. Therefore, if we know the position of the source S, the direction ζ_0 of the seismic impulse and the media structure $\Omega=\{B_{i,j,k}\}$ we can compute the intersecting point s_0 of the bicharacteristic curve and the surface $z=0$. The point s_0 is in fact the centre of the surface waves in the plane $z=0$ generated by the section of the wave front and the plane $z=0$. When actual measurement of the seismic waves is done, the coordinates of the point s_0 can be triangulated using the data from several stations. This way verification of the media model $\Omega=\{B_{i,j,k}\}$ is done. Exact coordinates of the epicenter of an earthquake and the center of the surface waves r_0 are computed using different and quite reliable techniques, like time - frequency analysis, based on the data from seismic stations. Given a certain 3D media model $\Omega=\{B_{i,j,k}\}$, we compute the point s_0. If the points r_0 and s_0 coincide within the error of the computations, then the media model is plausible.

For practical purposes 3D models $\Omega=\{B_{i,j,k}\}$ are generated using Monte Carlo type methods. Of course, like any other inverse problem, this algorithm has multiple solutions in the sense that many models can fulfil the requirement $s_0 = r_0$.

Characteristic set and bicharacteristic strip in homogeneous block $B_{i,j,k}$

Let $L_{k,m}(x,D) = \sum_{|\alpha|\le 2} a_{\alpha}^{k,m}(x)D^{\alpha}$. Then the characteristic set of a linear strongly coupled system

$$\sum_{m=1}^{n} L_{k,m}(x,D)u_m = f_i, k = 1,\dots.n$$

is given by $det|p_{k,m}(x,\xi)| = 0$ where $p_{k,m}(x,\xi) = \sum_{|\alpha|=2} a_{\alpha}^{k,m}(x)\xi_m^{\alpha}$(see [6], p.40).

Each element of the characteristic matrix is the principal symbol of the corresponding equation with respect to the corresponding argument. The characteristic set of system (1) in every the block $B_{i,j,k}$ is given by the equation

$$0 = p(x,\xi) =$$

$$\begin{vmatrix} \alpha+{\xi_1}^2 & \xi_1\xi_2 & \xi_1\xi_3 \\ \xi_1\xi_2 & \alpha+{\xi_2}^2 & \xi_2\xi_3 \\ \xi_1\xi_3 & \xi_2\xi_3 & \alpha+{\xi_3}^2 \end{vmatrix}$$

$$= \alpha \begin{vmatrix} \alpha+{\xi_2}^2 & \xi_2\xi_3 \\ \xi_2\xi_3 & \alpha+{\xi_3}^2 \end{vmatrix}$$

$$+ \begin{vmatrix} {\xi_1}^2 & \xi_1\xi_2 & \xi_1\xi_3 \\ \xi_1\xi_2 & \alpha+{\xi_2}^2 & \xi_2\xi_3 \\ \xi_1\xi_3 & \xi_2\xi_3 & \alpha+{\xi_3}^2 \end{vmatrix} =$$

$$= \alpha^2\left(\alpha+{\xi_1}^2+{\xi_2}^2+{\xi_2}^2\right),$$

where $\alpha = -(\lambda+\mu)^{-1}\left[\rho\tau^2 - \mu\left({\xi_1}^2 + {\xi_2}^2 + {\xi_3}^2\right)\right]$.

If we denote by $\alpha = -(\lambda+\mu)^{-1}\left[\rho\tau^2 - \mu\left({\xi_1}^2 + {\xi_2}^2 + {\xi_3}^2\right)\right]$ we have

$$0 = \begin{vmatrix} \alpha + {\xi_1}^2 & \xi_1\xi_2 & \xi_1\xi_3 \\ \xi_1\xi_2 & \alpha + {\xi_2}^2 & \xi_2\xi_3 \\ \xi_1\xi_3 & \xi_2\xi_3 & \alpha + {\xi_3}^2 \end{vmatrix} = \alpha \begin{vmatrix} \alpha + {\xi_2}^2 & \xi_2\xi_3 \\ \xi_2\xi_3 & \alpha + {\xi_3}^2 \end{vmatrix} +$$

$$\begin{vmatrix} {\xi_1}^2 & \xi_1\xi_2 & \xi_1\xi_3 \\ \xi_1\xi_2 & \alpha + {\xi_2}^2 & \xi_2\xi_3 \\ \xi_1\xi_3 & \xi_2\xi_3 & \alpha + {\xi_3}^2 \end{vmatrix} =$$

$$= \alpha^2\left(\alpha + {\xi_1}^2 + {\xi_2}^2 + {\xi_2}^2\right).$$

These simple calculations show that the characteristic set of system (1) consists of two subsets

$$p_1(x,\xi) = \rho\tau^2 - \mu\left({\xi_1}^2 + {\xi_2}^2 + {\xi_3}^2\right) = 0$$
$$p_2(x,\xi) = \rho\tau^2 - (\lambda + 2\mu)\left({\xi_1}^2 + {\xi_2}^2 + {\xi_3}^2\right) = 0 \tag{4}$$

since $(\lambda + \mu) > 0$.

Therefore the wave propagating in homogeneous block $B_{i,j,k}$ is actually a composition of two waves. This result corresponds to the theory of P (primary) and S (secondary) body waves. The P-wave corresponds to the set defined by $p_2(x,\xi) = 0$, and the S-wave to the one defined by $p_1(x,\xi) = 0$.

Hence the following theorem holds:

Theorem: A body wave propagating in homogeneous block $B_{i,j,k}$ is the composition of two waves – the P-wave and the S-wave. There are no other components of the wave.

The characteristic set of an operator contains the wave front of the solution u (see [2], vol. I, Theorem 8.3.1, p.271). Roughly speaking, the wave front of u is a conic set where u is not smooth (see [2], Def. 8.1.2 p.254). In terms of physics the wave front describes the position of the wave at the moment.

Furthermore, the characteristic set of an operator with real principal part $p(x,\xi)$ and constant coefficients is invariant under the bicharacteristic flow (see [2], vol. I, Chapter 8). The restriction of the bicharacteristic strip into R^4 is named bicharacteristic curve. It is applicable to 3D modeling of the Earth, for, generally speaking, the singularities propagate over bicharacteristic curves. In other words, the singularity that is generated by an earthquake in block $B_{0,0,0}$ propagates over the bicharacteristic curve in $B_{0,0,0}$ until it intersects at point (x_1, y_1, z_1) the boundary to the neighbouring block, $B_{1,0,0}$ for instance. Continuous boundary conditions mean that at point (x_1, y_1, z_1) system (1) in the block $B_{1,0,0}$ has a singularity, which propagates over the bicharacteristic curves in $B_{1,0,0}$, etc.

By definition if $p(x^0,\xi^0) = 0$ then the bicharacteristic strip at point (x^0,ξ^0) is defined by the Hamilton equations

$$\frac{dx}{ds} = \frac{\partial p(x,\xi)}{\partial \xi}, \frac{d\xi}{ds} = \frac{\partial p(x,\xi)}{\partial x}$$

with initial data $(x,\xi) = (x_0, \xi_0)$ for $t = 0$. The corresponding bicharacteristic curve is

$$\begin{aligned} x_1 &= c^{-1}\xi_1^0.(t - t^0) + x_1^0 \\ x_2 &= c^{-1}\xi_2^0.(t - t^0) + x_2^0 \qquad (5) \\ x_3 &= c^{-1}\xi_3^0.(t - t^0) + x_3^0 \end{aligned}$$

since $t - t^0 = 2c\sqrt{(\xi_1^0)^2 + (\xi_2^0)^2 + (\xi_3^0)^2}.s = 2c|\xi^0|$ and without loss of generality we may assume $|\xi^0| = 1$. Constant $c = \sqrt{\mu/\rho}$ for bicharacteristics generated by $p = p_1(x,\xi)$ and $c = \sqrt{(\lambda + 2\mu)/\rho}$ for ones generated by $p = p_2(x,\xi)$.

The values of ξ_1^0, ξ_2^0 and ξ_3^0 are determined by the features of the seismic source. Without loss of generalization we can assume the source of the seismic wave to be a point one with direction of the impulse $\xi_1^0, \xi_2^0, \xi_3^0$.

Reflection and refraction

Equation (5) describes the bicharacteristic curves of (1) in each $B_{i,j,k}$, and their behavior on the boundary $\partial B_{i,j,k}$ is studied by geometrical optics and microlocal analysis.

Let b^{in} be a bicharacteristic curve in $B_{i,j,k}$ and $b^{in} \cup \{F_{i,j,k,l}(x, y, z) = 0\} = p_0$. At point p_0 b^{in} can be reflected or refracted. Let b_{rr} be the refracted curve and b_{rl} be the reflected one. Both b_{rr} and b_{rl} are bicharacteristics through point p_0: b_{rr} is in the block next to $B_{i,j,k}$ (in the sense of propagation of the singularity generated in S) and b_{rl} is in $B_{i,j,k}$. The singularity at p_0 propagates over the bicharacteristics as well, and in this way the well-known following formulae for reflection and refraction from geometrical optics are obtained.

If the incidental bicharacteristic curve b^{in} is reflected, the angle θ_{in} of incidence to the surface $F_{i,j,k,l}(x, y, z) = 0$ is equal to the angle of reflection θ_{rl}, since in the same block the equation (5) has the same coefficients. As for refraction at a surface, the match of the boundary conditions of the neighboring blocks at the two sides of the boundary leads to the well - known formula from geometric optics $v_1 \sin\theta_{rr} = v_2 \sin\theta_{in}$, where θ_{rr} is the angle of refraction, v_1 is the speed of the wave in the "incidence" block and v_2 is the one in "refraction" block.

Computation of reflected and refracted bicharacteristic curves is simple. Let $\vec{n} = (n_1, n_2, n_3) = \left[\left(\frac{\partial F}{\partial x}\right)^2 + \left(\frac{\partial F}{\partial y}\right)^2 + \left(\frac{\partial F}{\partial z}\right)^2\right]^{-1/2} \left(\frac{\partial F}{\partial x}, \frac{\partial F}{\partial y}, \frac{\partial F}{\partial z}\right)(p_b)$ be the normal unit vector to surface $F_{i,j,k,l} = 0$ at the point of incidence p_b, $\xi^{in} = (\xi_1^{in}, \xi_2^{in}, \xi_3^{in})$ be the unit vector along the incidental bicharacteristic curve, $\xi^{rr} = (\xi_1^{rr}, \xi_2^{rr}, \xi_3^{rr})$ be the unit vector along the refracted one, and $\xi^{rl} = (\xi_1^{rl}, \xi_2^{rl}, \xi_3^{rl})$ be the unit vector along the reflected one.

Wave speed is a physical feature of every material and is known in advance. For instance, the velocity of the P-wave in a homogeneous isotropic medium is $v_P = \sqrt{(\lambda + 2\mu)/\rho}$; for the S-wave it is $v_S = \sqrt{\mu/\rho}$.

Quantities $\sin\theta_{in} = \sin\theta_{rl}$ and $\sin\theta_{rr}$ are easy to compute using scalar, or dot product $\cos\theta = \xi \cdot \vec{n}$ of unit vectors ξ and the normal unit vector $\vec{n}$, for instance

$$\sin^2\theta_{in} = 1 - \left(\xi_1^{in} n_1 + \xi_2^{in} n_2 + \xi_3^{in} n_3\right)^2.$$

Then equations of refraction and reflection from geometrical optics yield

$$\xi_1^{rr} n_1 + \xi_2^{rr} n_2 + \xi_3^{rr} n_3 = \left[1 - \left(\frac{v_2}{v_1}\right)^2 \left(1 - [\xi_1^{in} n_1 + \xi_2^{in} n_2 + \xi_3^{in} n_3]^2\right)\right]^{1/2} \xi_1^{rl} n_1 + \xi_2^{rl} n_2 + \xi_3^{rl} n_3 = \left[1 - \left(1 - [\xi_1^{in} n_1 + \xi_2^{in} n_2 + \xi_3^{in} n_3]^2\right)\right]^{1/2}. \quad (6)$$

In addition, the incidental bicharacteristic curve, the refracted one and the normal-to-the-surface vector lie on the same plane and give us the relation

$$\begin{vmatrix} n_1 & n_2 & n_3 \\ \xi_1^{in} & \xi_2^{in} & \xi_3^{in} \\ \xi_1^{rr} & \xi_2^{rr} & \xi_3^{rr} \end{vmatrix} = 0. \quad (7)$$

The same relation is valid for vector ξ^{rl}. Finally, since we consider vectors ξ^{in}, ξ^{rr} and ξ^{rl} to be unit ones, we obtain

$$(\xi_1^{rr})^2 + (\xi_2^{rr})^2 + (\xi_3^{rr})^2 = 1 (\xi_1^{rl})^2 + (\xi_2^{rl})^2 + (\xi_3^{rl})^2 = 1. \quad (8)$$

Equations (6), (7) and (8) define uniquely the vectors of refraction ξ^{rr} and reflection ξ^{rl} up to the sign.

3D modeling of the Lithosphere

Using bicharacterstic curves, described in the previous section, it is possible to define the following criterion for a 3D model of the Earth's crust and upper mantle.

Definition: Let $\{B_{i,j,k}\}$ be a set of blocks and the source of a seismic wave be a point one, at the point S with direction along vector ξ^0. Let P be the point of the Earth's surface belonging to the bicharacteristic curves generated by system (1), the set of blocks $\{B_{i,j,k}\}$ and the source S. For a given set of blocks $B_{i,j,k}$ it is plausible that the point P coincides with the epicentre E of the surface waves generated by the earthquake.

Since seismic stations record both surface and body waves, point E is a subject of triangulation if there are enough sensors in the region. Computing the bi-characteristic curves in the whole set $\{B_{i,j,k}\}$ gives rise to an important question. At the boundary between two blocks – with surfaces $F_{i,j,k,l}(x,y,z) = 0$ - is the bicharacteristic curve reflected, refracted, or both? The answer comes from the so-called reflection and refraction index. It is a physical feature of the material that builds the block.

How to compute the refraction and reflection index is well described in Aki and Richards (2002), W.M. Ewing, W. S. Jardetzky, F. Press. (1957) or in W. L. Plant (1979).

Furthermore, the records of body waves are useful in determining the block structure of the blocks closest to the seismic stations. The wave front in a homogeneous block is a subset of the characteristic set of system (1), therefore it has constant speed by (4).

Using bi-characteristic curves and the characteristic set we can compute arrival time for P- and S- waves. In combination with the criteria from the Definition, we can generate and test plausible 3D models of the Earth's crust and upper mantle.

Bibliography

[1] Keiiti Aki, Paul G. Richards. Quantitative Seismology, Second edition, University Science Books (2002).

[2] Lars Hormander. The Analysis of Linear Partial Differential Operators I, Springer - Verlag (1983).

[3] W.Maurice Ewing, Wenceslas S. Jardetzky, Frank Press. Elastic Waves in Layered Media. McGraw-Hill series in the Geological Sciences (1957).

[4] Walter L Plant. Elastic waves in the Earth, Developments in Solid Earth Geophysics 11, Elsevier (1979).

[5] Giuliano F. Panza, Fabio Romanelli, Franco Vaccari. Seismic wave propagation in laterally heterogeneous anelastic media: theory and applications to seismic zonation, Advances in Geophysics, vol.43, pp 1- 95 (2001).

[6] G.N.Polozhii, Equations of the mathematical physics (in Russian), "High School" publishing House, Moscow (1964).

[7] Maarten de Hoop, Gunter Hormann, Michael Oberguggenberger. Evolution systems for paraxial wave equations of Schrodinger - type with non-smooth coefficients. J. Differential Equations 245 (2008), 1413-1432.

CHAPTER TWO

HIGH PERFORMANCE SCIENTIFIC COMPUTING

TUNING FOR SCALABILITY ON HYBRID HPC CLUSTER

E. ATANASSOV, T. GUROV, A. KARAIVANOVA, S. IVANOVSKA, M. DURCHOVA, D. GEORGIEV AND D. DIMITROV

Introduction

New developments in hardware for high-performance computing motivate significant changes in software technologies and a convergence between the services that are used for HPC and distributed computing. The European strategy for development of e-infrastructure for research purposes includes as one of its main building blocks the Grid infrastructure (see, e.g., [5, 6]), built by linking the national Grid initiatives in the European countries. Notably, the documents [10, 11] identify the following key elements of the European vision:

- the GEANT research network;
- Grids for e-science;
- storing and post-processing of scientific data;
- supercomputer e-infrastructures;
- global virtual research communities.

Following these tendencies, the Institute of Information and Communication Technologies of the Bulgarian Academy of Sciences (IICT-BAS) in Sofia has built a high-performance grid computing cluster (see Figure 1), which also includes data storage in a high-performance disk array. The cluster serves as the center of the Bulgarian Grid Infrastructure and ensures the participation of Bulgarian research groups in European virtual research communities. One of the main considerations for the procurement was that we believed that Bulgarian researchers would be more interested in tightly coupled parallel applications. That is why we put emphasis on fast interconnection, which is achieved with non-blocking InfiniBand fabrics. In this way our cluster is one of the few clusters in the European Grid

Infrastructure with such advanced capabilities. It is also the most powerful interdisciplinary Bulgarian Grid cluster in terms of raw computational power, and supports all core services necessary to support Bulgarian researchers - senior scientists or Ph.D. students.

Figure 1: HPC Grid Cluster in IICT-BAS

The main computational part of the cluster is the set of blade server nodes, which are deployed inside three HP Cluster Platform Express 7000 enclosures. There are 36 identical blades of the type HP BL 280c, equipped with dual Intel Xeon X5560 @ 2.8GHz and 24 GB RAM per blade. Interfacing with the storage are 8 server nodes HP DL 380 G6 with dual Intel X5560 @ 2.8Ghz and 32 GB RAM. These nodes perform various controlling functions. They are connected with the storage systems by redundant Fibre Channel connections. Currently the cluster has three SAN systems: MSA2312fc, P2000 G3 and the newest one, IBM Storwize V7000, offering access to a total of 132 TB of disk storage. Following the tendencies for achieving highly cost-efficient and energy-efficient computing, we have gradually added new types of computational capabilities. Namely, we have added two HP ProLiant SL390s G7 servers with Intel(R) Xeon(R) CPU E5649 @ 2.53GHz, which can be equipped with up to 8 NVIDIA Tesla M2090 cards each. One such card has 512 cores for GPU computing and achieves 1331 GigaFLOPS in single precision and 665 GigaFLOPS in double precision. The cards have 6 GB ECC GDDR5 RAM with 177 GBytes/sec memory bandwidth. Currently we have 9 cards in these two servers, providing computational power close to 10 TeraFLOPS in single precision. The procurement procedure for acquiring an additional 7 cards is under way. We note that these servers are also interconnected with non-blocking InfiniBand interconnection, thus keeping all servers under the same connectivity parameters and enabling the development of hybrid applications that combine CPU and GPU computing in an optimal way [1].

Configuration, testing, certification and inclusion into the European Grid Infrastructure

System Configuration

The configuration of the system was carried out in two stages. In the first stage, the physical installation was carried out by the supplier. The disk space of the storage systems was distributed, using mostly RAID 6 with only the most performance-critical data stored using RAID 10. In order to facilitate the automated installation of the operating system software we used the **Perceus** configuration management system [12]. Although it is mostly oriented towards "stateless" installations, we used the so-called "stateful" option, which needed some tweaking to support the RAID controllers at the blades. Most of the disk space from the storage systems is used as a ``/home" file system, while a substantial part is also allocated for scratch space. Users of the cluster have direct access to these two file systems. They were configured using the Lustre file system [13], which is open-source software and a popular choice among high-performance systems, providing in some cases petabytes of disk space for large supercomputer installations. In our case three of the controlling servers are dedicated to serving the Lustre file systems, so that one of them acts as a metadata server and the other two provide access to the actual disk space in which to spread the workload. The Lustre file systems span across the three storage systems, thus achieving higher total throughput.

In the second stage of the installation we configured the Grid software, using the **YAIM** method. The remaining 5 controlling nodes were used to install virtual machines, where the servers supporting the Grid functions are installed. We used **KVM** for the virtualization and we enabled live migration over the InfiniBand interconnection. In this way it was possible to achieve fast live migration, without dropping any network packets during the transition.

Although all the grid-related services were installed on virtual machines, the blade servers were installed as grid worker nodes with direct, bare-metal installation for maximum performance.

The configurations of the so-called "worker nodes" follow the release cycle of the EMI project, starting with the gLite distribution version 3.2 and then upgrading to the consecutive EMI releases. One of the blade nodes is designated as a Grid User Interface, enabling direct job submission to the cluster, submission of jobs to all grid sites, and data management. Since the cluster has 36 blade nodes, each with two 4-core processors, and since hyper-threading is enabled, the optimal number of parallel processes to be launched is 576. Although the controlling servers have the same computational capabilities and the same non-blocking interconnection as the blade servers, they are not normally used for computations by users. Control of the user tasks is performed by the TORQUE Resource Manager [14]. A virtual machine is set up to act as a TORQUE server and accepts tasks that are submitted from the gateway node or from the grid controlling nodes. Access through the grid is performed by the servers:

- cr1.ipp.acad.bg - Computing Element of type CREAM
- cr2.ipp.acad.bg - experimental Computing Element of type CREAM, providing access to the GPU computing nodes only.

Figure 2: Nagios monitoring system

Service Status Details For Host 'cr1.ipp.acad.bg'

Host	Service	Status	Last Check	Duration	Attempt	Status Information
cr1.ipp.acad.bg	hr.srce.CREAMCE-CertLifetime	OK	15-11-2010 14:00:15	6d 5h 33m 37s	1/2	CERT LIFETIME OK - Certificate will expire in 186.99 days (May 21 11:50:29 2011 GMT)
	org.sam.CREAMCE-JobState-/ops/Role=lcgadmin	OK	15-11-2010 15:27:06	2d 23h 17m 54s	1/2	OK: success.
	org.sam.CREAMCE-JobSubmit-/ops/Role=lcgadmin	OK	15-11-2010 15:27:06	2d 23h 2m 10s	1/2	OK: success.
	org.sam.WN-Bi-/ops/Role=lcgadmin	OK	15-11-2010 15:17:11	6d 5h 25m 28s	1/2	wn03.hpcg: OK: getCE: cr1.ipp.acad.bg:8443/cream-pbs-ops
	org.sam.WN-CAver-/ops/Role=lcgadmin	OK	15-11-2010 15:17:28	6d 5h 25m 24s	1/2	wn03.hpcg: OK
	org.sam.WN-Csh-/ops/Role=lcgadmin	OK	15-11-2010 15:17:14	6d 5h 25m 26s	1/2	wn03.hpcg: OK
	org.sam.WN-MPI-/ops/Role=lcgadmin	OK	15-11-2010 14:37:05	2d 20h 56m 44s	1/2	wn03.hpcg: OK: MPI Status: OK
	org.sam.WN-Rep-/ops/Role=lcgadmin	OK	15-11-2010 15:18:16	6d 5h 24m 27s	1/2	wn03.hpcg: OK: success
	org.sam.WN-RepCr-/ops/Role=lcgadmin	OK	15-11-2010 15:17:39	6d 5h 25m 4s	1/2	wn03.hpcg: OK: File was copied to SE se001.ipp.acad.bg and registered in LFC prod-lfc-shared-central.cern.ch. CLI
	org.sam.WN-RepDel-/ops/Role=lcgadmin	OK	15-11-2010 15:18:16	6d 5h 24m 27s	1/2	wn03.hpcg: OK: Replicas for [lfn:/grid/ops/SAM/sam-log-rm-cr-wn03.hpcg.101115131722.30146851] were deleted. CLI
	org.sam.WN-RepFree-/ops/Role=lcgadmin	OK	15-11-2010 15:17:10	6d 5h 25m 25s	1/2	wn03.hpcg: OK: ok
	org.sam.WN-RepGet-/ops/Role=lcgadmin	OK	15-11-2010 15:17:48	6d 5h 24m 56s	1/2	wn03.hpcg: OK: File was copied from SRM. Diff successful. CLI
	org.sam.WN-RepISenv-/ops/Role=lcgadmin	OK	15-11-2010 15:17:10	6d 5h 25m 26s	1/2	wn03.hpcg: OK: LCG_GFAL_INFOSYS is set to bdii.ipp.acad.bg:2170
	org.sam.WN-RepRep-/ops/Role=lcgadmin	OK	15-11-2010 15:18:08	6d 5h 24m 33s	1/2	wn03.hpcg: OK: File was replicated to SE se01.afroditi.hellasgrid.gr. Replicas listed successfully. CLI
	org.sam.WN-SoftVer-/ops/Role=lcgadmin	OK	15-11-2010 15:17:15	6d 5h 25m 24s	1/2	wn03.hpcg: OK: 3.2.0
	org.sam.mpi.CE-JobState-/ops/Role=lcgadmin	OK	15-11-2010 15:14:43	2d 20h 57m 17s	1/2	OK: success.
	org.sam.mpi.CE-JobSubmit-/ops/Role=lcgadmin	OK	15-11-2010 15:14:43	2d 20h 45m 17s	1/2	OK: success.

17 Matching Service Entries Displayed

Figure 3: Service status details

All installed Grid services are monitored by the Nagios monitoring system (Figure 2) which makes it possible to identify and resolve infrastructure and grid middleware problems before they affect critical processes (Figure 3). The HPC Grid cluster is involved in the EGI dashboard interface for a ticket system as in BG01-IPP, (Figure 4).

Currently, the cluster BG01-IPP supports European Virtual Organizations (VOs) such as Biomed, CMS, and some regional and national VOs such as env.see-grid-sci.eu, mm-comp-chem.grid.acad.bg and bg-edu.grid.acad.bg. The main grid services for the last VOs are running off virtual machines from the controlling nodes. It is notable that env.see-grid-sci.eu and mm-comp-chem.grid.acad.bg VOs have high percentages of parallel jobs.

Figure 4: EGI dashboard interface for ticket system

Testing

To certify the cluster for full-scale operations a series of tests were performed.

High Performance Linpack

The productivity of supercomputer systems is usually measured in billions of floating point operations in double precision per second and the results of the first 500 systems are published regularly on the site http://www.top500.org. It should be noted that this output is required to be achieved in solving a system of linear equations, whereas the choice of libraries, problem sizes and software configuration can be varied. In our tests we used the package High-Performance Linpack (http://www.netlib.org/benchmark/hpl/), which is based on MPI. Since the peak performance of our Intel CPUs cannot be increased by turning hyper-threading on, we used only the available real cores, i.e. 288. In order to achieve the best performing configuration of the HPL software, the following website http://hpl-calculator.sourceforge.net/ provided useful information with regards to the optimal choice of parameters for the file HPL.dat. The peak performance of the cluster is equal to 3.225 billion operations per second, according to the specification. In the tests we achieved more than 3 billion operations per second, i.e. the ratio between attained performance and peak performance is more than 93%. This is an excellent result that highlights the capabilities of the system for parallel computing. Some of the tests were carried out in the Grid environment, proving that access to the high-performance capabilities of the cluster is also possible using Grid middleware.

It should be emphasized that in these tests the processor is almost fully loaded and there is no practical gain from the presence of hyper-threading, which allows for launching twice as many parallel processes. In real-time problems hyper-threading technology can offer significant advantages, since it allows an "overlap" of the waiting time when accessing memory or logic transitions. Some of our Monte Carlo applications achieve between 30 and 60 percent speedup from hyper-threading. Since some applications cannot benefit from hyper-threading, users can choose to use only the physical processors, but they still have to reserve all the available (logical) processors in order to obtain full control of the machine.

MPI tests with InfiniBand

One of the most popular standardized tests for high-performance MPI communications are the tests **osu_latency** and **osu_bw**, which measure the latency and bandwidth of the connection between two servers of the cluster. Significant differences were noted between the basic distribution of OFED and the later versions that we installed - around 10% in some benchmarks. The results in Table 1 were obtained when we used the following test:

```
/usr/mpi/gcc/openmpi-1.3.3/bin/mpirun -H wn02.hpcg,wn03 -n 2
/usr/mpi/gcc/openmpi-1.3.3/tests/osu_benchmarks-3.1.1/osu_latency
```

Table 1: OSU MPI Latency Test v3.1.1

Size	Latency (us)	Size	Latency (us)
0	2.45	2048	8.75
1	2.54	4096	10.73
2	2.46	8192	14.74
4	2.17	16384	20.84
8	2.14	32768	33.15
16	2.18	65536	53.76
32	2.24	131072	100.41
64	2.44	262144	178.19
128	4.10	524288	333.10
256	4.45	1048576	625.75
512	5.01	2097152	1249.96
1024	6.22	4194304	2462.30

One can see that the latency of short communications is at times less than 2.5μ. With regard to the bandwidth, (see Table 2) we used the following test:

/usr/mpi/gcc/openmpi-1.3.3/bin/mpirun -H wn02.hpcg,wn03 -n 2 /usr/mpi/gcc/openmpi-1.3.3/tests/osu_benchmarks-3.1.1/osu_bw

Table 2: OSU MPI Bandwidth Test v3.1.1

Size	Bandwidth (MB/s)	Size	Bandwidth MB/s)
0	-	2048	936.44
1	1.13	4096	1211.40
2	2.32	8192	1528.57
4	4.68	16384	1471.90
8	8.82	32768	1572.26
16	18.51	65536	1680.17
32	35.68	131072	1727.01
64	67.71	262144	1727.80
128	124.40	524288	1731.80
256	228.00	1048576	1733.65
512	427.28	2097152	1734.14
1024	698.39	4194304	1734.69

In these tests one can see that when the size of the message grows to more than 131072 bytes the maximum throughput of around 1734 megabytes per second is achieved, which is close to the theoretical maximum. For comparison we conducted the same tests on the same server from the cluster. Thus we found that the difference between the results is relatively small and that access to the memory of another server is comparable in speed to memory access from the same server. We note that the use of adjacent or distant blade servers does not significantly affect these results, due to the non-blocking interconnection.

Tests of the file system

One of the most common tests for file system performance is **bonnie++**, which is available as a standard OS package. The cluster file system has two types of Lustre file systems, one of which is used to securely store data - /home, and the other is used for temporary files and broadband access. In tests with **bonnie++** the same order of magnitude of results was achieved for both, for example reading speeds of around 436 MB/s.

Following the certification process the cluster was included in the European Grid Infrastructure, managed by the project EGI-InSPIRE, [15]. In addition to that, the cluster was included in the infrastructure of the project HP-SEE (see [16]) to create a network of high-performance clusters and supercomputers in the region.

Continuous operation of the cluster: monitoring, problem solving and statistical data collection

As part of the regional and European Grid Infrastructure the cluster is monitored continuously and passes certification tests every hour. This requires the project team to ensure regular monitoring to eliminate possible problems and to update software. In order to enable detailed accounting of the resources used, the team of the HP-SEE project developed and deployed a system which gathers and analyses more data than the regular Grid accounting software [3].

Figure 5 shows how the Bulgarian HPCG cluster in IICT-BAS supports virtual research communities related to computational physics, computational chemistry and life sciences in the SEE region [4], and how their accounting data are distributed on different computing centers in the region, including our center named HPCG, http://gserv4.ipp.acad.bg/HPSEEAccounting/.

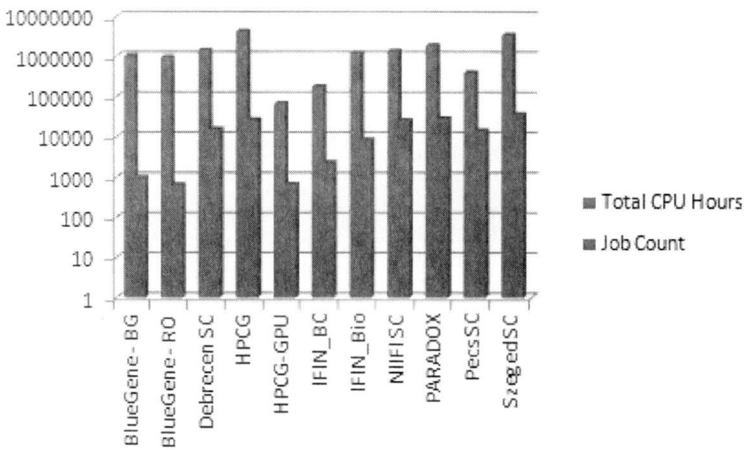

Figure 5: HP-SEE usage for the last two project years over different HPC centers in SEE region (log scale).

It should be noted that the development of Grid applications requires good understanding of the dependencies of the executable code on different versions of system libraries, and developers should be ready for the inherent heterogeneity of the Grid environment. In the Bulgarian Grid infrastructure the goal is to maintain maximum compatibility between operating systems installed on Bulgarian Grid clusters, allowing a number of codes to be prepared initially on smaller clusters and then ported to the high-performance grid cluster with minimal modification.

The main Grid applications using the cluster are: a system dealing with the modeling of transport of hazardous substances into the atmosphere for the purpose of early warning and rapid response; multi-scale modeling of the atmosphere,[8]; Monte Carlo sensitivity analysis [2]; and computer simulations of gas flow in micro-channels [9]. The cluster also has many international users, mostly obtaining accounts through the HP-SEE project.

There is substantial diversity of applications using the cluster, varying from single-CPU applications to tightly coupled parallel applications using a high percentage of the total available CPU power. The distribution of jobs can be seen from two angles in Figures 6 and 7. Although single-CPU jobs dominate in terms of number of jobs, as seen in the first figure, the total CPU usage reaches a peak in the range between 81 and 160 parallel processes. This observation motivated us to make some configuration changes and enhancements in the deployed versions of maui and torque in order to facilitate faster execution of such jobs. When such jobs are submitted using the Grid job submission mechanisms, there is the undesirable consequence that the job can be launched on any combination of job slots, leading to uneven distribution of processes among worker nodes. We deployed filtering scripts that prevent such a possibility, thus ensuring dedicated worker nodes for optimal quality of service.

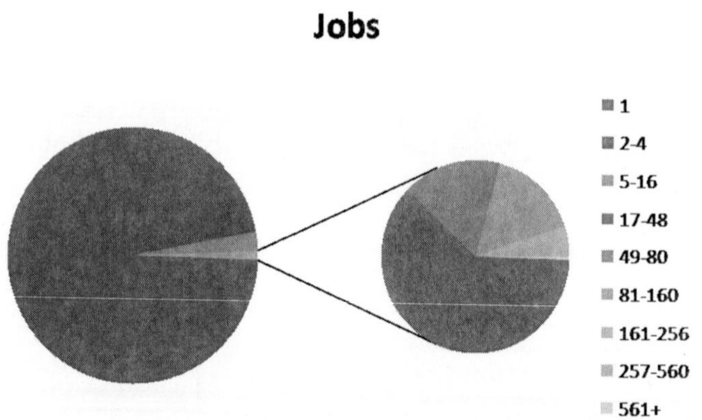

Figure 6: Distribution of number of jobs by number of parallel processes

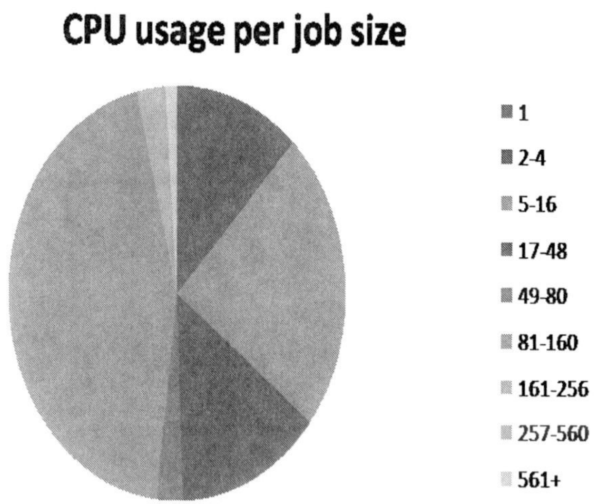

Figure 7: Distribution of CPU usage by number of parallel processes

Conclusion

The cluster was certified and put into operation within one month after completion of the procurement process, and is used continuously by scientists from various scientific organizations from Bulgaria, European Grid Virtual Organizations and scientists from the regional HP-SEE project. It provides a strong basis for Bulgarian participation in the largest European Grid project and the construction of a new high-performance computing regional infrastructure. The substantial expansion of the user base provides further justification of the need to expand the cluster using state-of-the-art high-performance computing technologies, for example, using accelerator cards like NVIDIA Tesla and Intel Xeon Phi at a larger scale.

The European Grid Initiative is developing new services, based on Cloud technologies, [7]. Efforts to integrate the cluster using OpenStack middleware are under way and production operations are planned to start in April 2014. Due to the distinguishing features of Grid and Cloud technologies we expect that the data-intensive applications will be using the Cloud services more intensively, while the applications that stress the parallel performance of the cluster will continue using the cluster through the Grid interfaces.

Acknowledgment

This work was supported by the National Science Fund of Bulgaria under Grants DCVP 02/1 (SuperCA++) and the European Commission under EU FP7 project EGI-InSPIRE (contract number RI-261323).

Bibliography

[1] E. Atanassov and T. Gurov and A. Karaivanova. Security issues of the combined usage of Grid and Cloud resources. *MIPRO 2012/DC-VIS*: 417-420, 2012.

[2] E. Atanassov and S. Ivanovska. Sensitivity Study of Heston Stochastic Volatility Model Using GPGPU. In Lirkov, Ivan and Margenov, Svetozar and Wasniewski, Jerzy, editors, *LSSC* in Lecture Notes in Computer Science, pages 439-446, 2011. Springer.

[3] D. Dimitrov and E. Atanassov and T. Gurov. SEE-GRID-SCI Accounting Portal. *SEE-GRID-SCI User Forum*, :73-76, 2012.

[4] M. Dulea and A. Karaivanova and A. Oulas and I. Liabotis and D. Stojiljkovic and O. Prnjat. *High-Performance Computing Infrastructure for South East Europe's Research Communities: Results of the HP-SEE User Forum 2012*. Springer Publishing Company, 2014.

[5] Foster, Ian. What is the Grid? A Three Point Checklist. 2002.

[6] Foster, Ian and Kesselman, Carl and Tuecke, Steven. The Anatomy of the Grid: Enabling Scalable Virtual Organizations. *Int. J. High Perform. Comput. Appl.*, 15(3):200--222, 2001.

[7] Ian T. Foster and Yong Zhao and Ioan Raicu and Shiyong Lu. Cloud Computing and Grid Computing 360-Degree Compared. *CoRR*, abs/0901.0131, 2009.

[8] G. K. Gadzhev and K. G. Ganev and M. Prodanov and D. E. Syrakov and N. G. Miloshev and G. J. Georgiev. Some numerical studies of the atmospheric composition climate of Bulgaria. *AIP Conf. Proc., AIP Publishing*, 1561:100-111, 2013.

[9] K. S. Shterev and S. K. Stefanov and E. I. Atanassov. A Parallel Algorithm with Improved Performance of Finite Volume Method (SIMPLE-TS). In Lirkov, Ivan and Margenov, Svetozar and Wasniewski, Jerzy, editors, *LSSC* in Lecture Notes in Computer Science, pages 351-358, 2011. Springer.

[10] Communication from the commission to the European parliament, the council, the European economic and social committee and the committee of the regions, A Reinforced European Research Area Partnership for Excellence and Growth. 2012.
http://ec.europa.eu/research/era/pdf/era-communication/era-communication_en.pdf.
[11] Communication from the commission to the European parliament, the council, the European economic and social committee and the committee of the regions, ICT infrastructures for e-SCIENCE. 2009.
http://www.parliament.bg/pub/ECD/79030COM_2009_108_EN_ACTE_f.pdf.
[12] http://moo.nac.uci.edu/ hjm/Perceus-Report.html.
[13] http://www.lustre.org.
[14] http://www.adaptivecomputing.com/products/open-source/torque.
[15] http://www.egi.eu.
[16] http://www.hp-see.eu.

PARALLEL ALGORITHM FOR FIELD FIRE SIMULATION

STEFKA FIDANOVA AND PENCHO MARINOV

Introduction

Every year South European countries have vast areas burned and changes to the vegetation coverage caused by wild-land fires. In recent decades, with the consequences of human influence on nature and with climate change, this part of the Europe has become drier. That has increased the occurrence of wild-land fires. The USA, Mexico, Australia and Brazil have the same problem. The Bulgarian Ministry of Forests and Agriculture officially announced in 2006 that Bulgaria has had a significant increase in wild-land fires since 1976.

Rothermel [23] started field fire modeling forty years ago in the USA. He shows how the energy freed during pyrolysis is distributed in nature. Both USA models, WRF-Fire and FARSITE, use as their basis the Rothermel rate of spread equation [12,17]. In the beginning of the eighties Russian researchers from the University of Tomsk developed their own wildfire model, very different from the American one [14]. The Russian model is very specific for the Siberian forest and is difficult to use for any other region. The Canadian model PROMETHEUS [25] is designed to work in Canadian fuel complexes. Most of these models are very slow and unusable for front fire development. They are used for training firemen and to estimate the potential spread of a wildfire and the damage which it can cause [9]. There are some groups from Spain [11] and France [7, 18] who work in the field of forest fire modeling, but more theoretically. Having in mind all the listed statistics a team from the Bulgarian Academy of Sciences has started working on modeling field fire using a game method. It is a cellular automate with cells that can represent the spread of a fire for flat areas where the vegetation can be diverse. We have tested the cells with different shapes, and have chosen the hexagon as the optimal. This shape gives better contact with the neighbor cells and the fire spread is more realistic. The software in use is meant to provide predictions of different scenarios of the fire spread, which can help the firefighting teams optimize their work. The firemen need a fast model which can calculate in minutes the fire's acceleration for several hours. When the area is large and the front of the fire changes very fast the model does many calculations. That is why we have created a parallel version of our algorithm.

The rest of the paper is organized as follows: in Section 2 we describe the Game Method; in Section 3 the numerical simulation is presented; in Section 4 we report computational results and some comparisons; in Section 5 we draw some conclusions and directions for future work.

Game Method

The Game Method for Modeling (GMM) is developed by Prof. Krassimir Atanassov [4]. The idea for GMM comes from Conoway's Game of Life [8, 15]. It uses an orthogonal grid of square cells, each with one of the two possible states, alive or dead. Their first application is for describing astronomical processes [24]. They are used in the field of combinatorial geometrics [6] and field fire modeling [10]. Movements of objects and their quantitative and qualitative changes can be described by this method.

In GMM the considered area is described by a mesh. The cells of the mesh can be triangular, square (more often), hexagonal or other. There are initial parameters, called initial state, which will be changed during the time steps. There are transition rules which define the way the cell's parameters will change according to the previous stage and according to the stage of the neighbor cells. Thus in every time step we can know the stage of every one of the cells in the area under consideration. The initial configuration can be every set of eligible parameters. The final configuration is a set of parameters which are the result of a certain number of applications of the transition rules. A single application of the rules is called an elementary step. The stopping criteria can vary according to the problem modeled. One such criterion, often used, is a pre-defined number of time steps. Other criteria for stopping could be when the cells form a pre-defined configuration, or when the process oscillates.

The GMM can be described as follows:

$$L = A(K) = A_1(A_1(\times A_1(K) \times))$$

where:

A_1	--	is the set of transition rules
K	--	is the initial stage
L	--	is the final configuration

Numerical Simulation

Forest fire models are usually classified as follows:

- stochastic and deterministic models [16, 22];
- empirical, semi-empirical models [3];
- physical models [20, 21].

Stochastic models are based on the observation of experimental and wildland fires, from which the parameters of burning are determined. Empirical and semi-empirical models are based on the assumption that the energy transferred to the unburned fuel is proportional to the energy released by the combustion of the fuel. Physical models lead to differential equation systems which describe the energy transfer in the burning area. These models require time-consuming numerical calculations [2, 13, 19]. Normally they are run on parallel computers, because they are very slow and inefficient on a single processor, even for a small area.

In our previous works [10, 26] we applied GMM with square cells to the modeling of the spread of forest fires. Later we decided to use cells with a hexagonal shape. They are more appropriate for two reasons: the hexagon is closer to the circle, the shape of a fire without wind; there is only one kind of neighbor cells, and there are no corner neighbors (see Figure 1).

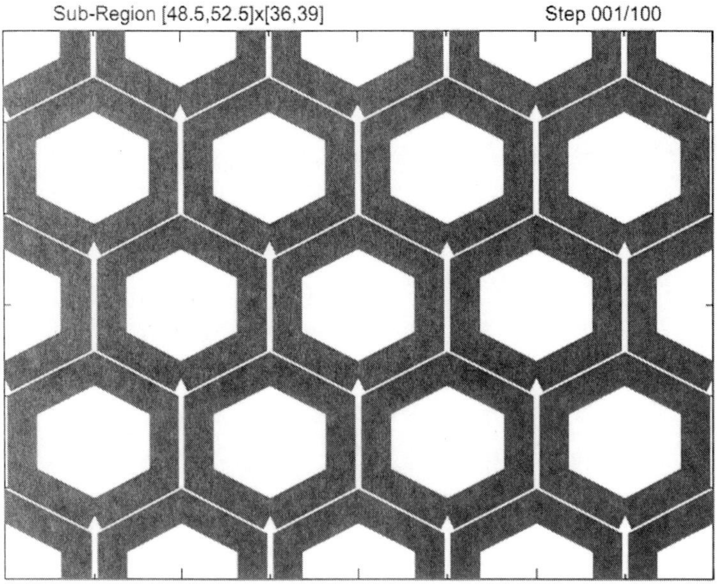

Figure 1: Hexagonal cells

The parameters for our GMM model are:

- Burning time;
- Time to start to burn;
- Wind parameters, force and direction;
- Surface.

The burning time shows how many time steps the material inside the cell needs in order to be totally burned, or a burning duration. The time to start to burn shows the ignition speed if one neighbor cell burns. For example: if the speed for ignition is 2 time steps and the burning duration is 5 time steps, the cell will start to burn if during 2 time steps some neighbor cells burn, and the cell will be totally burned 5 time steps after ignition. We suppose that the size of the parameters and the size of the cells are fixed in advance. If the material in the cell is unburned, then the burning duration and the speed for ignition are equal to 0.

The rules of our fire model are as follows:

- At the initial time step the modeling starts from the initial state of the area where one or more cells are burning;
- Every time step the burning duration of burning cells decreases by 1 till it becomes 0 (totally burned);
- If a cell is burning, the speed of ignition of closer cells is changed, depending of the force and direction of the wind;
- When the speed of ignition becomes 0 the cell starts to burn;
- The process continues until no other change of the parameters is possible or the number of the time steps is equal to the pre-defined time steps; otherwise go to 2.

One of the advantages of our model is that it can start from any stage of the area in question, which is realistic. Normally, a forest fire is only discerned after some acceleration.

Parallel Algorithm

Our model is much faster than the American ones. In the case when the area of concern is very large and we need to know what to expect about the development of the fire front for a short time, the parallel version of the algorithm can be useful. We prepare a version of our algorithm which is convenient for the IBM Blue Gene/P supercomputer. For parallelization we use MPI (Message Passing Interface). Our parallel algorithm can easily be adjusted for other parallel computers too.

In June 2007, IBM unveiled Blue Gene/P, the second generation of the Blue Gene series of supercomputers. Each Blue Gene/P compute chip contains four PowerPC 450 processor cores, running at 850 MHz. The cores are cache coherent and the chip can operate as a 4-way symmetric multiprocessor. The memory subsystem on the chip consists of small private L2 caches, a central shared 8 MB L3 cache, and dual DDR2 memory controllers. The chip also integrates the logic for node-to-node communication. A compute card contains a Blue Gene/P chip with 4 GB DRAM, comprising a "compute node". 32 compute cards are plugged into an air-cooled node board. A rack contains 32 node boards (thus 1024 nodes, 4096 processor cores). By using many small, low-power, densely packaged chips, Blue Gene/P exceeds the power efficiency of other supercomputers of its generation, and at 371 MFLOPS/W Blue Gene/P installations rank at or near the top of the Green500 lists in 2007-2008 [1].

In our parallel algorithm we divide the area under consideration into overlapped bands. The bands have the same width (number of cells). The width of the overlapped area depends on the force of the wind. If there is no wind, the overlapping is one row. Thus we minimize data transfer, which is very important in parallel computing because it is the slowest part of parallel computing. The used cores are numbered. The core number 0 receives all the data, the area and cell parameters. It divides the area and sends the parts to other cores. Other cores calculate the changes of cell parameters in their parts and after every time step they exchange information for the overlapping parts. Thus they unify the information. If we divide the area into N bands, the core with number i, $i = 2, \dots, N-1$ exchanges information with cores with numbers $i-1$ and $i+1$. Core 1 exchanges information with core 2 and core N exchanges information with core $N-1$. After the final time step all cores with number greater than 0 send their areas to the core 0. Core 0 assembles the information coming from other cores and constructs the final configuration of the area concerned.

We prepare a test problem where the area is flat, with uniform vegetation (same burning time and ignition speed for every cell), to test the parallelization of our algorithm. Our tests consist of 100×100 and 1000×1000 hexagonal cells. The number of time steps is fixed to be 100. Every processor of our supercomputer has 4 cores with common memory. When the cores used are on the same processor they use common memory and the transfer of data is fast. When the cores used are on different processors, the transfer of data is slower. We run the algorithm on 4 cores on the same processor, 2 cores from 2 different processors, 4 cores from 4 different processors, 3 cores on the same processor and 2 cores on the same processor.

When 2 cores are used, the algorithm proceeds in a sequential way, because the core 0 only divides the data and at the end assembles them. In the case where the two cores are on the same processor, they use a common memory; thus there is no parallelism and the transfer of data is fast. We will denote the running time in this case with 1 and will use it to estimate the running time of other cases. When the two cores are on different processors there is transfer of data between processors. Thus in this case we estimate data transfer. When 4 cores are used, the area is divided into 3 bands. When 3 cores are used the area is divided into 2 bands.

Table 1: Running time on various number of cells

2 cores	2 cores	3 cores	4 cores	4 cores
1 processor	2 processors	1 processor	1 processor	4 processors
1	+2%	-6%	-10%	+10 %

Table 1 shows the increase and decrease of the running time of the algorithm. When the algorithm is run on several cores on the same processor the running time decreases according to the sequential variant, because of the fast transfer of data between cores on the same processor. When the algorithm is run on cores on different processors, the running time increases, because of the slow transfer of data between the processors.

Let us run the algorithm on 5 cores. The cores from 1 to 4 are to be on the same processor and core 0 is to be on another processor. In this case there is slow data transfer from core 0 to other cores, which occurs only at the beginning and at the end of the calculations, and fast data transfer between other cores. Thus the running time decreases by 8% according to the sequential variant.

We verified if the final configuration achieved by the parallel algorithm is the same as the configuration achieved by the sequential one, or if the parallel algorithm performs in a proper way.

Conclusion

In this paper we develop an algorithm for field fire modeling. The algorithm is based on a game method for modeling. We prepare a parallel version of our algorithm for the IBM Blue Gene supercomputer using MPI. We study the influence of the number of cores and the architecture used on the algorithm's performance. We can conclude that the minimal running time is when the algorithm is run on cores on the same processors. Another possibility can be to have only the core 0 on the different processor and the others to be on the same. During the algorithm run there is data transfer from core 0 at the beginning and back to core 0 at the end of the run. There is data transfer between other cores at every time step. Thus it is better for them to be on the same processor.

Acknowledgments

This work is supported by the Bulgarian NSF under the grant I01/0006 and by the PRACE project funded in part by the EU's 7th FP, grant agreement no. RI-211528 and FP7-261557.

Bibliography

[1] "Overview of the IBM Blue Gene/P project". IBM Journal of Research and Development. Jan 2008.

[2] Anderson D.H., E.A. Catchpole, N.J. DeMestre, T. Parkes.1982. Modeling the spread of grass fires, J. Aust. Math. Soc. (B) 23, 451-466.

[3] Andre J.C.S., D.X. Viegas. 1994. A strategy to model the average fire line movement of a light to medium intensity surface forest fire, in: Proceedings of the 2nd International Conference on Forest Fire Research, Coimbra, pp. 221-242.

[4] Atanassov K., On a combinatorial game-method for modeling, Advances in Modeling & Analysis, AMSE Press, Vol. 19, 1994, No. 2, 41-47

[5] Atanassov K., Application of a combinatorial game-method in combinatorial geometry. Part 1: Combinatorial algorithms for solving variants of the Steiner-Rosenbaum's problem, Advances in Modeling & Analysis, Vol. 2, 1998, Vol. 2, No. 1-2, 23-29.

[6] Atanassov K., Application of a combinatorial game-method in combinatorial geometry. Part 2: Algorithm for grouping and transferring of points and a general algorithm, Advances in Modeling & Analysis, Vol. 2, 1998, Vol. 2, No. 1-2, 31-36.

[7] Asensio M. I., Ferragut L., Simon J.. High definition local adjustment model of 3D wind fields performing only 2D computations. Comm. in Numer. Methods in Eng., Serie A. Mat., (2009).

[8] Deutsch A., Dormann S., Cellular Automaton Modeling Biological Pattern Formation, Birkh¨auser, Boston, 2005.

[9] Dillon, G., Morgan P., Holden Z., Mapping the potential for high severity wildfire in the western United States. Fire Management Today. 71(2), 2011, 25-28.

[10] Dobrinkova N., Fidanova S. and Atanasov K., Game-Method Model for Field Fires, Large Scale Scientific Computing, Lecture Notes in Computer Science No 5910, ISSN 0302-9743, Springer, Germany, 2010, 173-179

[11] Ferragut L., Asensio M. I., Simon, Forest fire simulation: mathematical models and numerical methods, Monografias de la Real Academia de Ciencias de Zaragoza 34, 2010, 51 -- 71.

[12] Finney, M. A. 1998. FARSITE: Fire Area Simulator model development and evaluation. Res. Pap. RMRS-RP-4. Ft. Collins, CO: U.S. Department of Agriculture, Forest Service, Rocky Mountain Research Station. 47 pages.

[13] Green D.G., A.M. Gill, I.R. Noble, 1983. Fire shapes and the adequacy of fire spread models, Ecol. Mod. 20, 33-45.

[14] Grishin A., Mathematical models of forest fire, Pub University of Tomsk, 1981, (in Russian).

[15] Gros C., Complex and Adaptive Dynamical Systems. Springer, Berlin, 2011.

[16] Halada L., P. Weisenpacher. 2005. Principles of forest fire spread models and their simulation, J. Appl. Math., Stat. Informat. 1 (1), 3-13.

[17] Mandel, J., Beezley, J.D., Coen, J.L., Kim, M.: Data assimilation for wild-land fires: Ensemble Kalman filters in coupled atmosphere-surface models. IEEE Control Systems Magazine 29 (2009) 47--65.

[18] Sero-Guillaume O., Margerit J., Modeling forest fires. Part I: a complete set of equations derived by extended irreversible thermodynamics, Int. J. of Heat and Mass Transfer 45, 2002, 1705 -- 1722.

[19] Oldehoeft, R. 2000. Taming complexity in high-performance computing, Math. Comp. Simul. 54, 314-357.

[20] Pastor E., L. Zarate, E. Planas, J. Arnaldos. 2003. Mathematical models and calculation systems for the study of wild land fire behavior, Prog. Energy Comb. Sci. 29,139-153.

[21] Perry G.L.W. 1998. Current approaches to modeling the spread of wild land fire: a review, Prog. Phys. Geog. 2 (2) 222-245.

[22] Pitts W.M. 1991. Wind effects on fires, Prog. Energy Combust. Sci. 17 (1991) 83-134.

[23] Rothermel R. C. A mathematical model for predicting fire spread in wild land fires. USDA Forest Service Research Paper INT-115, 1972.

[24] Sasselov D., Atanassov K., On the generalized nets realization of a combinatorial game-method for modeling in astronomy, Advances in Modeling & Analysis, AMSE Press, Vol. 23, 1995, No. 4, 59-64.

[25] Tymstra, C., Bryce, R.W., Wotton, B.M., Armitage, O.B. Development and structure of Prometheus: the Canadian wild land fire growth simulation Model. Inf. Rep. NOR-X-417. Nat. Resour. Can., Can. For. Serv., North. For. Cent., Edmonton, 2009.

[26] Velizarova E., Sotirova E., Atanassov K., Vassilev P., Fidanova S., On the Game Method for Forest Fire Spread Modeling with Consideration of the Wind Effect. In proc. of IEEE Conf. on Intelligent Systems, Sofia, Bulgaria, September, 2012, ISBN 978-1-4673-2277-5, 216 - 220.

A MULTISCALE MULTILEVEL ALGORITHM USING ANALYTICAL COARSE OPERATOR APPLIED TO BONE TISSUE MODELING

R. SVIERCOSKI AND S. MARGENOV

Introduction

Trabecular bone tissue is an example of a deformable medium that has a complex hierarchical morphology, in the sense that essential features need to be considered - from nanometer to millimeter scales. These features modeled at various scales determine how well the bone tissue meets conflicting mechanical and mass-transport needs. However, the modeling needed to predict the flow and mechanical behavior in such systems with their hierarchical structures and multiple, often poorly separated length-scales is very computationally demanding. It is thus impractical to make simulations of the everyday mechanical behavior and flow of bone tissue.

The main goal of this paper is to propose a new, multilevel, multiscale method, at low computational cost and easily implementable, for modeling a linear elasticity system which has an isotropic heterogeneous bulk modulus and a spatially varying Poisson ratio close to the incompressibility limit, which we will make more precise further on. This new procedure uses a multigrid combined with an analytical approximation of the homogenized bulk modulus, as well as an average of the variable Poisson's ratio for the fluid phase, at coarser levels. The numerical results will be presented in a 2D version of the problem.

The use of homogenization tools incorporated into multigrid schemes for solving problems in heterogeneous media is justified, since obtaining optimal performance of the multigrid alone is a challenging task, particularly when the heterogeneous media are described by step functions or jumping coefficients. Among the issues related to using the multigrid to solve multiscale systems is mesh anisotropy. This is caused by elements having a very large aspect ratio, typically appearing as a factor in the condition number of the stiffness matrix, which can easily generate a highly ill-conditioned problem [7]. In this particular case, the other difficulty is that the system becomes ill-conditioned also because the Poisson ratio of the fluid phase approaches the incompressibility limit, $\nu \to 0.5$.

Even though both the Poisson's ratio $\nu(x)$ and the bulk modulus $K(x)$ are upscaled, the term upscaled refers to the bulk modulus only, since it is a tensor that is a multiplicative term in the compliance matrix $\mathbf{C}(x)$ ahead. The proposed upscaling of ν does not follow from classical homogenization theory. The general framework of the multiscale multigrid method corresponds to defining coarse-grid operators at each level of the V-cycle by using an upscaled value for ν and for $K(x)$. The upscaled bulk modulus is derived from an analytical approximation K^* of the true upscaled, or effective, tensor $\overline{\mathbf{K}}$. The existence of $\overline{\mathbf{K}}$ is a classical result in homogenization theory from [9]. The proposed multilevel setting can be considered as an iterative homogenization procedure.

Let us recall that finding the true upscaled tensor, $\overline{\mathbf{K}}$, in a simplified way has been the subject of much research, generally aiming to reduce computational effort. Given its importance in various fields of application, there is a large literature on the subject and the reader is referred to [8] for a review and other general applications.

Since the numerical computation of $\overline{\mathbf{K}}$ is usually computationally demanding, the underlying hypothesis is that the effective tensor at each level can be replaced by suitable approximations. The use of approximations to replace $\overline{\mathbf{K}}$ is not new in the literature, particularly when built into a numerical scheme. The most popular schemes use the arithmetic and harmonic average, or a combination of both [1, 3, 5]. However, these averages are generally too far from the true upscaled tensor, especially when high aspect ratios are present; thus better approximations to the true upscaled tensor can lead to more reliable algorithms. Here, the Analytical Coarse Operator (ACO) proposed in [12, 13] will be used, together with the averaging of the variable Poisson's ratio $\nu(x)$.

In the next sections, the algorithm will be presented followed by numerical convergence results that indicate the reliability of the procedure when compared with well-known upscaled tensors given by the arithmetic, harmonic and geometric averages, and the respective averaging of ν approaching the incompressible limit $\nu \approx 0.5$. First, the general multiscale problem is outlined, together with the theoretical basis for its upscaled approximation at coarser levels. This work is a continuation of a previous work where only the bulk modulus was variable, see Sviercoski and Margenov [12].

The Problem at the Fine-scale and the Homogenization

Without loss of generality, let $\Omega \subset R^2$ be a bounded domain with boundary $\Gamma = \partial\Omega$ and $\mathbf{u} = (u_1, u_2)$, the displacement in Ω. The pure displacement deformation of a body under the influence of applied forces, $\mathbf{f}$, (and considering only first order terms in the displacement) is described by:

$$\begin{cases} -\nabla \cdot \boldsymbol{\tau}(x) = \mathbf{f} & x \in \Omega \\ \mathbf{u}(x) = 0 & x \in \partial\Omega \end{cases} \tag{1}$$

where $\boldsymbol{\tau}(x)$ is the stress tensor, with components, $\tau_{ij}(x)$, given by Hooke's law:

$$\tau_{ij}(x) = \Sigma_{k,l=1}^{2} c_{ijkl}(x)\varepsilon_{kl}(\mathbf{u}), \quad 1 \leq i,j, \leq 2. \tag{2}$$

The components of the strain-displacement tensor are given by:

$$\varepsilon_{ij}(\mathbf{u}) = \frac{1}{2}\left(\frac{\partial u_i}{\partial x_j} + \frac{\partial u_j}{\partial x_i}\right), \quad 1 \leq i,j, \leq 2, \tag{3}$$

and $c_{ijkl}(x)$ are the spatially dependent properties describing the behavior of the material. These properties are related to Lamé's coefficients $(\lambda(x), \mu(x))$:

$$\lambda(x) = \frac{3K(x)\nu(x)}{(1+\nu(x))} = K(x)\Lambda(x), \qquad \mu(x) = \frac{3K(x)(1-2\nu(x))}{2(1+\nu(x))} = K(x)\kappa(x) \tag{4}$$

where they have been used as the relationship for the Young's module $E(x) = 3K(x)(1 - 2\nu(x))$, as a function of the spatially dependent, bulk modulus $K(x)$ and of the Poisson ratio $\nu(x) \in [0, \frac{1}{2})$. The case when the spatially variable $\nu(x) = \frac{1}{2} - \delta$ ($\delta > 0$ is a small parameter) leads to the notion of *almost incompressible* material. We observe that eq. (1) becomes ill-posed at the incompressibility limit, when $\nu(x) \to \frac{1}{2}$.

For $\mathbf{f} = (f_1, f_2)^T \in (L_2(\Omega))^2$, the weak formulation of (1) reads as finding $\mathbf{u} \in (H_0^1(\Omega))^2 = \{\mathbf{u} \in (H^1(\Omega))^2 | \mathbf{u}_{\partial\Omega} = 0\}$ such that for all $\mathbf{v} \in (H_0^1(\Omega))^2$:

$$A(\mathbf{u}, \mathbf{v}) = \int_\Omega \lambda div(\mathbf{u}) div(\mathbf{v}) + 2\mu \Sigma_{k,l=1}^2 \varepsilon_{ij}(\mathbf{u}) \varepsilon_{ij}(\mathbf{v}) = \int_\Omega \mathbf{f}^T \mathbf{v} dx \quad (5)$$

The bilinear form $A(\mathbf{u}, \mathbf{v})$ can be written, from [7], as:

$$A(\mathbf{u}, \mathbf{v}) = \int_\Omega < C(x) \mathbf{d}(\mathbf{u}), \mathbf{d}(\mathbf{v}) > dx, \quad (6)$$

where,

$$\mathbf{C}(x) = \mathbf{K}(x) \begin{bmatrix} (\Lambda(x) + 2\kappa(x)) & 0 & 0 & \Lambda(x) \\ 0 & \kappa(x) & \kappa(x) & 0 \\ 0 & \kappa(x) & \kappa(x) & 0 \\ \Lambda(x) & 0 & 0 & (\Lambda(x) + 2\kappa(x)) \end{bmatrix} \quad (7)$$

and $\mathbf{d}(\mathbf{u}) = \left[\frac{\partial u_1}{\partial x_1}, \frac{\partial u_1}{\partial x_2}, \frac{\partial u_2}{\partial x_1}, \frac{\partial u_2}{\partial x_2}\right]$. In the 2D case, $\mathbf{K}(x)$ is a 4×4 isotropic diagonal tensor. Note also that the formulation of the compliance matrix $\mathbf{C}(x)$ is used in a general setting, unlike the work from [12] where a modified $\mathbf{C}(x)$ was used for the particular case of a pure displacement problem.

An upscaled or homogenized version of (6) means finding an upscaled tensor representation of the $\mathbf{K}(x)$ and of the scalar $\nu(x)$, here denoted by the tensor $\overline{\mathbf{K}}$ and the scalar $\overline{\nu}$. Therefore, the respective upscaled form of the bilinear system is considered as $\overline{A}(\overline{\mathbf{u}}, \overline{\mathbf{v}})$, such that:

$$\overline{A}(\overline{\mathbf{u}}, \overline{\mathbf{v}}) \to_H A(\mathbf{u}, \mathbf{v}) \quad (8)$$

converges to the fine-scale operator in the homogenized (or averaged) sense, where $\overline{\mathbf{u}}$ is the upscaled approximation of the solution $\mathbf{u}$. When only the homogenization of $\mathbf{K}(x)$ is considered, then the convergence is called H −convergence (not to be confused with H as coarse mesh parameter) and its mathematical detailed explanation can be found in classical literature, such as [6, 9]. Here, however, besides the upscaled coefficient for $\mathbf{K}(x)$, the averaging also applies to the variable $\nu(x)$. These will lead us to define at the coarse level the upscaled diagonal tensor with entries $\overline{\Lambda} + 2\overline{\kappa}$ and $\overline{\kappa}$. While the averaging of $\mathbf{K}(x)$ follows from homogenization theory, the averaging of $\nu(x)$ does not follow from the same theory. It is used here only as an alternative.

The derivation of K^*, which is usually a full tensor, starts by using an approximation to the solution of the cell-problem from [10], leading to the lower and upper bounds of the upscaled tensor $\overline{\mathbf{K}}$, known as the generalized Voigt-Reuss inequality (GVR) [6]:

$$\mathbf{K}_h \leq \widetilde{\mathbf{K}} \leq \overline{\mathbf{K}} \leq \widehat{\mathbf{K}} \leq \mathbf{K}_a \tag{9}$$

The lower bound, $\widetilde{\mathbf{K}}$, is the arithmetic average (say in the x_2 variable) of the harmonic average (in the x_1 variable) of $\mathbf{K}(x)$, whereas the upper bound, $\widehat{\mathbf{K}}$, is the harmonic average of the arithmetic average of $\mathbf{K}(x)$. These are much stricter bounds than the classical Voigt-Reuss inequality given by the harmonic, $\mathbf{K}_h$ and arithmetic, $\mathbf{K}_a$, averages.

The diagonal entries of K^* are given by the average between the geometric and arithmetic averages of the (GVR) bounds (9) at each direction. The off-diagonal terms are derived by using a rotation of a diagonal matrix $\mathbf{D}$ by an angle θ related to the center of mass of the reference cell. In 2D, it is given by:

$$K^* = \frac{1}{2}\begin{pmatrix} K_{11}^a + K_{11}^g & -b\sin(2\theta) \\ -b\sin(2\theta) & K_{22}^a + K_{22}^g \end{pmatrix} = \mathbf{P}D^*\mathbf{P}^{-1}, \tag{10}$$

where D^* is diagonal representation of K^* with 2 eigenvalues and $\mathbf{P}$ is the eigenvector matrix. There are two possible rotations, when the main diagonals of K^* are equal or not. In each case, one can choose appropriately $\mathbf{P}$, and the coefficient b will correspond to:

$$b = \begin{cases} \frac{K_{ii}^* - K_{jj}^*}{\cos(2\theta)} & if \quad K_{ii}^* \neq K_{jj}^* \\ \frac{\hat{K}_{ii} + \hat{K}_{jj}}{2} - \frac{\tilde{K}_{ii} + \tilde{K}_{jj}}{2} & if \quad K_{ii}^* = K_{jj}^* \end{cases} \tag{11}$$

where $\tilde{K}_{ii}$ and $\hat{K}_{ii}$ come from (9) above. In the cited literature, there are conditions on the angle θ to ensure that K^* is positive definite. In this particular application, an upscaled full tensor may render a non-symmetric operator. Thus, in order to ensure symmetry, and that at the same time the heterogeneities are accounted for, the eigenvalues λ_1 and λ_2 of K^* will be used to define:

$$\mathbf{K}^{\#} = \frac{\lambda_1 + \lambda_2}{2} \tag{12}$$

It can be proved that each eigenvalue satisfies the classical Voigt-Reuss inequality (9) above.

The next step to obtain the coarse operator is to average the Poisson ratio $\nu(x)$. Here, the same averaging used for the bulk modulus $K(x)$ is used for $\nu(x)$ to obtain ν^* and $\nu^{\#}$, which are then used to compute $\overline{\Lambda} + 2\overline{\kappa}$ and $\overline{\kappa}$ presented below.

The Analytical Coarse Operator

In this paper, the system defined through (5)-(6) is written using the Displacement Decomposition (DD) form. Under this setting, and since $\mathbf{K}(x)$ is a diagonal isotropic tensor, the coefficient matrix $\mathbf{C}(x)$ can be written as four 2×2 block diagonal matrices C_i, $i = 1,..,4$ as:

$$\mathbf{C}(x) = K(x) \begin{pmatrix} \Lambda + 2\kappa & 0 & 0 & \Lambda \\ 0 & \kappa & \kappa & 0 \\ 0 & \kappa & \kappa & 0 \\ \Lambda & 0 & 0 & \Lambda + 2\kappa \end{pmatrix} = \begin{pmatrix} C_1(x) & C_2(x) \\ C_3(x) & C_4(x) \end{pmatrix} \tag{13}$$

To solve the resulting discrete system $\mathbf{A}(x)\mathbf{u}_h = \mathbf{b}$ from (5), at the fine mesh h, a multiscale multilevel composite block iterative method is applied to the related coupled Finite Element (FE) system, where at the coarse mesh, H, the approximations $\mathbf{K}^{\#}$ and $\nu^{\#}$ are used to represent the fine-scale features.

At the coarse level H, the upscaled representation of $\mathbf{C}(x)$ will be given as two 2×2 block diagonals $C^* = [C_1^*, 0; 0, C_4^*]$, where $\mathbf{C}_1^* = K^{\#}\begin{pmatrix} (\overline{\Lambda} + 2\overline{\kappa}) & 0 \\ 0 & \overline{\kappa} \end{pmatrix}$, and $C_4^* = K^{\#}\begin{pmatrix} \overline{\kappa} & 0 \\ 0 & (\overline{\Lambda} + 2\overline{\kappa}) \end{pmatrix}$ where $K^{\#}$ and $\nu^{\#}$ are the averages between the eigenvalues of K^* as (12) and the respective ν^*. Then from (4), $\overline{\Lambda} + 2\overline{\kappa} = \frac{3\nu^{\#}}{(1+\nu^{\#})} + \frac{3(1-2\nu^{\#})}{(1+\nu^{\#})}$. The theoretical justification for using the block diagonal displacement decomposition (DD) instead of the full matrix at coarser levels is provided by the second Korn's inequality [2].

The multigrid algorithm follows by first defining a sequence of grids: $\Omega_0 \subset \Omega_1 \subset \ldots \subset \Omega_H \subset \ldots \subset \Omega_h = \Omega$ together with the corresponding sequence of equations being discretized. The approximations $K_0^*, K_1^*, \ldots$ and $\upsilon_0^*, \upsilon_1^*, \ldots$ at each grid level lead to a sequence of linear operators from the coarse to the fine level $L_0^* U_0 = f_0, \ldots, L_H^* U_H = f_H, L_h^* U_h = f_h$, where a coarse-grid operator at a given level H, is:

$$(L_H^*)_{ij} = \int_\Omega \nabla \varphi_H^j C^* \nabla \psi_H^i dx \tag{14}$$

where $\{\varphi_H^j\}_{j=1}^{n_k}$ and $\{\psi_H^j\}_{j=1}^{n_k}$ are the test functions at the corresponding level.

The application of a two-level grid to the initial fine-grid operator, $C_h U_h = f_h$, at the nth iteration, is performed following the steps described below.

1. Compute K^* and υ^* at each level at the finest grid resolution, in order to obtain L^* for each subgrid.

2. Pre-smoothing. Compute an approximation to the initial value U_h^n by applying η steps of the smoothing iteration, S, to the system at the h −level. This can be formalized as: $U_h^\eta = S^{\eta_{pre}}(U_h^n, L_h, f_h)$.

3. Coarse-grid correction. Define a coarse grid, H level, and its operator L^*_H. (3.1) - Restrict the residual to the H grid, by solving $r_H = I_h^H(f_h - L_h U_h^\eta)$. (3.2) - Solve for the corrector K_H, with $L^*_H K_H = r_H$. (3.3) - Compute the new approximation $U^å$ by interpolating the correction back to the grid h, I_H^h, to give: $U^* = U_h^\eta + I_H^h K_H$.

4. Post-smoothing. Use $U^å$ as the initial value to give: $U_h^{n+1} = S^{\eta_{pos}}(U^*, L_h, f_h)$.

The resulting two-level multigrid method is the operator:

$$G = S^{\eta_{pos}}(I - I_H^h (L_H^å)^{-1} I_h^H L_h) S^{\eta_{pre}} \tag{15}$$

Here the classical point-wise Gauss-Seidel (GS) smoothing is used, and the intergrid operators are the standard full weight restriction and bilinear interpolation, as in early work [1]. Moreover, unless otherwise indicated, it is assumed that $K(x)$ is completely resolved on the finest grid; that is, the jumps of discontinuity coincide with grid lines. At each level, the corresponding homogenized coefficients, K* and υ*, and the related diagonal entries of C_1^* and C_4^*, are computed within each subdomain, implying that the derived coarse-operator L^* has a coefficient still varying from one cell to another.

This same procedure can be applied with other approximations of the upscaled tensor and other averaging of ν, such as the ones using the arithmetic, harmonic or geometric averages [3]. These operators here will be identified in the tables below as ARCO, HCO and GCO, respectively. The results will be presented next.

Numerical Results

In this section, the multilevel algorithm described above for the system (6) will be used to demonstrate numerically the reliability of the ACO operator, which will be measured by comparing the number of V-cycles with other well known analytical operators. The results described in the tables demonstrate that the operators given by the arithmetic average, ARCO, and the harmonic HCO average have their limitations because neither of them can be reliable for a range of contrast ratios within the same medium.

The tests were selected illustrating that the bulk moduli $K(x)$ and $\nu(x)$ are given by step functions describing non-periodic media. This is emphasized here because classical homogenization results focus primarily on periodic media, but the results are also known to follow for the non-periodic case. Test 1 uses checkerboard geometry. The choice of such a medium is the usual (and sometimes extreme) example of a non-periodic medium used in the literature to illustrate the generalization of the classical theory. It is also known that the upscaled tensor for this case is the diagonal isotropic tensor given by the geometric average of $K(x)$ [6]; that is the reason why the geometric coarse operator, GCO, appears in Table 1. For Test 2, the geometry is an example close to the realistic micro structure of trabecular bone tissue.

All the results were obtained on the unit square domain, $\Omega = [0,1]^2$, with spatially varying Poisson's ratio approaching the value $\nu = 0.5$, with stress force equal to $\mathbf{f} = (1,1)$. At the boundary $\partial\Omega$ there is no displacement, meaning $\mathbf{u} = (0,0)$. The accuracy at the iterative stopping criterion was set to 10^{-7}. The results are for two grids, where the finer has 32×32 nodes and the coarser has 16×16 nodes. Each grid is resolved by linear rectangular finite elements. Even though the application here is for pure displacement, the bilinear form used applies to more general settings.

Tests 1 and 2 have bulk moduli $K(x)$ and $\nu(x)$ respectively, defined by a step function like:

$$k(x) = \begin{cases} \alpha & at \quad \Omega_{Black} \\ 1 & at \quad \Omega_{White} \end{cases} \tag{16}$$

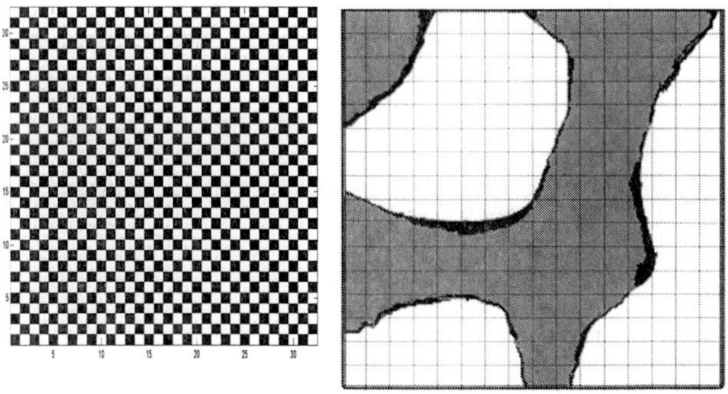

Figure 1: (Left): Checkerboard geometry used to obtain Table 1, for various values of α and ν. (Right): Illustration of the geometry of a trabecular bone[2]. The dark area has bulk modulus $K = 14\ GPa$ and Poisson ratio, $\nu = 0.325$. The light area is the fluid with $K = 2.3\ GPa$ and varying Poisson ration $\nu \to 0.5$. The grid corresponds to the coarse grid used in the simulation, with results presented in Table 2. The finer grid, 32×32, is assumed to resolve the geometry.

Test 1 The medium is the checkerboard geometry illustrated in Fig. 1 (left), where the values change from one element to another from 1 to α, for $K(x)$ and from 0.325 to $\nu \to 0.5$ for the Poisson ratio, $\nu(x)$. Observe that, given its geometrical description, the upscaled tensor will be isotropic and constant. Note that the Gauss-Seidel smoothing parameter η changes with changes in α. However, for fixed α and a range of ν, the tests ACO, ARCO, HCO and GCO use the same η. The results in Table 1 suggest that the ACO is a more reliable operator compared to the others. The reliability here is meant across the heterogeneity ratio for α and the incompressibility limit ν. Although the ARCO performed better than the ACO in some cases, it did not for the case of $\alpha = 10^2$ and neither did it for a more realistic tissue geometry as illustrated in Test 2. The HCO and GCO did not converge for high contrast ratios; that is the reason why they do not appear in the Table 1 when $\alpha = 10^{-2}$ and $\alpha = 10^2$.

[2]adapted from http://www.umich.edu/ bme332/ch9bone/bme332bone.htm

Table 1: Test 1- Comparison between the number of V-cycle iterations to solve (6) using ACO, ARCO, HCO and GCO. The medium is illustrated in Fig. 1 (left), where the binary colored values are 1 and α and 0.325 and ν. The sign $*$ indicates that the method did not converge. For the case $\alpha = 10^2$ and $\alpha = 10^{-2}$ neither HCO nor GCO converged. The ACO method performed better than all the others because it converged with fewer iterations, on average, than the ARCO.

$\alpha = 10^{-2}$ & $\eta = 20$	$\nu = 0.49$	$\nu = 0.495$	$\nu = 0.498$	$\nu = 0.499$
ACO	69	76	77	77
ARCO	26	28	30	30
$\alpha = 10^{-1}$ & $\eta = 10$	$\nu = 0.49$	$\nu = 0.495$	$\nu = 0.498$	$\nu = 0.499$
ACO	24	34	46	53
ARCO	24	33	46	52
HCO	*	*	*	*
GCO	29	34	46	53
$\alpha = 10$ & $\eta = 10$	$\nu = 0.49$	$\nu = 0.495$	$\nu = 0.498$	$\nu = 0.499$
ACO	34	39	44	48
ARCO	45	53	59	62
HCO	30	28	29	32
GCO	32	37	42	45
$\alpha = 10^2$ & $\eta = 10$	$\nu = 0.49$	$\nu = 0.495$	$\nu = 0.498$	$\nu = 0.499$
ACO	47	76	126	163
ARCO	79	126	207	267

Test 2 The geometry depicted in Fig. 1 (right) is an example of a micro structure which is a close-to-realistic representation of trabecular bone tissue. The values considered for both bulk modulus and Poisson ratio are taken from the literature on the subject. The dark area has bulk modulus equal $K = 14\ GPa$ and Poisson ratio, $\nu = 0.325$. The light area is the fluid with $K = 2.3\ GPa$ and varying Poisson ratio $\nu \to 0.5$. The illustrated grid is the coarse grid used in the simulation, with the assumption that the finer grid, with 32×32, resolves the geometry. The number of v-cycles

for each averaging procedure presented in Table 2 indicates that the ACO is the best among them. Note that while the ARCO performed better than the HCO for Test 1, it did not for this case.

Table 2: Test 2 - Number of V-cycle iterations using ACO, ARCO and HCO for the bone geometry illustrated in Fig. 1 (right), using $\eta = 20$. The sign $*$ means that the algorithm did not converge. Observe that the ACO performed best among them.

ν	0.49	0.495	0.498	0.499
ACO	39	45	65	81
ARCO	*	*	*	*
HCO	33	43	64	80

Discussion of the Results and Conclusions

In the literature, there are estimates relating the number of V-cycle iterations for resolving the DD system, N_{DD}, to the number of V-cycle iterations for the scalar elliptic equation, N_E. For instance, the inequality $N_{DD} \leq C(1-2\nu)^{-1/2} N_E$ holds true. However, this result follows from the second Korn's inequality, which concerns the case of isotropic homogeneous media (see, e.g., [2]). The conclusion here is that for the case when the coefficients are heterogeneous, the estimate is not uniform with respect to the coefficient jumps, particularly for contrasts higher than 1 order of magnitude. These representative numerical results illustrate the robustness of ACO across geometries and contrast ratios at the incompressibility limit. They also suggest that the HCO is not a robust operator as sometimes claimed in the literature, e.g. [3], because it failed to converge in many cases. Further improvements could be expected by using an analytical prolongation operator [13], instead of the one used here, together with an aggressive coarsening by considering a coarser mesh than the one considered here as a precondition. The application of this procedure to a more realistic 3D bone structure is also ongoing work, as are further improvements to the procedure.

Acknowledgments The authors were fully supported by the European Commission, under FP7 Grant FastMM, No PIIF - 236913.

Bibliography

[1] Alcouffe, R. E., Brandt, A., Dendy, J. E., Painter, J. W. *The Multi-grid Method for the Diffusion Equation with Strongly Discontinuous Coefficient*, SIAM J. Sci. Stat. Comput. (2), 4, pp. 430-454 (1981).

[2] Axelsson, O. and Gustafsson, I. *Iterative Methods for the Navier Equations of Elasticity*, Comp. Meth. Appl. Mech. Engin., 15; pp. 241–258, (1978).

[3] Dendy, J. E. and Moulton, D. *Black Box Multigrid with Coarsening by a Factor of Three*, Numer. Linear Algebra Appl.(17); pp. 577–598, (2010).

[4] Flaig, C. and Arbenz, P. *A Highly Scalable Matrix-free Multigrid Solver for μFE Analysis Based on a Pointer-less Cctree*. Proceedings LSSC 2011. Springer Lecture Notes in Computer Science 7116, pp. 498-506 (2011).

[5] Jikov, V. V., Kozlov, S.M., Oleinik, O.A. *Homogenization of Differential Operators and Integral Functionals*. Springer Verlag (1994).

[6] Krauss J., Margenov, S. *Robust Algebraic Multilevel Methods and Algorithms* Radon Series on Comput. and Appl. Mathematics (5), De Gruyter, Berlin (2009).

[7] Milton, G. *The Theory of Composites* volume 6 - Cambridge Monographs on Applied and Computational Mathematics. Cambridge University Press, Cambridge, UK, (2002).

[8] Sanchez-Palencia, E. *Non-Homogeneous Media and Vibration Theory*, Lecture Notes in Physics, 127, Springer-Verlag, Berlin. (1980).

[9] Sviercoski, R. F., Winter, C. L., Warrick, A.W. *Analytical Approximation for the Generalized Laplace's Equation with Step Function Coefficient*. SIAM Journal of Applied Mathematics vol. 68 (5) - pp. 1268-1281, (2008).

[10] Sviercoski, R. F. *An Analytical Effective Tensor for Flows through Generalized Composites*. Advances in Water Resources, 33(7), pp. 728-739, (2010).

[11] Sviercoski, R. F., Margenov, S. *Displacement Decomposition ACO based preconditioning of FEM elasticity systems*, in APPLICATION OF MATHEMATICS IN TECHNICAL AND NATURAL SCIENCES. Editor: Michail D. Todorov, AIP Conf. Proc. 1561, 112 (2013); http://dx.doi.org/10.1063/1.4827220.

[12] Sviercoski, R. F., Popov, P., Margenov, S. *An Analytical Coarse Grid Operator Applied to Multiscale Multigrid Method,* Fast Multiscale Multigrid Methods - FastMM. Submitted.

IMPROVED IMPLEMENTATION OF A LARGE-SCALE AIR POLLUTION MODEL

TZVETAN OSTROMSKY

The Danish Eulerian Model

In this section we reveal the main features of the Danish Eulerian Model (DEM) [12, 13] and its current high-performance implementation UNI-DEM [11].

The Danish Eulerian Model (DEM) is a powerful air pollution model, designed to calculate the concentrations of various dangerous pollutants and other chemical species over Europe. In fact, its spatial domain covers a larger geographical region (4800 × 4800 km), including the whole of Europe, the Mediterranean and some parts of Asia, Africa and the adjacent oceans. It takes into account the main physical, chemical and photochemical processes between the species under consideration, the emissions, and the quickly-changing meteorological conditions.

Mathematical representation of UNI-DEM

The Danish Eulerian Model (DEM) is mathematically represented by the following system of partial differential equations, in which the unknown concentrations of a large number of chemical species (pollutants and other chemically active components) take part. The main physical and chemical processes (advection, diffusion, chemical reactions, emissions and deposition) are represented in that system.

$$\frac{\partial c_s}{\partial t} = -\frac{\partial(uc_s)}{\partial x} - \frac{\partial(vc_s)}{\partial y} - \frac{\partial(wc_s)}{\partial z} +$$

$$+\frac{\partial}{\partial x}\left(K_x \frac{\partial c_s}{\partial x}\right) + \frac{\partial}{\partial y}\left(K_y \frac{\partial c_s}{\partial y}\right) + \frac{\partial}{\partial z}\left(K_z \frac{\partial c_s}{\partial z}\right) + \qquad (1)$$

$$+E_s + Q_s(c_1, c_2, \dots c_q) - (k_{1s} + k_{2s})c_s, \quad s = 1,2, \dots q \ .$$

where

- c_s -- the concentrations of the chemical species;
- $u,\ v,\ w$ -- the wind components along the coordinate axes;
- $K_x,\ K_y,\ K_z$ -- diffusion coefficients;
- E_s -- the emissions;
- $k_{1s},\ k_{2s}$ -- dry / wet deposition coefficients;
- $Q_s(c_1, c_2, \dots c_q)$ -- non-linear functions describing the chemical reactions between species under consideration.

Splitting of the system

The above large and rather complex system (1) is not suitable for direct numerical treatment. For the purpose of numerical solution it is split into sub-models, which represent the main physical and chemical processes. The most straightforward sequential splitting [6] is used in the current production version of the model, although other splitting methods have also been considered and implemented in some experimental versions [1, 4]. In this way, the following 3 sub-models are formed:

$$\frac{\partial c_s^{(1)}}{\partial t} = -\frac{\partial(u c_s^{(1)})}{\partial x} - \frac{\partial(v c_s^{(1)})}{\partial y} + \frac{\partial}{\partial x}\left(K_x \frac{\partial c_s^{(1)}}{\partial x}\right) + \frac{\partial}{\partial y}\left(K_y \frac{\partial c_s^{(1)}}{\partial y}\right) = A_1 c_s^{(1)}(t)$$

horizontal advection & diffusion

$$\frac{\partial c_s^{(2)}}{\partial t} = E_s + Q_s(c_1^{(2)}, c_2^{(2)}, \dots c_q^{(2)}) - (k_{1s} + k_{2s})c_s^{(2)} = A_2 c_s^{(2)}(t)$$

chemistry, emissions & deposition

$$\frac{\partial c_s^{(3)}}{\partial t} = -\frac{\partial(w c_s^{(3)})}{\partial z} + \frac{\partial}{\partial z}\left(K_z \frac{\partial c_s^{(3)}}{\partial z}\right) = A_3 c_s^{(3)}(t)$$

vertical transport.

The EMEP grid or its refinements (see Table 1) are used for the spatial discretization of the domain (EMEP is abbreviation of the *European*

Monitoring and Evaluation Programme). Spatial and time discretization of the above sub-models makes each of them a huge computational task, challenging for the most powerful supercomputers available nowadays. High performance and parallel computing has become vital for the real-time numerical solution of the model. Therefore parallelization has been a crucial point in the software implementation of DEM since its very early stages. A coarse-grain parallelization strategy based on partitioning of the spatial domain appears to be the most efficient and well-balanced way, using the widest class of today's parallel machines (with not too many processors), although some restrictions apply. Other parallelizations have also been developed and tried on certain classes of supercomputers [9, 10].

Various numerical methods are used in the numerical solution of the sub-models, as follows:

- On the **advection-diffusion part** -- Finite elements, followed by predictor-corrector schemes with several different correctors.
- On the **chemistry-deposition part** -- An improved version of the QSSA (Quasi Steady-State Approximation) [5].
- On the **vertical transport** -- Finite elements, followed by theta-methods.

UNI-DEM package -- contents and properties

The advances in the development of DEM during the last two decades resulted in a variety of different versions with respect to the grid-size / resolution, the dimension and the number of layers [14], as well as the number of species in the chemical scheme. The most advanced of them have been united under a common driver routine in a package called UNI-DEM. It provides a uniform and easy access to the available up-to-date versions of the model with a rather intuitive way of selecting the desired parameters (appropriate default values can be used as well).

The versions incorporated in UNI-DEM and the user-defined parameters used to choose between them are shown in Table 1 below.

Table 1: User-defined parameters and their optional values

Choosable parameters for selecting an optional UNI-DEM version				
Parameter	Description	Optional values		
NX = NY	Grid size	96 × 96	288 × 288	480 × 480
	Grid step	50 km.	16.7 km.	10 km.
NZ	# layers (2D/3D)	1		10
NEQUAT	#chem. species	35	56	168
NSIZE	chunk size	integer divisor of (NX*NY)		

Basic parallelization algorithm (distributed memory model)

The traditional parallelization strategy followed in UNI-DEM is based on partitioning of the spatial domain (the horizontal grid, in particular) into strips, and distributed memory processing of the sub-domains obtained in separate processors. This strategy requires communications (point-to-point) on each time step. To implement it with maximal portability, the MPI library of standard communication routines is used. MPI was originally developed as a package of communication subroutines for distributed memory parallel computers. Later, gaining popularity among parallel programmers and having proved to be efficient and easy to use, it became one of the most popular and portable tools for parallel programming. Now it can be used on a much wider class of parallel systems, including shared-memory supercomputers and clusters with a hierarchical structure. For optimal load-balance, **non-blocking MPI communications** are used in this so-called **communication stage**.

Each sub-domain is assigned to a separate MPI task. As there is no data dependency between the MPI tasks on both the chemistry and the vertical exchange stages, there is no need of data exchange on each time step. On the advection-diffusion stage, however, the boundary conditions treatment in the separate MPI tasks requires overlapping of the inner boundaries and exchange of certain boundary values on the neighboring sub grids on each time step. The main consequences are as follows:

- Computational and storage overhead due to the extended-boundary sub-domains on the advection-diffusion stage, which grows with increases in the number of MPI tasks;
- A certain load imbalance on the advection-diffusion stage due to the unequal size of the sub-domains (extended along the inner boundaries only);

New meteorological data sets for each new month of the modeled period must be loaded from local files. This leads to interruption of the smoothing of the calculation process and often causes heavy I/O overload. Furthermore, this overload can easily become a performance bottleneck with increases in the number of parallel tasks, as will be shown by experiments in the next section.

A huge amount of output data is written in large **local temporary files**, which also causes heavy I/O overload and increases the disk storage requirements of the system. These must be read again at the end (**gathering stage**) and post-processed in order to extract the output results. Another performance bottleneck appears here.

Additional pre-processing and post-processing stages are needed for **scattering** the input data and **gathering** the results. This is also a significant overhead, increasing with the number of parallel tasks.

Increasing of the data locality and the reuse of data in the faster cache memory is ruled by the parameter NSIZE. It determines how to group in **chunks** of proper size the numerous small tasks in the chemistry-deposition stage for more efficient cache utilization. Different parallel systems need tuning of the parameter NSIZE for optimal performance.

Scalability results for the basic algorithm

In the table below some scalability results of UNI-DEM runs are given (including execution times, speed-ups and parallel efficiency). The corresponding experiments have been performed on the IBM MareNostrum III platform at BSC - Barcelona, the most powerful Spanish supercomputer and one of the most powerful in Europe. It has a theoretical peak performance of about 1 PetaFLOPS. The machine consists of 3028 IBM iDataPlex dx360 M4 compute nodes, placed in 84 compute racks. Each compute node is composed of two 8-core Intel Xeon processors (16 cores per node or in total more than 48000 Intel SandyBridge-EP E5-2670 cores). The cores are working at a frequency of 2.6 GHz and are capable of executing 8 flops per cycle (i.e. 20.8 GFLOPS per core). There are 32 GB RAM per node and 20 MB local cache memory. The total disk storage is 1,9 PB. The machine has two interconnection networks: Infiniband and Gigabit Ethernet.

The finest-grid 2D version of UNI-DEM ((480 × 480)) has been used in the experiments. The time steps in both Advection and Chemistry sub-models are equal -- 90 seconds (small enough to ensure stability of the results, which should be correlated with the spatial grid step). The cache utilization parameter NSIZE is equal to 32. The times shown in Table 2 are for a period of one year.

Table 2: Time (T) in seconds and speed-up (*Sp*) of UNI-DEM (MPI parallelism with the basic algorithm) on IBM MareNostrum III at BSC - Barcelona. Times for one-year runs are given in the table.

Time and *speed-up* of UNI-DEM (basic version) on the IBM MareNostrum III (480 × 480 × 1) grid, 35 species, CHUNKSIZE=32									
#	Advection		Chemistry		I/O	Com.	TOTAL		
CPU	T[s]	(Sp)	T [s]	(Sp)	T [s]	T[s]	T [s]	(Sp)	*E*[%]
16	8976	(16)	6672	(16)	4730	1865	22387	(16)	100%
32	5085	(28)	3372	(32)	5943	2307	16790	(21)	67 %
64	2630	(55)	1563	(68)	7258	1090	12613	(28)	44 %
96	2095	(69)	1365	(78)	11114	695	15302	(23)	24 %
120	1774	(81)	850	(126)	12417	619	15684	(23)	19 %
160	1481	(97)	701	(152)	14412	595	17230	(21)	13 %

The results of one-year experiments with the 2D UNI-DEM on the fine-resolution spatial grid (480 × 480 × 1) executed on the IBM MareNostrum III are presented in Table 2. Times T (in seconds), **speed-up (Sp)** and parallel efficiency *E* (in percent) are given in the corresponding columns of the table. The chemistry stage has almost linear speed-up, the advection scales as well, taking into account the overlapping subdomains overhead, but the I/O time grows quickly with increases in the number of processes. As a result the total efficiency drops down to 13%.

Improved data management parallelization algorithm

Main features and specifics

An improved data distribution mechanism has been developed and implemented in UNI-DEM. The I/O operations from/to local temporary files were replaced by MPI non-blocking communications, reducing significantly the non-scalable time overhead, as well as the local storage requirements. Here there are the main features of the new algorithm.

- Certain processors (a fixed number) are used only for I/O procedures (global file transfer and scatter/gather operations), as well as exchanging (via MPI) with the rest of the processors the corresponding data chunks. In particular, 11 processors are reserved for the 11 meteorological input data sets and 5 for the output data sets, (16 in total).
- The rest of the processors are acting as in the basic algorithm but without the above (overhead-causing) operations. Receiving and sending of (local) I/O data is done via MPI instead of reading/writing temporary files (normally much faster on most modern supercomputers).
- The number of opened files in a time is kept constant (independent of the number of MPI processes).
- Part of the non-scalable overhead (producing and using local temporary files) is fully avoided.

Scalability results for the improved data management UNI-DEM

In Table 3 similar scalability results of runs of the improved data management UNI-DEM are given (similar to those in the previous table). The experiments have been performed on the same platform (IBM MareNostrum III at BSC - Barcelona), with the same input parameters.

Table 3: Time (T) in seconds and speed-up (*Sp*) of UNI-DEM (MPI parallelism with the improved data management algorithm) on IBM MareNostrum III at BSC, Barcelona. Times for one-year runs are given in the table.

Time and *speed-up* of UNI-DEM (improved version) on the IBM MareNostrum III ($\mathbf{480} \times \mathbf{480} \times \mathbf{1}$) grid, 35 species, CHUNKSIZE=32									
#	Advection		Chemistry		I/O	Com.	TOTAL		
CPU	T [s]	(Sp)	T [s]	(Sp)	T [s]	T [s]	T [s]	(Sp)	*E* [%]
16	8976	(16)	6672	(16)	4730	1865	22387	(16)	100 %
32	8911	(16)	6761	(16)	1934	3538	21303	(17)	53 %
48	5025	(29)	3422	(31)	1829	3386	13758	(26)	54 %
80	2705	(53)	1584	(67)	1773	2975	9120	(39)	49 %
112	2039	(70)	1334	(80)	1678	2375	7464	(48)	43 %
136	1829	(79)	876	(122)	1481	2007	6223	(58)	42 %
176	1521	(94)	711	(150)	1529	1863	5674	(63)	36 %

It can be seen from the table that the I/O time now is much smaller and scales better (does not increase with increasing the parallelism, as it used to be with the original algorithm). In fact, all the I/O overhead due to file processing is taken now by the 16 I/O processors and pipelined with the computations performed by the rest of the processors. The time for MPI communications is higher, as expected (more data is communicated), but in general does not increase with increasing the number of parallel tasks. As a result, the total speed-up and efficiency are considerably better.

Concluding remarks and plans for future work

Essential improvements were done in the data distribution mechanism and the data management strategy in the distributed-memory parallel implementation of the Danish Eulerian Model. These are connected with the I/O data management and transfer of data between the parallel MPI processes. The I/O operations from/to local temporary files were replaced by MPI non-blocking communications, reducing significantly the non-scalable time overhead, as well as the local storage requirements.

Improvements were done in the parallel MPI implementation of UNI-DEM on the IBM MareNostrum III at BSC - Barcelona. They increase significantly the total speed-up and efficiency of the code, as shown by experiments.

For the near future, plans are as follows:

- The new data management strategy should be refined and optimized, reducing still further the non-scalable overhead.
- It can be expected that these improvements will be particularly efficient on parallel cluster supercomputers with a relatively low number of I/O devices (with respect to the cluster's computational power). These improvements may also help where the cluster is not capable of ensuring enough disk storage per node for the original UNI-DEM model in its full range of options. Therefore, the version of UNI-DEM with these improvements will be implemented on the Bulgarian IBM Blue Gene/P supercomputer.
- The improvements will also be implemented and used in the specialized version SA-DEM [2, 3, 7, 8] for producing sensitivity analysis data, in order to increase its scalability and capability to exploit efficiently the computational power of the large cluster supercomputers like IBM Blue Gene/P and IBM MareNostrum III.

Acknowledgments

This research is partly supported by the following grants:

- the Bulgarian NSF Grant # DCVP02/1 /2010 - SuperCA++;
- the European FP7-REGPOT Grant # 316087 /2012 - AComIn.

I would like to acknowledge also the support of the Barcelona Supercomputing Centre (BSC) for granting me access to their most powerful supercomputers.

Bibliography

[1] I. Dimov, I. Faragó, Á. Havasi, Z. Zlatev, Operator splitting and commutativity analysis in the Danish Eulerian Model, *Math. Comp. Sim.* **67** (2004), pp. 217-233.

[2] I. Dimov, R. Georgieva, Tz. Ostromsky, Z. Zlatev, Sensitivity Studies of Pollutant Concentrations Calculated by UNI-DEM with Respect to the Input Emissions, *Central European Journal of Mathematics* **11** *(8)*, "Numerical Methods for Large Scale Scientific Computing" , 2013, pp. 1531- 1545.

[3] I. Dimov, R. Georgieva, Tz. Ostromsky, Monte Carlo Sensitivity Analysis of an Eulerian Large-scale Air Pollution Model, Reliability Engineering & System Safety, to appear. E-version: *http://dx.doi.org/10.1016/j.ress.2011.06.007*

[4] I. Dimov, Tz. Ostromsky, Z. Zlatev, Challenges in using splitting techniques for large-scale environmental modeling, in: Advances in Air Pollution Modeling for Environmental Security (Faragó, I., Georgiev, K., Havasi, Á. - eds.) *NATO Science Series* **54**, (2005), Springer, pp. 115-132.

[5] E. Hesstvedt, Ø. Hov and I. A. Isaksen, Quasi-steady-state approximations in air pollution modeling: comparison of two numerical schemes for oxidant prediction, Int. Journal of Chemical Kinetics 10 (1978), pp. 971-994.

[6] G. I. Marchuk, Mathematical modeling for the problems of the environment, Studies in Mathematics and Applications, No. 16, North-Holland, Amsterdam, 1985.

[7] Tz. Ostromsky, I. Dimov, R. Georgieva, Z. Zlatev, Air pollution modeling, sensitivity analysis and parallel implementation, International Journal of Environment and Pollution, Vol. 46, No. 1-2 (2011), pp. 83-96.

[8] Tz. Ostromsky, I. Dimov, P. Marinov, R. Georgieva, Z. Zlatev, Advanced Sensitivity Analysis of the Danish Eulerian Model in Parallel and Grid Environment, AIP Conf. Proc. 1404 (2012), pp. 225 - 232. E-version: *http://link.aip.org/link/doi/10.1063/1.3659924*

[9] Tz. Ostromsky, Z. Zlatev, Parallel Implementation of a Large-scale 3D Air Pollution Model, Large Scale Scientific Computing (S. Margenov, J. Wasniewski, P. Yalamov, Eds.), LNCS-2179, Springer (2001), pp. 309-316.

[10] Tz. Ostromsky, Z. Zlatev, Flexible Two-level Parallel Implementations of a Large Air Pollution Model, in: Numerical Methods and Applications

(I.Dimov, I.Lirkov, S. Margenov, Z. Zlatev, Eds.), LNCS-2542, Springer (2002), pp. 545-554.

[11] WEB-site of the Danish Eulerian Model, available at: *http://www.dmu.dk/AtmosphericEnvironment/DEM*

[12] Z. Zlatev, Computer treatment of large air pollution models, Kluwer, 1995.

[13] Z. Zlatev, I. Dimov, Computational and Numerical Challenges in Environmental Modeling, Elsevier, Amsterdam (2006).

[14] Z. Zlatev, I. Dimov, K. Georgiev, Three-dimensional version of the Danish Eulerian Model, *Zeitschrift für Angewandte Mathematik und Mechanik,* **76**, *S4* (1996), pp. 473-476.

CHAPTER THREE

CHAOTIC SYSTEMS AND THEIR APPLICATION IN INDUSTRY

CHAOTIC SYSTEMS AND THEIR APPLICATION IN INDUSTRY

GREGORY AGRANOVICH, ELENA LITSYN AND ANGELA SLAVOVA

Introduction

Pattern formation in dynamic systems is found in many areas of science and engineering, such as biology, fluids, plasmas, neurobiology, and nonlinear optics [1]. Turing's seminal paper [14] on pattern formation sought to account for morphogenesis in biological organisms, e.g. the tentacle patterns in Hydra. The dynamics of spatial pattern formation have also been used to model the development of patterns on seashells and animal coasts. In engineering, pattern formation has been suggested as a mechanism for fingerprint enhancement. A system of reaction-diffusion equations is the prototypical system for modeling pattern formation.

In this chapter we propose an explanation of pattern formation behavior in a variety of complex systems, e.g. biological, neuronal etc. We describe Cellular Nonlinear Networks (CNN) models [2,12] that exhibit spatially organized patterns of activity which are formed by reaction-diffusion. Partial differential equations of diffusion type have long served as models for regulatory feedbacks and pattern formation in aggregates of living cells. We propose new receptor-based models for pattern formation and regulation in multi-cellular biological systems. The idea is that patterns are controlled by specific cell-surface receptors, which transmit to the cells signals responsible for their differentiation. The main aim of this work is to check which aspects of self-organization and regeneration can be explained within the framework of CNNs.

The simplest model describing receptor-ligands is given in the form of three equations. It takes into consideration the density of free receptors, of the bound receptors and of the ligands. We use a representation of this simplest receptor-based model that is as generic as possible and based on the scheme shown in Fig.1.

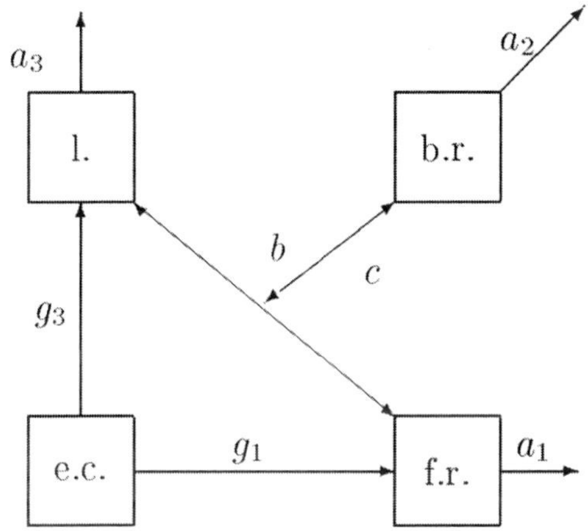

Fig.1. General scheme of the simplest receptor-based model.

The abbreviations in Figure 1 are as follows: l. -ligands, b.r. - bound receptors, e.c. - epithelial cells, f.r. - free receptors. This model is based on the idea that epithelial cells secrete ligands (a regulatory biochemical), which diffuse locally within the interstitial space and bind to free receptors on the cell surface. We assume that new ligands and new free receptors are produced on the cell surface through a combination of recycling (dissociation of bound receptors) and *de novo* production within the cell. Then a ligand binds to a free receptor reversibly, which results in a bound receptor that is internal to the cell. Bound receptors also dissociate. Both ligands and free receptors undergo natural decay.

Hysteresis seems to be important in modeling biological development, since according to observations inductive signals are present only in the specific time interval of the development [9]. It triggers the changes in the cell's nucleus and evokes differentiation, which does not revert when the signal is stopped. The developmental process is irreversible. Hysteresis results from multiple steady states. A reaction-diffusion model involving a hysteretic function was proposed by Hoppensteadt and Jäger [8]. They assumed that the cell's growth had a hysteretic dependence on the amount of nutrients and acid present. Pattern formation in this model is caused by the initial instability of ordinary differential equations (ODEs).

Receptor-Based Model

We consider a one-dimensional epithelial sheet of length L. We denote the density of ligands by $v(t,x)$, where x and t are space and time coordinates, with x increasing from 0 to L along the body column. The density of free receptors is denoted by $u(x,t)$. Consider a system of one reaction-diffusion equation and one ordinary differential equation (ODE):

$$\left| \begin{array}{l} u_t = \Delta u + f(u,v) \\ v_t = g(u,v), \end{array} \right. \tag{1}$$

where the functions $f(u,v)$ and $g(u,v)$ represent the rate of production of new free receptors and ligands respectively, and they are given by:

$$\left| \begin{array}{l} f(u,v) = -c_1 \frac{u}{1+u^2} + \frac{b_1 u}{(1+u^2-u)(1+v)} \\ g(u,v) = -c_2 \frac{v}{1+v^2} + \frac{b_2 v}{(1+v^2-v)(1+u)}, \end{array} \right. \tag{2}$$

c_1 is the rate of decay of free receptors, c_2 is the rate of decay of ligands, $b_i > 0$, $i = 1,2$ are constants.

Systems composed of both diffusion-type and ordinary differential equations cause some difficulties, since both the existence and the behavior of the solutions are more difficult to establish. Many aspects of qualitative behavior have to be investigated numerically. For this purpose we shall apply a Cellular Nonlinear Network (CNN) approach [2] to studying the system (1), (2).

The case when $\partial_v f(u,v) \leq 0$ and $\partial_u g(u,v) \leq 0$ hold has been studied by Heinze and Schweizer [6]. They prove the existence of stationary and travelling fronts, and they also investigate the stability of these solutions.

System (1), (2) describes a receptor-based model in which the production of new receptors and ligands in a steady state has a hysteretic dependence on the amount of new receptors and ligands, in the sense that in a steady state the density of the new receptors and ligands is a third order polynomial divided by a second order polynomial. Periodic solutions of (1), (2) have been found, and the existence of a stationary standing wave or a spatio-temporal solution oscillating in time have been obtained numerically.

We shall study the dynamic behavior of the CNN model of (1), (2) and the emergence or complexity of the model will be proved. The edge of chaos phenomena will be presented. Then we shall design a discrete-continuous regulator of the CNN model in order to stabilize the chaotic motion to an admissible solution which is connected in some way to the original behavior of the system (1), (2).

Cellular Nonlinear Network model

We can present the dynamic systems for a general CNN whose cells are made of time-invariant circuit elements for each cell $C(ij)$ in the following way [2]:

$$\dot{x}_{ij} = -g(x_{ij}, u_{ij}, I_{ij}^{s}), \tag{3}$$

where $x_{ij} \in R^m$, u_{ij} is usually a scalar. In most cases, the interactions (spatial coupling) with the neighbor cell $C(i+k, j+l)$ are specified by a CNN synaptic law:

$$I_{ij}^{s} = A_{ij,kl}x_{i+k,j+l} + \tilde{A}_{ij,kl} * f_{kl}(x_{ij}, x_{i+k,j+l}) + \tag{4}$$

$$+\tilde{B}_{ij,kl} * u_{i+k,j+l}(t).$$

The first term $A_{ij,kl}x_{i+k,j+l}$ of (4) is simply a linear feedback of the states of the neighborhood nodes. The second term provides an arbitrary nonlinear coupling, and the third term accounts for the contributions from the external inputs of each neighbor cell that is located in the N_r neighborhood.

Complete stability, i.e. convergence of each trajectory towards some stationary state, is a fundamental dynamic property used to design CNNs for accomplishing important tasks in applications. The most basic result on complete stability is certainly the one which requires that the CNN interconnection matrix $\tilde{A}$ be symmetric. Also some special classes of nonsymmetric CNNs, such as cooperative CNNs, were shown to be completely stable. In the general case, however, competitive (inhibitory) CNNs may exhibit stable nonlinear oscillations.

In our particular case for the model with hysteresis (1), (2) we shall take a one-dimensional discretized Laplacian template:

$$A: (1, -2, 1)$$

Therefore the CNN representation for our hysteresis model (1), (2) will be the following:

$$\frac{du_j}{dt} = (u_{j-1} - 2u_j + u_{j+1}) + \tag{5}$$

$$+f(u_j, v_j)$$

$$\frac{dv_j}{dt} = g(u_j, v_j), 1 \leq j \leq N.$$

The above system is actually a system of ODE which is identified as the state equation of an autonomous CNN made of N cells. For the output of our CNN model (5) we will take the standard sigmoid function.

Edge of chaos in the hysteresis CNN model

The theory of local activity [3,4] provides a definitive answer to the fundamental question: what are the values of the cell parameter for which the interconnected system may exhibit complexity? The answer is: the necessary condition for a non-conservative system to exhibit complexity is to have its cell locally active. The theory which will be presented below offers a constructive analytical method for uncovering local activity. In particular, for a diffusion CNN model, one can determine the domain of the cell parameters in order for the cells to be locally active, and thus potentially capable of exhibiting complexity. This precisely defined parameter domain is called the edge of chaos.

We develop the following constructive algorithm for determining the edge of chaos:

1. Map hysteresis model (1), (2) into the following associated discrete-space version which we shall call the hysteresis CNN model:

$$\frac{du_j}{dt} = (u_{j-1} - 2u_j + u_{j+1}) + \tag{6}$$

$$+f(u_j, v_j) = U + f$$

$$\frac{dv_j}{dt} = g(u_j, v_j) = g, 1 \leq j \leq N.$$

2. Find the equilibrium points of (6). According to the theory of dynamic systems, the equilibrium points (u^e, v^e) of (6) are those for which:

$$U + f(u^e, v^e) = 0, \tag{7}$$

$$g(u^e, v^e) = 0.$$

If we substitute (2) into (7), we obtain:

$$U - c_1 \frac{u^e}{1+(u^e)^2} + \frac{b_1 u^e}{(1+(u^e)^2 - u^e)(1+v^e)} = 0 \tag{8}$$

$$-c_2 \frac{v}{1+(v^e)^2} + \frac{b_2 v^e}{(1+(v^e)^2 - v^e)(1+u^e)} = 0.$$

After solving (8), we obtain that it has one, two or three real roots, $E_1 = (u_1^e, v_1^e)$, $E_2 = (u_2^e, v_2^e)$, $E_3 = (u_3^e, v_3^e)$, respectively. In general, these roots are functions of the cell parameters $c_{1,2}$ and $b_{1,2}$.

3. We calculate now the four cell coefficients of the Jacobian matrix of (8) about each system equilibrium point E_i, $i = 1,2,3$:

$$\begin{bmatrix} \frac{\partial f(u,v)}{\partial u} & \frac{\partial f(u,v)}{\partial v} \\ \frac{\partial g(u,v)}{\partial u} & \frac{\partial g(u,v)}{\partial v} \end{bmatrix} |_{(u,v)=E_i, i=1,2,3} = \begin{bmatrix} f_1^e & f_2^e \\ g_1^e & g_2^e \end{bmatrix}. \tag{9}$$

4. Calculate the trace $Tr(E_i)$ and the determinant $\Delta(E_i)$ of the Jacobian matrix (9) for each equilibrium point:

$$Tr(E_i) = f_1^e(E_i) + g_2^e(E_i), \tag{10}$$

$$\Delta(E_i) = f_1^e(E_i)g_2^e(E_i) - f_2^e(E_i)g_1^e(E_i).$$

5. We shall identify the cell state variables u_j and v_j as follows: u_j is associated with the node-to-datum voltage at node j of a grid G_1 of linear resistors; v_j is associated with the node-to-datum voltage at node j of a second grid G_2. We identify the coupling input U with the current leaving node j, i.e. entering the cell connected to node j. In this case, the associated electronic circuit has three terminals with two node-to-datum voltages (u_j, v_j) and one terminal current U. The importance of the circuit model is not only in the fact that we have a convenient physical implementation, but also in the fact that well-known results from classic circuit theory can be used to justify the cells' local activity. In this sense, if there is at least one equilibrium point for which the circuit model of the cell acts like a source of a "small signal", i.e. if the cell is capable of injecting a net small-signal average power into the passive resistive grids, then the cell is said to be locally active [3].

Definition 2.1 *A diffusion cell is locally active at an equilibrium point E_i, if the matrix:*

$$L_{E_i} = \begin{bmatrix} -2f_1^e & -(f_2^e + g_1^e) \\ -(f_2^e + g_1^e) & -2g_2^e \end{bmatrix} \tag{11}$$

is not semi-definite at the equilibrium point E_i, $i = 1,2,3$.

Definition 2.2 *Local activity region $LAR(E_i)$ is defined as follows:*

$$LAR(E_i): g_2^e > 0 \quad 4f_1^e g_2^e < (f_2^e + g_1^e)^2, i = 1,2,3. \tag{12}$$

Definition 2.3 *Stable and locally active region $SLAR(E_i)$ at the equilibrium point E_i for the hysteresis CNN model (6) is such that $Tr < 0$ and $\Delta > 0$.*

6. Edge of chaos

In the literature [4], the so-called edge of chaos (EC) means a region in the parameter space of a dynamic system where complex phenomena and information processing can emerge. We shall try to define more precisely these phenomena till now known only via empirical examples. Moreover, we shall present an algorithm for determining the edge of chaos for diffusion CNN models as for our hysteresis CNN model (6). Let us set $U = 0$ in the equilibrium equations:

$$U + f(u^e, v^e) = 0, \tag{13}$$

$$g(u^e, v^e) = 0.$$

After solving the above system we get that it can have one, two or three real solutions and therefore we have three equilibrium points $E_i(u_i^e, v_i^e)$, $i = 1,2,3$.

Our next step is to calculate the local cell coefficients f_1^e, f_2^e, g_1^e, g_2^e from (9) about each equilibrium point E_i, $i = 1,2,3$. We determine LAR and SLAR for each point in the cell parameter space and we find that there is at least one equilibrium point $E(0,0)$ for which these conditions hold. We shall identify the edge of chaos domain EC in the cell parameter space by using the following definition:

Definition 2.4 *A hysteresis CNN is said to be operating on the edge of chaos EC if there is at least one equilibrium point E_i, $i = 1,2,3$ which is both locally active and stable when $U = 0$.*

The following theorem then holds:

Theorem 2.5 *A hysteresis CNN model for the system (1), (2) is operating in the edge of chaos regime if and only if $c_1 > b_1 > 0$, $c_2 > b_2 > 0$. For these parameter values there is at least one equilibrium point which is both locally active and stable.*

Proof: After solving (13) we have three equilibrium points $E_1 = (0,0)$, $E_2 = (-1,-1)$ and $E_3 = (-1, \frac{-2b_1-3c_1}{3c_1})$. Then we check the conditions for local activity and stability given by Definitions 2.2 and 2.3. The results show that the equilibrium point $E_1 = (0,0)$ satisfies these conditions for the following parameter set $c_1 > b_1 > 0$ and $c_2 > b_2 > 0$. Therefore, there is at least one equilibrium point which is both locally active and stable. According to Definition 2.4, this means that the hysteresis CNN model (6) is operating in the edge of chaos regime. Theorem is proved.

According to the simulations for the hysteresis CNN model (6) we obtain the following figures:

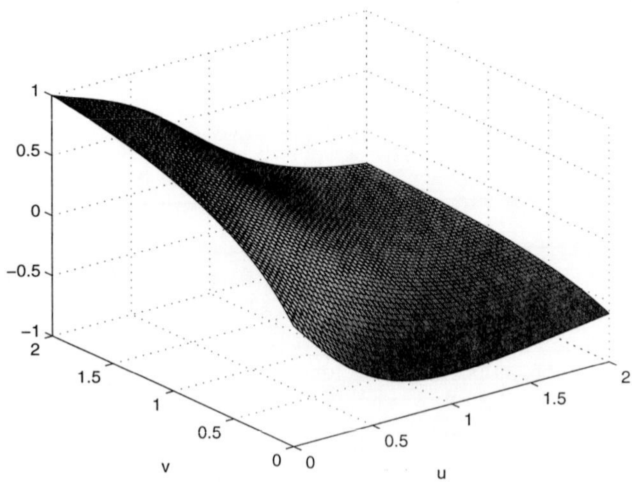

Fig.2. Stationary wave solution of the receptor-based CNN model.

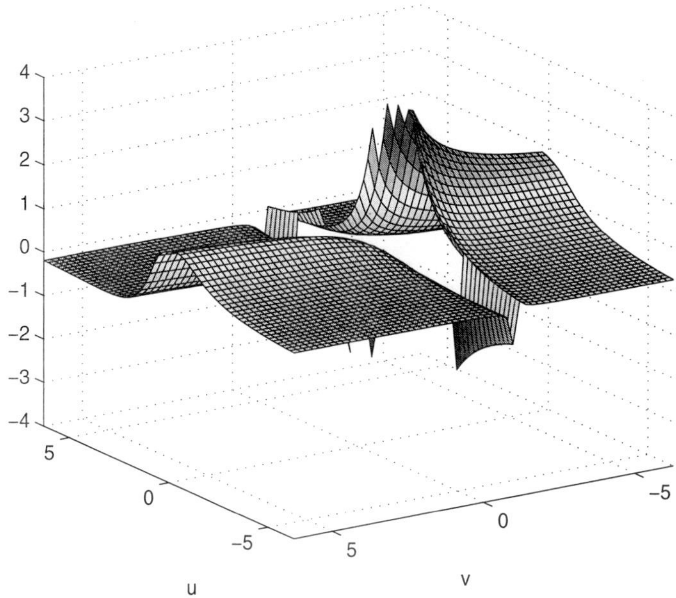

Fig.3. Spatio-temporal solution of the receptor-based CNN model.

Remark 2.6 Simulations show that for the model (6) we can have a gradient-like solution for the density of free receptors (standing wave) which is stationary in time (see Fig.2) or a spatio-temporal solution oscillating in time (see Fig. 3). The formation and persistence of the several peaks in Fig.3 are the result of the bi-stability of the reaction term. For such a model we can have various stable solutions which are transitions between the stable steady states.

Continuous feedback control of the CNN model

Let us rewrite the hysteresis CNN model (6) by the following simultaneous $2 * N$ ordinary differential equations:

$$\frac{du_j}{dt} = (u_{j-1} - 2u_j + u_{j+1}) + \tag{14}$$

$$+f(u_j, v_j) + z_{uj}, j = 1 \dots N,$$

$$\frac{dv_j}{dt} = g(u_j + z_{vj}, j = 1 \dots N,$$

where z_{uj}, z_{vj} are controls and

$$\left| \begin{array}{l} f(u,v) = -c_1 \frac{u}{1+u^2} + \frac{b_1 u}{(1+u^2-u)(1+v)} \\ g(u,v) = -c_2 \frac{v}{1+v^2} + \frac{b_2 v}{(1+v^2-v)(1+u)}, \end{array} \right. . \quad (15)$$

Number of cells N lies in bounds $1 \leq N \leq 25$. Constant coefficients

$$c_j \in [0,1], b_j \in [1,2]. \quad (16)$$

Boundary conditions for (14) are

$$u(t,-1) = u(t, N+1) = 0$$

and initial conditions are in the intervals

$$u(0,j) \in [0,2], v(0,j) \in [0,2].$$

The matrix form of (14) is

$$\frac{dU}{dt} = AU + F(U,V) + Z_U, \quad (17)$$

$$\frac{dV}{dt} = G(U,V) + Z_V$$

where the tridiagonal $N \times N$ matrix

$$A = \begin{bmatrix} -2 & 1 & 0 & \dots & \dots & \dots & 0 \\ 1 & -2 & 1 & 0 & \dots & \dots & 0 \\ \dots & \dots & \dots & \dots & \dots & \dots & \dots \\ 0 & \dots & \dots & 0 & 1 & -2 & 1 \\ 0 & \dots & \dots & \dots & 0 & 1 & -2 \end{bmatrix}. \quad (18)$$

The nonlinear state model of (14) is in the following matrix form

$$\frac{dX}{dt} = A_e X + F_e(X) + Z \tag{19}$$

in which the block matrices are

$$X = \begin{bmatrix} U \\ V \end{bmatrix}, F_e = \begin{bmatrix} F \\ G \end{bmatrix}, Z = \begin{bmatrix} Z_U \\ Z_V \end{bmatrix}, A_e = \begin{bmatrix} A & 0_{N\times N} \\ 0_{N\times N} & 0_{N\times N} \end{bmatrix}. \tag{20}$$

The linearized model of (19) in the neighborhood of the equilibrium point X_s is

$$\frac{dX}{dt} = (A_e + F_{Xe}(X_s))X + Z, \tag{21}$$

where

$$F_{Xe}(X_s) = \frac{\partial F_e(X_s)}{dX} = \begin{bmatrix} \frac{\partial F(X_s)}{\partial U} & \frac{\partial F(X_s)}{\partial V} \\ \frac{\partial G(X_s)}{\partial U} & \frac{\partial G(X_s)}{\partial V} \end{bmatrix}. \tag{22}$$

At the equilibrium point $E_1 = (0,0)$ the coefficient matrix of the linearized system (21) takes the form

$$A_e + F_{Xe}(X_s) = \begin{bmatrix} A + (b_1 - c_1)E & 0_{N\times N} \\ 0_{N\times N} & (b_2 - c_2)E \end{bmatrix}. \tag{23}$$

It follows from (23) that the eigenvalues $\{\lambda_j^0, j = 1 \dots 2N\}$ of the linearized system (21) are

$$\lambda_j^0 = \lambda_j + b_1 - c_1, j = 1 \dots N; \tag{24}$$

$$\lambda_{N+1}^0 = \cdots = \lambda_{2N}^0 = \mu^0 = b_2 - c_2,$$

where λ_j, $j = 1 \dots N$ are the eigenvalues of the matrix A (18).

We shall seek stabilized controls for (21), (23) as follows

$$Z_U = k_u V, Z_V = k_v U + k_w V, \tag{25}$$

where the values of the scalar control coefficients k_u, k_v, k_w are to be found. The close-loop system (21), (23), (25) has the matrix representation

$$\frac{dX}{dt} = (A_e + F_{Xe}(X_s) + R)X, \tag{26}$$

where

$$R = \begin{bmatrix} 0_{N\times N} & k_u E \\ k_v E & k_w E \end{bmatrix}.$$

So we obtain the following close-loop system's dynamical matrix

$$A_{cl} = \begin{bmatrix} A + (b_1 - c_1)E & k_u E \\ k_v E & (k_w - b_2 - c_2)E \end{bmatrix}. \tag{27}$$

The characteristic polynomial of (27) is

$$det(sE - A_{cl}) = det\begin{bmatrix} (s - b_1 + c_1)E - A & -k_u E \\ -k_v E & (s - k_w - b_2 + c_2)E \end{bmatrix}$$

$$= det(\begin{bmatrix} (s - b_1 + c_1)E - A & -k_u E \\ -k_v E & (s - k_w - b_2 + c_2)E \end{bmatrix} \times$$

$$\times \begin{bmatrix} E & 0_{N\times N} \\ \frac{k_v}{s-k_w-b_2+c_2} E & E \end{bmatrix})$$

$$= det\begin{bmatrix} (s - b_1 + c_1 - \frac{k_u k_v}{s-b_2+c_2})E - A & -k_u E \\ 0_{N\times N} & (s - k_w - b_2 + c_2)E \end{bmatrix}$$

$$= (s - k_w - b_2 + c_2)^N \cdot det[(s - b_1 + c_1 - \frac{k_u k_v}{s-k_w-b_2+c_2})E - A]$$

$$= (s - k_w - b_2 + c_2)^N \cdot det[\frac{(s-b_1+c_1)(s-k_w-b_2+c_2)-k_u k_v}{s-k_w-b_2+c_2})E - A].$$

Let matrix A have m eigenvalues λ_j of order m_j, $j = 1 \dots m$. So the characteristic polynomial of A is

$$det[sE - A] = \prod_{j=1}^{m} (s - \lambda_j)^{m_j}.$$

Then the characteristic polynomial of (27) can be represented as

$$det(sE - A_{cl}) = \prod_{j=1}^{m} [(s - b_1 + c_1 - \lambda_j)(s - k_w - b_2 + c_2) - k_u k_v]^{m_j} \tag{28}$$

As it follows from (24), (26) the eigenvalues $\{\lambda_j^{cl}, j = 1 \dots 2N\}$ of the close-loop system (21), (23), (25) are solutions of the equations

$$(s - \lambda_j^0)(s - k_w - \mu^0) - k_u k_v = 0, j = 1 \dots N,$$

or, after some trivial algebraic calculations,

$$s^2 - (k_w + \mu^0 + \lambda_j^0)s + [\lambda_j^0(k_w + \mu^0) - k_u k_v] = 0, j = 1 \dots N. \tag{29}$$

The next theorem gives the opportunity to find feedback coefficients (25) for the stabilizing close-loop system (26) and in addition ensures the designed rate of convergence.

Theorem 3.1 *If the parameters of the close-loop system (26) with open-loop eigenvalues (24) satisfy the following inequalities*

$$k_w \leq -2\sigma - \mu^0 - max_j(\lambda_j^0) \tag{30}$$

$$k_u k_v \leq \sigma^2 + \sigma(k_w + \mu^0) + min_j \lambda_j^0(\sigma + k_w + \mu^0)$$

for some $\sigma \geq 0$, then the eigenvalues of the close-loop system satisfy the inequality $\lambda_j^{cl} \leq -\sigma$. For each value of $\sigma \geq 0$ there exist values of control feedback parameters (25), for which the inequalities (30) are satisfied.

Proof: Let us perform in the characteristic equations (29) of the close-loop system a change of variable s to $s - \sigma$:

$$s^2 - (2\sigma + k_w + \mu^0 + \lambda_j^0)s + [\sigma^2 + \sigma(k_w + \mu^0 + \lambda_j^0) + \lambda_j^0(k_w + \mu^0) - k_u k_v] = 0.$$

It is well known, [16], that positivity of coefficients is the necessary and sufficient condition for a second - order polynomial to be Hurwitz. Let us shift a polynomial argument s by $\sigma > 0$ to the left. It is obvious, that the Hurwitz property of the shifted polynomial implies that the roots s_i of the initial polynomial satisfy inequality $s_i < \sigma$.

Then the conditions $\lambda^{cl} \leq -\sigma$ are satisfied if and only if the following inequalities

$$k_w \leq -2\sigma - \mu^0 - \lambda_j^0, k_u k_v \leq \sigma^2 + \sigma(k_w + \mu^0) + \lambda_j^0) + \lambda_j^0(k_w + \mu^0)$$

hold for every eigenvalue λ_j^0, $j = 1 \dots N(24)$ of the open-loop system. Replacing the right-hand parts of these inequalities by the least value according to j, we obtain (30). The rest of the proof follows immediately from the form of the inequalities.

As can be seen from figures 4 and 5, the proposed method allows us to stabilize the system's dynamics so as to assign its rate of convergence.

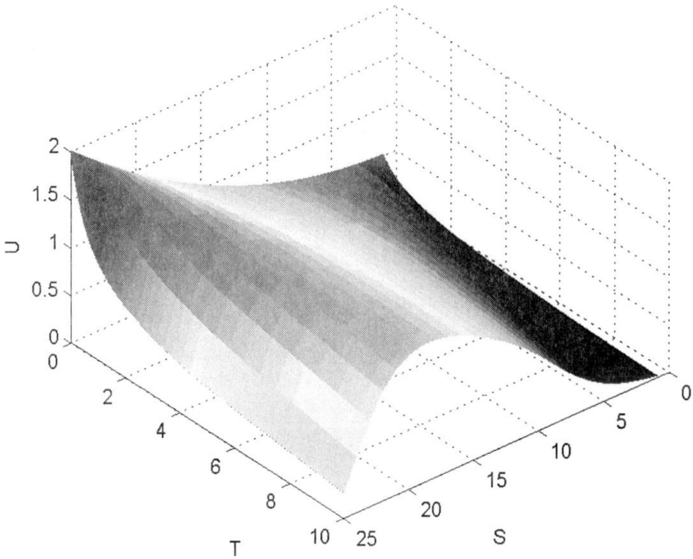

Fig.4. Spatio-temporal solution of the un-stabilized receptor-based CNN model.

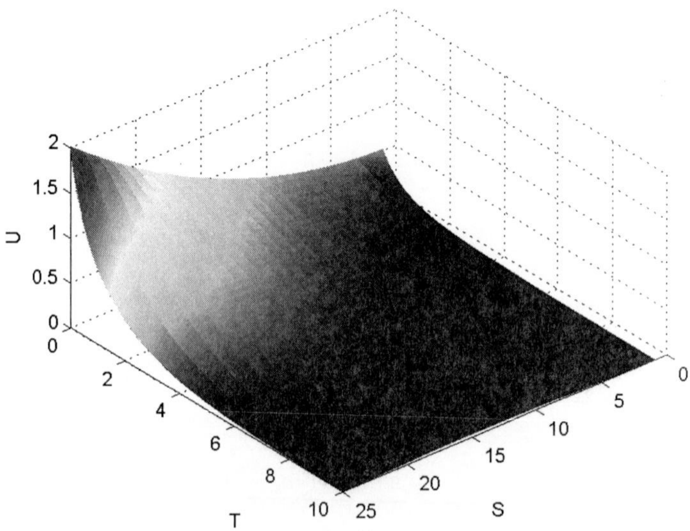

Fig.5. Spatio-temporal solution of the stabilized receptor-based CNN model, $\sigma = 0.5$.

Coupled FitzHugh-Nagumo neural system

Nonlinear reaction-diffusion type of equations are widely used to describe phenomena in different fields, as in biology: the Fisher model, the Hodgkin-Huxley model and its simplification, the FitzHugh-Nagumo nerve conduction model, etc. [1,5,10]. In this section we shall study a coupled FitzHugh-Nagumo neural system and the phenomenon "edge of chaos", as well as the feedback stabilization of the system.

Hodgkin injected a DC-current of varying amplitude and discovered that some systems could exhibit repetitive spiking with arbitrary low frequencies, while others discharged in a narrow frequency band. In the seminal paper by Rinzel and Ermentrout it was shown that the difference in behavior is due to different bifurcation mechanisms of excitability. For dynamic systems in neuroscience, the type of bifurcation determines the computational properties of neurons. Neuronal models can be excitable for some values of parameters, and fire spikes periodically for other values.

These two types of dynamics correspond to a stable equilibrium and a limit cycle attractor, respectively. When the parameters change, the models can exhibit a transition from one qualitative type of dynamics to another. Thus, the stability and bifurcation of neural network systems attract a lot of attention. At the same time, the transmission of information among neurons is carried out through synapses, and therefore the coupling among neurons is also achieved through synapses. Coupling among neurons can be classified into a gap junction and a chemical synapse coupling. Chaos and bifurcations can occur even in the most simple system, and, moreover, coupled neurons could synchronize and exhibit collective behavior.

The famous Hodgkin-Huxley neuron model [7] is the first mathematical model describing neural excitation transmission derived from the angle of physics and lays down the basis of electrical neurophysiology. The FitzHugh-Nagumo equation, which is a simplification of the Hodgkin-Huxley model, describes the generation and propagation of the nerve impulse along the giant axon of the squid. The FitzHugh-Nagumo systems [5,10] are of fundamental importance for understanding the qualitative nature of nerve impulse propagation. Based on the finite propagating speed in the transmission of signals between the neurons, the following coupled FitzHugh-Nagumo neural system is proposed:

$$\left|\begin{array}{l} \dot{u}_1 = -u_1(u_1-1)(u_1-a) - u_2 + cf(u_3) \\ \dot{u}_2 = b(u_1 - \gamma u_2) \\ \dot{u}_3 = -u_3(u_3-1)(u_3-a) - u_4 + cf(u_1) \\ \dot{u}_4 = b(u_3 - \gamma u_4), \end{array}\right. \qquad (31)$$

where a, b, γ are positive constants, $u_{1,2}$ represent transmission variables, and $u_{3,4}$ are receiving variables; c measures the coupling strength, $f \in C^3$, $f(0) = 0$, $f'(0) = 1$. We shall take $f(x) = tanh(x)$ in our investigation. System (31) is symmetric. Thus it is interesting to consider the existence, spatio-temporal patterns and stability of its Hopf bifurcating periodic solutions.

Edge of chaos of coupled FitzHugh-Nagumo CNN model

For a coupled FitzHugh-Nagumo system (31), the CNN model will be the following:

$$\left|\begin{aligned} \frac{du_j^1}{dt} &= -u_j^1(u_j^1-1)(u_j^1-\mathrm{a}) - u_j^2 + cf(u_j^3) \\ \frac{du_j^2}{dt} &= b(u_j^1 - \gamma u_j^2) \\ \frac{du_j^3}{dt} &= -u_j^3(u_j^3-1)(u_j^3-a) - u_j^4 + cf(u_j^1) \\ \frac{du_j^4}{dt} &= b(u_j^3 - \gamma u_j^4), j = 1, \dots, n. \end{aligned}\right. \tag{32}$$

The system is transformed into a system of ordinary differential equations which is identified as the state equations of a CNN with appropriate templates. We map the variables u_1, u_2, u_3 and u_4 into CNN layers such that the state voltage of a CNN cell at a grid point is $u_j^i, i = 1,2,3,4$, $n = M.M$; M is number of the cells.

Simulations of the above CNN model for different parameter sets are given in Fig.6:

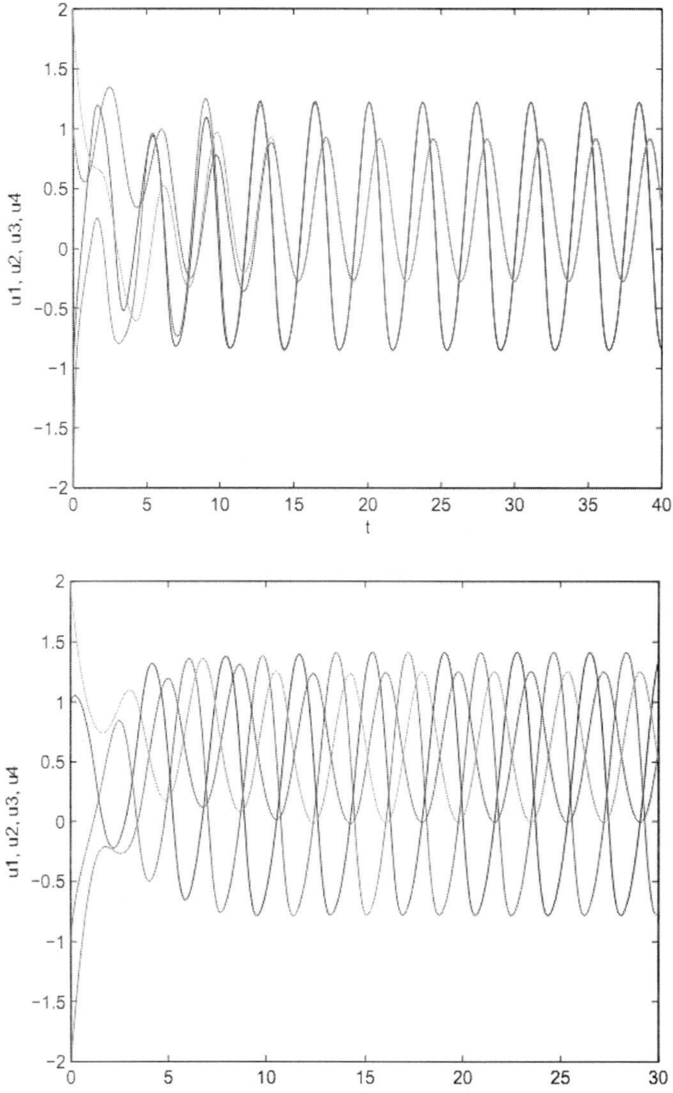

Fig.6. Simulations of the CNN model (32).

We shall present an algorithm for determining the edge of chaos for reaction-diffusion CNN models as coupled FitzHugh-Nagumo CNN models (32).

We apply the following constructive algorithm:

1. Map a coupled FitzHugh-Nagumo system (31) into the associated discrete-space version (32), which we shall call the coupled FitzHugh-Nagumo CNN model.

2. Find the equilibrium points of (32). According to the theory of dynamic systems the equilibrium points $\hat{u}_j^i$ of (32) are those for which:

$$\begin{cases} -u_j^1(u_j^1-1)(u_j^1-a)-u_j^2+c\ tanh(u_j^3)=0 \\ b(u_j^1-\gamma u_j^2)=0 \\ -u_j^3(u_j^3-1)(u_j^3-a)-u_j^4+c\ tanh(u_j^1)=0 \\ b(u_j^3-\gamma u_j^4)=0. \end{cases} \tag{33}$$

Equation (33) may have one, two, three or four real roots $\hat{u}_j^1$, $\hat{u}_j^2$, $\hat{u}_j^3$, $\hat{u}_j^4$ respectively. In general, these roots are functions of the cell parameters a, b, c, γ. In other words, we have $\hat{u}_j^i = \hat{u}_j^i(a, b, c, \gamma)$, $i = 1,2,3,4$. We shall consider only the equilibrium point $E_0 = (0,0,0,0)$.

3. Calculate now the Jacobian matrix of (33) about equilibrium point E_0. In our particular case the associate linear system in a sufficient small neighborhood of the equilibrium point E_0 can be given by

$$\frac{dz}{dt} = DF(E_0)z$$

$DF(E_0) = J$ is the Jacobian matrix of each of the equilibrium points and can be computed by:

$$J_{p,s} = \frac{\partial F_p}{\partial u_s}|_{u=E_0}, 1 \le p, s \le n. \tag{34}$$

In our particular case the Jacobian matrix in the equilibrium point E_0 is:

$$J = \begin{bmatrix} -a & -1 & c & 0 \\ b & -b\gamma & 0 & 0 \\ c & 0 & -a & -1 \\ 0 & 0 & b & -b\gamma \end{bmatrix}.$$

4. Calculate the trace $Tr(E_0) = \sum_{q=1}^{N} \lambda_q$. In the equilibrium point $E_0 = (0,0,0,0)$ trace is $Tr(0,0,0,0) = -a - b\gamma - a - b\gamma = -2(a + b\gamma)$.

5. We shall identify the cell state variables u_j as follows: u_j is associated with the node-to-datum voltage at node (j) of a two-dimensional grid G of linear resistors. The importance of the circuit model is not only in the fact that we have a convenient means of physical implementation, but also in the fact that well-known results from classic circuit theory can be used to justify the cells' local activity. In this sense, if there is at least one equilibrium point for which the circuit model of the cell acts like a source of "small signal" power, in a precise sense defined in Defn. 4.1, i.e. if the cell is capable of injecting a net small-signal average power into the passive resistive grids then the cell is said to be locally active [3,4].

Definition 4.1 *Stable and Locally Active Region $SLAR(E)$ at the equilibrium point E_0 for coupled FitzHugh-Nagumo CNN model (32) is such that $Tr < 0$.*

In our particular case we have: $Tr(0,0,0,0) = -2(a + b\gamma) < 0$ for all a, b, γ positive. Therefore in the equilibrium point $E_0 = (0,0,0,0)$ we have a stable and locally active region.

6. Edge of chaos.

We shall identify the edge of chaos domain (EC) in the cell parameter space by using the following definition:

Definition 4.2 *Coupled FitzHugh-Nagumo CNN model is said to be operating on the edge of chaos EC if and only if there is at least one equilibrium point E_0, which belongs to $SLAR(E)$.*

The following theorem then holds:

Theorem 4.3 *CNN model of coupled FitzHugh-Nagumo system (31) is operating in the edge of chaos regime for all* a, b *and* γ *positive. For values of this parameter there is at least one equilibrium point which belongs to* $SLAR(E)$.

Proof: After solving (32) we find that one of the equilibrium points is $E_0 = (0,0,0,0)$. Then we check the conditions for local activity and stability given by Definition 4.1. The results show that the equilibrium point $E_0 = (0,0,0,0)$ satisfies these conditions for the parameter set. Therefore, there is at least one equilibrium point which is both locally active and stable. According to Definition 4.2, this means that the FitzHugh-Nagumo CNN model (32) is operating in the edge of chaos regime. Theorem is proved.

The edge of chaos EC in which the coupled FitzHugh-Nagumo CNN model (32) is operating is given in Fig.7 for different parameter sets:

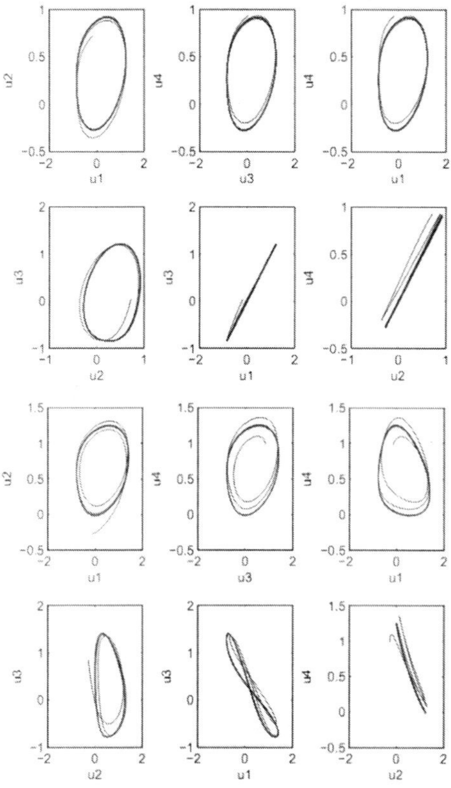

Fig.7. EC

Remark 4.4 In the literature [11,15] spatio-temporal patterns are derived by the following result: the trivial solution of the system (31) undergoes a Hopf bifurcation, that is, when $c = l$ (respectively, $c = -l$), where l is some constant, giving rise to one branch of synchronous (respectively, anti-phase) periodic solutions. So in Fig. 7 we present this synchronization for $c = -2$, $c = 2$, respectively. The parameter set is the following: $a = 0.33, \gamma = 0.47, b = 1$.

Stabilizing feedback control for the coupled FitzHugh-Nagumo CNN model

Let us extend the model (32) by adding to each cell the local linear feedback:

$$\left|\begin{aligned}&\frac{du_j^1}{dt}=-u_j^1(u_j^1-1)(u_j^1-a)-u_j^2+cf(u_j^3)-ku_j^1\\&\frac{du_j^2}{dt}=b(u_j^1-\gamma u_j^2)\\&\frac{du_j^3}{dt}=-u_j^3(u_j^3-1)(u_j^3-a)-u_j^4+cf(u_j^1)-ku_j^3\\&\frac{du_j^4}{dt}=b(u_j^3-\gamma u_j^4),j=1,\dots,n.\end{aligned}\right. \tag{35}$$

where k is the feedback controls coefficient, which is assumed to be equal for all cells.

The problem is to prove that this feedback, which is simple and available for implementation, can stabilize the coupled FitzHugh-Nagumo CNN model. In the following we present a proof of this statement and give sufficient condition on the feedback coefficient values which provide the stability of the CNN nonlinear model (35).

As a first step, we examine the stability conditions of the system (35), linearized in the neighborhood of the zero equilibrium point E_0. This system in a vector-matrix form is given by

$$\frac{dz}{dt}=J(k)z$$

where $J(k)$ is the Jacobian matrix of the controlled CNN in E_0:

$$J(k)=\begin{bmatrix}-(a+k) & -1 & c & 0\\ b & -b\gamma & 0 & 0\\ c & 0 & -(a+k) & -1\\ 0 & 0 & b & -b\gamma\end{bmatrix}. \tag{36}$$

Theorem 5.1 *Let the parameters a, b and γ of a coupled FitzHugh-Nagumo CNN system and feedback coefficient k (35) have positive values. Then its linearized in E_0 model (36) is assymptotically stable for all*

$$k > \sqrt{\left(\frac{(b-1)^2}{8b\gamma}\right)^2 + c^2} + \frac{(b-1)^2}{8b\gamma} - a \tag{37}$$

Proof: Define the quadratic Lyapunov function candidate $L(z) = \frac{1}{2}z^T z$. Then its derivative along the linearized controlled CNN is

$$\frac{dL(z)}{dt} = \frac{1}{2}z^T(J^T(k) + J(k))z = -z^T Q(k)z$$

where

$$Q(k) = \begin{bmatrix} (a+k) & -\frac{b-1}{2} & -c & 0 \\ -\frac{b-1}{2} & b\gamma & 0 & 0 \\ -c & 0 & (a+k) & -\frac{b-1}{2} \\ 0 & 0 & -\frac{b-1}{2} & b\gamma \end{bmatrix}. \tag{38}$$

Therefore $\frac{dL(z)}{dt} < 0$ implies a positive definiteness of $Q(k)$. It can be shown that $Q(k)$ (38) positive definiteness implies

$$\begin{aligned} &a + k > 0 \\ &(a+k)b\gamma - \frac{(b-1)^2}{4} > 0 \\ &b\gamma(a+k)^2 - \frac{(b-1)^2}{4}(a+k) - c^2 b\gamma > 0 \end{aligned}$$

or

$$\begin{aligned} &k > -a \\ &k > \frac{(b-1)^2}{4b\gamma} - a \\ &(a+k)^2 - 2\frac{(b-1)^2}{8b\gamma}(a+k) + \left(\frac{(b-1)^2}{8b\gamma}\right)^2 > \left(\frac{(b-1)^2}{8b\gamma}\right)^2 + c^2 \end{aligned}.$$

The last inequality has equivalent representation

$$a + k - \frac{(b-1)^2}{8b\gamma} > \sqrt{(\frac{(b-1)^2}{8b\gamma})^2 + c^2}$$

from which the theorem inequality follows. Theorem is proved.

For verification of Theorem 2 the eigenvalues of $J(k)$ (36) were calculated related to the values вЂ‹вЂ‹ of feedback coefficient k. Stability of the linear system requires that the eigenvalues $\lambda_j^i, i = 1, \ldots, 4$ of (36) вЂ‹вЂ‹ satisfy the inequality $max_i Re\lambda_j^i < 0$. Dependence of the $max_i Re\lambda_j^i$ on k for the parameter set, which is defined in Remark 1 and $c = -2$, is represented in Figure 8.

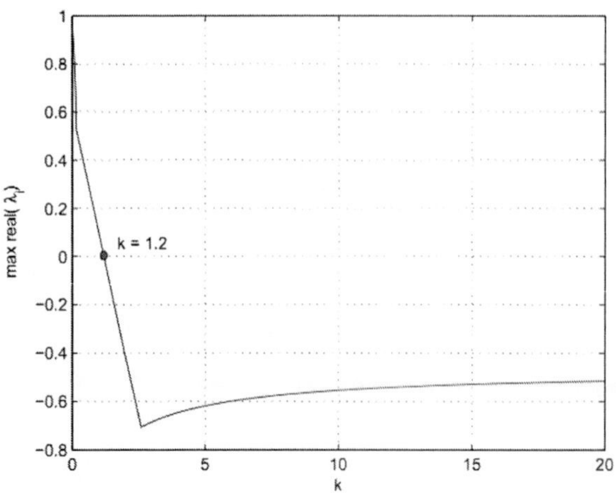

Fig.8. Dependence of real part value of dominant eigenvalue on the feedback coefficient CNN model (35).

The critical value of $k = 1.2$, for which the $max_i Re\lambda_j^i = 0$, is marked in the figure. For this parameter set the inequality (37) gives the critical value $k = 1.67$.

The following theorem provides a sufficient condition to stabilize the nonlinear controlled FitzHugh-Nagumo model (35).

Theorem 5.2 *Let the parameters a, b, γ and feedback coefficient k of the coupled controlled FitzHugh-Nagumo CNN (35) have positive values. Then it is globally asymptotically stable in the zero equilibrium point E_0 for all*

$$k > \sqrt{(\frac{(b-1)^2}{8b\gamma})^2 + c^2} + \frac{(1-a)^2}{4} + \frac{(b-1)^2}{8b\gamma}. \qquad (39)$$

Proof: Let us define the quadratic Lyapunov function candidate

$$L(u_j) = \frac{1}{2}\sum_{i=1}^{4} (u_j^i)^2 .$$

It is obvious that it is a positive definite, decreasing and radially unbounded Lyapunov function. Then its derivative along the model (35) of controlled CNN is

$$\frac{dL(u_j)}{dt} = -(u_j^1)^4 + (1+a)(u_j^1)^3 - a(u_j^1)^2 + (b-1)u_j^1u_j^2 - k(u_j^1)^2$$

$$+cu_j^1 f(u_j^3) - b\gamma(u_j^2)^2$$

$$-(u_j^3)^4 + (1+a)(u_j^3)^3 - a(u_j^3)^2 + (b-1)u_j^3u_j^4 - k(u_j^3)^2$$

$$+cf(u_j^1)u_j^3 - b\gamma(u_j^4)^2 .$$

Selecting the perfect square

$$-(u_j^1)^4 + 2\frac{1+a}{2}(u_j^1)^3 - \frac{(1+a)^2}{4}(u_j^1)^2 = -(u_j^1)^2(u_j^1 - \frac{1+a}{2})^2$$

in the first line of the expression and the same one in the third line, we have the inequality

$$\frac{dL(u_j)}{dt} \le (\frac{(1-a)^2}{4} - k)(u_j^1)^2 + (b-1)u_j^1u_j^2 + cu_j^1 f(u_j^3) - b\gamma(u_j^2)^2$$

$$(\frac{(1-a)^2}{4} - k)(u_j^3)^2 + (b-1)u_j^3 u_j^4 + cf(u_j^1)u_j^3 - b\gamma(u_j^4)^2 .$$

Now we strengthen the inequality, using

$$(b-1)u_j^1 u_j^2 \le |(b-1)| \cdot |u_j^1| \cdot |u_j^2|,$$

$$cu_j^1 f(u_j^3) \le |c| \cdot |u_j^1| \cdot |f(u_j^3)| \le |c| \cdot |u_j^1| \cdot |u_j^3|,$$

and obtain

$$\frac{dL(u_j)}{dt} \le (\frac{(1-a)^2}{4} - k)(u_j^1)^2 + |(b-1)| \cdot |u_j^1| \cdot |u_j^2|$$

$$+(\frac{(1-a)^2}{4} - k)(u_j^3)^2 + |(b-1)| \cdot |u_j^3| \cdot |u_j^4|$$

$$+2|c| \cdot |u_j^1| \cdot |u_j^3| - b\gamma(u_j^2)^2 - b\gamma(u_j^4)^2 .$$

The right-hand part of the last inequality can be represented as quadratic form $-|u_j|^T Q_{nl}(k)|u_j|$, where $|u_j| = [|u_j^1|,.., |u_j^4|]^T$ is the vector of absolute values of the CNN coordinates, and

$$Q_{nl}(k) = \begin{bmatrix} k - \frac{(1-a)^2}{4} & -\frac{|b-1|}{2} & -|c| & 0 \\ -\frac{|b-1|}{2} & b\gamma & 0 & 0 \\ -|c| & 0 & k - \frac{(1-a)^2}{4} & -\frac{|b-1|}{2} \\ 0 & 0 & -\frac{|b-1|}{2} & b\gamma \end{bmatrix}. \quad (40)$$

If the matrix $Q_{nl}(k)$ is positive definite then the nonlinear coupled FitzHugh-Nagumo model (35) is assymptotically stable in the neighborhood of the zero equilibrium point E_0.

It can be shown that $Q_{nl}(k)$ (40) positive definiteness implies

$$k - \frac{(1-a)^2}{4} > 0$$
$$(k - \frac{(1-a)^2}{4})b\gamma - \frac{(b-1)^2}{4} > 0$$
$$b\gamma(k - \frac{(1-a)^2}{4})^2 - \frac{(b-1)^2}{4}(k - \frac{(1-a)^2}{4}) - c^2 b\gamma > 0$$

or

$$k > \frac{(1-a)^2}{4}$$
$$k > \frac{(b-1)^2}{4b\gamma} + \frac{(1-a)^2}{4}$$
$$(k - \frac{(1-a)^2}{4})^2 - 2\frac{(b-1)^2}{8b\gamma}(k - \frac{(1-a)^2}{4}) + (\frac{(b-1)^2}{8b\gamma})^2 > (\frac{(b-1)^2}{8b\gamma})^2 + c^2 .$$

The last inequality has equivalent representation

$$k - \frac{(1-a)^2}{4} - \frac{(b-1)^2}{8b\gamma} > \sqrt{(\frac{(b-1)^2}{8b\gamma})^2 + c^2}$$

from which the theorem inequality follows. Theorem is proved.

For the parameter set, which is defined in Remark 1 and c = - 2, the inequality (37) gives the critical value $k = 2.11$.

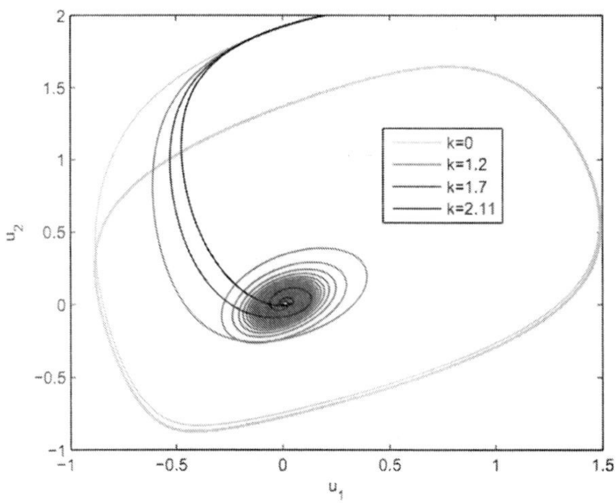

Fig.9. Phase trajectory $u_1 - u_2$ for different values of the feedback coefficient CNN model (35).

Acknowedgements

The authors are partially supported by the bilateral grant between the Israel Academy of Sciences and the Bulgarian Academy of Sciences.

Bibliography

[1] Britton N.F., Reaction-Diffusion Equations and Their Applications to Biology, New York: Academic, 1986.

[2] Chua L.O., CNN:A Paradigm for Complexity, World Scientific Series on Nonlinear Science, Series A - Vol. 31, World Scientific, 1998.

[3] Chua L.O., Local activity is the origin of complexity, Int.J. of Bifurcations and Chaos, Nov. 2005.

[4] Dogaru R., Chua L.O., Edge of chaos and local activity domain of FitzHugh-Nagumo equation, International Journal of Bifurcation and Chaos, vol. 8 (2), pp. 211-257, 1998.

[5] FitzHugh R., Impulses and physiological states in theoretical models of nerve membrane, Biophys. J., 1, 1961, pp.445-466.

[6] Heinze S., Schweizer B., Creeping fronts in degenerate reaction-diffusion systems,

[7] Hodgkin A.L., Huxley A.F., A quantitative description of membrane current and its application to conduction and excitation in nerve, J. Physiology, vol. 117, pp. 500-544, 1952.

[8] Hoppensteadt F., Jäger W., Pattern formation by bacteria, In S. Levin ed., *Lecture Notes in Biomathematics: Biological Growth and Spread,* pp.69-81, Heidelberg, Springer-Verlag, 1980.

[9] Macki J., Nistri P., Zecca P., Mathematical Models for Hysteresis, *SIAM Review,* 53:1:94-123, 1993.

[10] J.Nagumo, S.Arimoto, S.Yoshizawa, An active pulse transmission line simulating nerve axon, Proc.IRE 50, 1962, 2061-2070

[11] J.Rinzel, G.B.Ermentrout, Analysis of Neural Excitability and Oscillations, Methods in Neuronal Modeling, MIT, Cambridge, 1989.

[12] Slavova A., Applications of some mathematical methods in the analysis of Cellular Neural Networks, J.Comp.Appl.Math., 114, pp. 387-404, 2000.

[13] Slavova A., Zecca P., CNN model for studying dynamics and travelling wave solutions of FitzHugh-Nagumo equation, Journ. Comp. and Appl. Math., 151, pp. 13-24, 2003.

[14] Turing A.M., The chemical basis of morphogenesis, *Phil Trans.Roy.Soc. B,*237:37-72,1952.

[15] J. Wu, Symmetric functional differential equations and neural networks with memory, Trans. Am.Math.Soc. 350, 1998, 4799-4838.
[16] M. Vidyasagar, Nonlinear systems analysis, 2nd ed., Philadelphia: Society for Industrial and Applied Mathematics, 2002.

CHAPTER FOUR

INDUSTRIAL APPLICATIONS IN MECHANICS

ON 2D FINITE ELEMENT SIMULATION OF A THERMODYNAMICALLY CONSISTENT LI-ION BATTERY MICROSCALE MODEL

M. TARALOV, V. TARALOVA, P. POPOV, O. ILIEV, A. LATZ AND J. ZAUSCH

Introduction

Li-ion batteries are one of the most popular types of rechargeable batteries for portable electronics as well as for electromobility applications because they have one of the best energy-to-weight ratios, no memory effect, and one of the longest calendar life times, i.e. a slow loss of charge when not in use. Right now progress is driven mainly by laboratory experiments and experience, in contrast to other industries where extensive computer simulations help to improve the production process. Mathematical modeling aims at a better understanding of the complex electrochemical processes in the batteries and prediction of the influence of different factors on their longevity and performance. Furthermore, with the right tools, potentially dangerous scenarios could be identified thus ensuring safe working conditions.

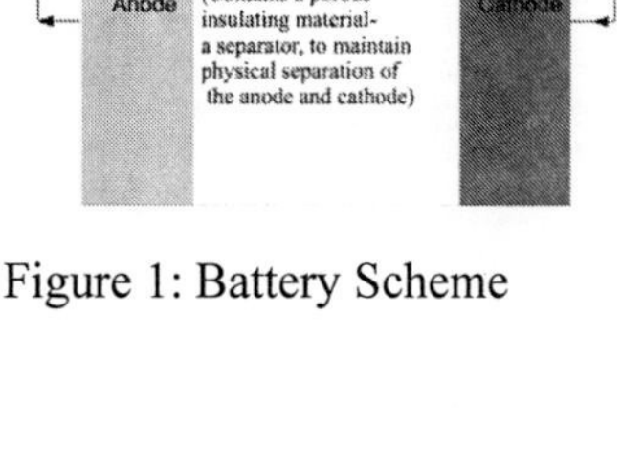

Figure 1: Battery Scheme

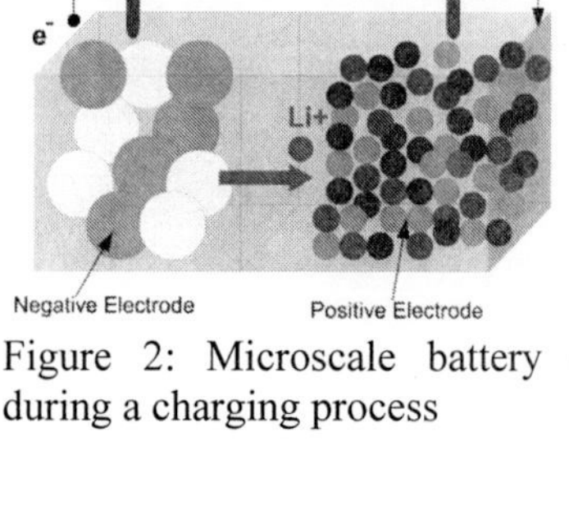

Figure 2: Microscale battery cell during a charging process

A Li-ion battery contains several electrically connected electrochemical cells. Each cell has at least one positive and one negative electrode, which are referred to as cathode and anode, and a separator, c.f. Figure 1. We are considering the micro length scale where the grain structure of the porous electrodes of these cells is resolved. This porous structure consists of an active particle skeleton filled with electrolyte. There are also other additives which however are not dealt with here. The separator also has a porous structure. The size of a single active particle varies between nanometers and micro meters and their number can reach thousands in each electrode, depending on the type of the battery. The active particles are made of solid materials in which lithium can enter and exit due to electrochemical reactions. The main process inside the battery is diffusion and migration, i.e. a potential-driven flux of lithium ions. The process of a lithium ion entering into the active particles is called intercalation. The inverse process of a lithium ion leaving the active particles material is called de-intercalation. These electrochemical reactions of intercalation and de-intercalation happen on the interface between the solid particles and the electrolyte. During discharge of the battery, i.e. when a power-consuming device is connected to the battery, the lithium ions diffuse to the surface of the active particles of the anode, de-intercalate into the electrolyte phase and then carry the electric current from the negative to the positive electrode through the pores of the separator. There the ions are intercalated into the active particles and diffuse into the particles. The electrical current within the active particles is mainly carried by electrons. The electrons do not enter the electrolyte phase. Only the Li ions can move through the electrolyte. Therefore the particles in the electrodes have to be connected to each other to guarantee the electric conductivity of the electrode as a whole. The electrical current, i.e. the electrons, leave the cell through current collectors, which are connected to the active material. This is shown in Figure 2. When the battery is being charged a higher voltage than the one produced by the battery is applied on the cathode thus forcing the current to pass from the cathode to the anode.

During the operation of the battery the temperature may rise significantly. Heat is one of the major contributors to the degradation of the materials inside the battery [13, 1]. It can also lead to unsafe working conditions caused by a local hotspot somewhere inside the cell. As such the full mathematical model described in [6] will be used in this work. This model considers separately the complex transport phenomena in the active particles of the electrodes and in the electrolyte on the microscale as well as the contribution of heat generators on the interfaces. To the best of our knowledge, there is little literature on systematic derivation of microscale Li-ion battery models which are thermodynamically consistent. Consequently there are also very few numerical experiments at such a scale.

Many of the existing simulations are based on the pseudo 2D model of Newman et. al. and use finite differences [9]. The early FORTRAN codes are supplied by Newman himself in his BAND routine. This model is widely used since due to its 1D+1D nature it is both easy to implement and also runs very fast. It is not thermodynamic, but there are thermal extensions based on global energy conservation, i.e. usually without considering the microstructure of the battery and also using a prescribed heat generator [3, 4, 10]. In order to account for the interface conditions more naturally there are also Finite Volume or Finite Element discretizations. If one wants also to observe local phenomena, a cell-resolved model could be used. They are far more expensive to solve however (especially in 3D), so simulating many charge/discharge cycles can take much time. For an example of finite volume discretization of a cell-resolved isothermal model [7], see [11, 8]. A cell-resolved model was solved in [2] with finite elements, using Comsol.

The main contributions of this paper are:

- carefully developing a FEM algorithm for solving the highly nonlinear problem with discontinuous coefficients and solutions, and performing a feasibility study;
- using the developed algorithm and software to study the temperature behavior of porous electrodes at different operating regimes.

Although the developed FEM algorithm in general contains known basic elements, we did not find in the literature FEM algorithms for simulation of pore scale processes in Li-ion batteries. We therefore had to carefully put together and test all the components. We had several reasons to choose FEM. Compared to the voxel-based FVM in the above-cited papers, FEM allows us to resolve more accurately the microstructure of the porous electrodes. Compared to DG, FEM uses less memory and is better studied. Memory use is an important issue when one performs simulations on 3D images of the complicated microstructure of electrodes - such simulations being our final goal.

Mathematical model

We use the model derived in [6]. The equations of the model differ in the particles and the electrolyte. The unknowns in these equations are the concentration of lithium ions, the electrical (or electrochemical in the electrolyte) potential and the temperature. They are denoted by c, Φ and T respectively. The functions $c(\mathbf{x}, t)$ and $\Phi(\mathbf{x}, t)$ are discontinuous across the interfaces.

Governing equations in the electrolyte

Let us introduce the flux of ions and the flux of charges:

$$\mathbf{N}_{+,e} = -\left(D_e \nabla c_e - \frac{t_+}{F}\mathbf{j}_e + \frac{D_e c_e k_{T,e}}{T}\nabla T\right) \tag{1}$$

$$\mathbf{j}_e = -\left(\kappa \nabla \varphi_e - \kappa \frac{1-t_+}{F}\left(\frac{\partial \mu_e}{\partial c_e}\right)\nabla c_e - \kappa \frac{1}{F}\left(\frac{\partial \mu_e}{\partial T}\right)\nabla T\right). \tag{2}$$

In the above fluxes F is the Faraday number, R is the universal gas constant, D is the interdiffusion coefficient (strictly positive), t_+ is the transference number of Li-ions, μ is the chemical potential, κ is the electrical conductivity (strictly positive) and k_T is the Soret coefficient. Also as per [6] in the electrolyte we actually solve for the electrochemical potential φ_e rather than the electric potential Φ_e. Then in the electrolyte the system of equations has the following form:

$$\frac{\partial c_e}{\partial t} = -\nabla \cdot \mathbf{N}_{+,e} \tag{3a}$$

$$0 = -\nabla \cdot \mathbf{j}_e \tag{3b}$$

$$c_{p,e}\rho \frac{\partial T}{\partial t} = \nabla \cdot (\lambda_e \nabla T) + \frac{|\mathbf{j}_e|^2}{\kappa} + \frac{\partial \mu_e}{\partial c_e}\frac{(\mathbf{N}_{+,e} - \frac{t_+}{F}\mathbf{j}_e)^2}{D_e} -$$

$$T\nabla \cdot \left(c_e \frac{\partial \mu_e}{\partial c_e}\frac{k_{T,e}}{T}\left(\mathbf{N}_{+,e} - \frac{t_+}{F}\mathbf{j}_e\right)\right) \tag{3c}$$

for $\mathbf{x} \in \Omega_e$, where Ω_e is the domain of the electrolyte, Ω is the whole domain and c_p, ρ, and λ are specific heat capacity per unit mass, density and heat conductivity (stricly positive) respectively. We use for boundary conditions

$$\mathbf{N}_{+,e} \cdot \mathbf{n} = 0 \tag{4a}$$

$$\mathbf{j}_e \cdot \mathbf{n} = 0 \tag{4b}$$

$$\lambda_e \nabla T \cdot \mathbf{n} = \alpha(T - T_{ext}) \tag{4c}$$

for $\mathbf{x} \in \partial\Omega \cap \partial\Omega_e$.

Governing equations in the solid particles

In the solid particles the ion transference number t_+ is approximately zero, since the current is mainly carried by electrons. The fluxes there have the following form:

$$\mathbf{N}_{+,s} = -\left(D_s \nabla c_s + \frac{D_s c_s k_{T,s}}{T} \nabla T\right) \tag{5}$$

$$\mathbf{j}_s = -\sigma \nabla \Phi_s \tag{6}$$

and the equations are:

$$\frac{\partial c_s}{\partial t} = -\nabla \cdot \mathbf{N}_{+,s} \tag{7a}$$

$$0 = -\nabla \cdot \mathbf{j}_s \tag{7b}$$

$$c_{p,s}\rho \frac{\partial T}{\partial t} = \nabla \cdot (\lambda_s \nabla T) + \frac{|\mathbf{j}_s|^2}{\kappa} - F \frac{\partial U_0}{\partial c_s} \frac{|\mathbf{N}_{+,s}|^2}{D_s} +$$

$$TF\nabla \cdot \left(c_s \frac{\partial U_0}{\partial c_s} \frac{k_{T,s}}{T} \mathbf{N}_{+,s}\right) \tag{7c}$$

for $\mathbf{x} \in \Omega_s$ where Ω_s is the domain of the active material of an electrode, so $s = anode, cathode$. With σ is denoted the average electronic conductivity and with U_0 - the half-cell open circuit potential. The boundary conditions are

$$\mathbf{N}_{+,s} \cdot \mathbf{n} = 0, \quad \mathbf{x} \in \partial\Omega \cap \partial\Omega_s \tag{8a}$$

$$\mathbf{j}_s \cdot \mathbf{n} = const, \quad \mathbf{x} \in \partial\Omega \cap \partial\Omega_{cathode} \tag{8b}$$

$$\Phi_s = const, \quad \mathbf{x} \in \partial\Omega \cap \partial\Omega_{anode} \tag{8c}$$

$$\lambda_s \nabla T \cdot \mathbf{n} = \alpha(T - T_{ext}), \quad \mathbf{x} \in \partial\Omega \cap \partial\Omega_s. \tag{8d}$$

Interface conditions

Let us denote the current density across the interface with

$$i_{se} = i_0 \left(\exp(\tfrac{\alpha_a F}{RT} \eta_s) - \exp(\tfrac{-\alpha_c F}{RT} \eta_s) \right), \tag{9}$$

where the pre-factor i_0 (exchange current density) is

$$i_0 = k c_e^{\alpha_a} c_s^{\alpha_a} (c_{s,max} - c_s)^{\alpha_c} \tag{10}$$

and the overpotential η_s is

$$\eta_s := \Phi_s - \varphi_e - U_0 \left(\frac{c_s}{c_{s,max}} \right). \tag{11}$$

Then for the first two equations of the model we have the following interface conditions

$$\mathbf{j}_s \cdot \mathbf{n}_s = \mathbf{j}_e \cdot \mathbf{n}_s = i_{se} = \mathcal{J}(c_e, c_s, \varphi_e, \Phi_s, T), \quad \mathbf{x} \in \Gamma \tag{12a}$$

$$\mathbf{N}_{+,s} \cdot \mathbf{n}_s = \mathbf{N}_{+,e} \cdot \mathbf{n}_s = \tfrac{i_{se}}{F} = \mathcal{N}(c_e, c_s, \varphi_e, \Phi_s, T), \quad \mathbf{x} \in \Gamma \tag{12b}$$

where $\mathbf{n}_s$ is the unit normal vector going out of the solid into the electrolyte and Γ is the interface boundary. The thermal interface conditions are given by

$$-\lambda_s \mathbf{n_s} \cdot \nabla T_s + \lambda_e \mathbf{n_s} \cdot \nabla T_e =$$

$$-i_{se}\eta_s - i_{se}\Pi + i_{se}\left(c_s \frac{\partial U_0}{\partial c} k_{T,s} + c_e \frac{\partial \mu_e}{\partial c} \frac{k_{T,e}(1-t_+)}{F}\right) \quad (13)$$

with $\mathbf{x} \in \Gamma$ and Π being the Peltier coefficient:

$$\Pi = T(\beta_s - \beta_e) + T\frac{\partial(U_0 + \frac{\mu_e}{F})}{\partial T}$$

where β is the Seebek coefficient.

Discretization

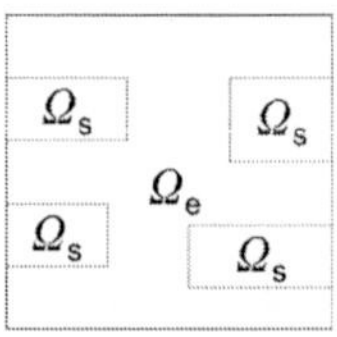

Figure 3: Sample domain for a battery. The whole domain is Ω. The domains for the electrodes are Ω_s. The domain of the electrolyte is $\Omega_e = \Omega \backslash \Omega_s$.

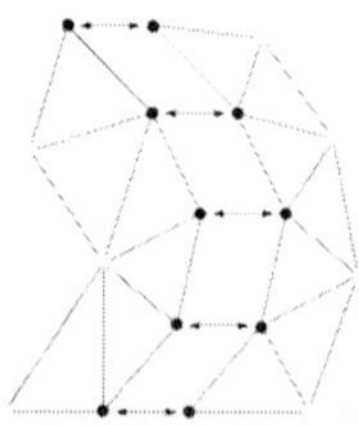

Figure 4: Example of splitting of the nodes on the interface. The split nodes are in black and the others are in white.

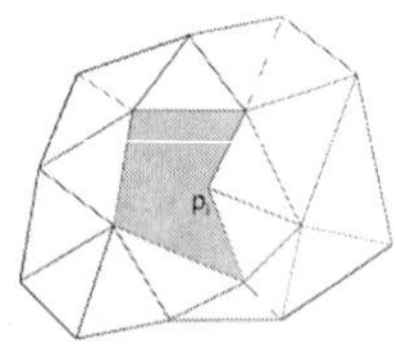

Figure 5: Support of a basis function on a split node. The interface is the dashed line.

For time discretization we use the Backward Euler method. For space disretization we use the finite element method with P1 conforming elements. Since we have two discontinuous functions we mesh the whole region (Fig. 3), and we split the nodes on the interface (Fig. 4). The resulting nodes are treated as two separate entities and the support of the nodal basis functions associated with them is only on one side of the interface (Fig. 5). The interface conditions are treated as nonlinear Neumann-type boundary conditions and accounted for naturally in the weak formulation of the problem. For the continuous temperature we just use the original nodes and the basis functions associated with it have support on both sides of the interface. Now let n be the total number of vertices in the mesh, n_e be the vertices in the electrolyte and n_s be the number of vertices in the solids. We denote the basis functions in the electrolyte with $\{\varphi_{e,i}\}_{i=1}^{n_e}$, in the particles with $\{\varphi_{s,i}\}_{i=1}^{n_s}$ and the ones continuous across the interfaces with $\{\psi_i\}_{i=1}^{n}$, and the discretrized concentrations, potentials and temperature with

$$C_{r,h} = \sum_{j=1}^{n_r} C_{r,i}(t)\varphi_{r,j}(\mathbf{x}) \quad P_{r,h} = \sum_{j=1}^{n_r} P_{r,i}(t)\varphi_{r,j}(\mathbf{x}) \quad T_h = \sum_{j=1}^{n} T_i(t)\psi_j(\mathbf{x})$$

where $r = e, s$. When we plug everything into the weak formulations of our system of equations we obtain the following system of nonlinear algebraic equations

$$\int_{\Omega_r} \frac{C_{r,h}^l - C_{r,h}^{l-1}}{\tau} \varphi_{r,j} \mathrm{d}\mathbf{x} - \int_{\Omega_r} \mathbf{N}_{r,h} \varphi_{r,j} \mathrm{d}\mathbf{x} + \delta_r \int_\Gamma \mathcal{N} \varphi_{r,j} \mathrm{d}s = 0, \quad (14a)$$

$$- \int_{\Omega_r} \mathbf{j}_{r,j} \cdot \nabla \varphi_{r,j} \mathrm{d}\mathbf{x} + \delta_r \int_\Gamma \mathcal{J} \varphi_{r,j} \mathrm{d}s = 0, \quad j = 1, \ldots, n_r \quad (14b)$$

$$\int_\Omega \psi_j c_p \rho \frac{T_h^l - T_h^{l-1}}{\tau} \mathrm{d}\mathbf{x} + \int_\Omega \nabla \psi_j \cdot (\lambda \nabla T_h^l) \mathrm{d}\mathbf{x} - \int_\Omega \psi_j \frac{|\mathbf{j}|^2}{\kappa} \mathrm{d}\mathbf{x} -$$

$$\int_\Omega \psi_j \frac{\partial \mu}{\partial c} \frac{(\mathbf{N}_+ - \frac{t_+}{F}\mathbf{j})^2}{D} \mathrm{d}\mathbf{x} - \int_\Omega \nabla(\psi_j T_h^l) \cdot \left(C_h^l \frac{\partial \mu}{\partial c} \frac{k_T}{T_h^l} \left(\mathbf{N}_+ - \frac{t_+}{F} \mathbf{j} \right) \right) \mathrm{d}\mathbf{x} -$$

$$\int_{\partial\Omega} \psi_j \alpha (T_h^l - T_{ext}) \mathrm{d}s - \int_\Gamma \psi_j \mathcal{T} \mathrm{d}s = 0, \quad j = 1, \ldots, \mathrm{n} \quad (14c)$$

where we have

$$\mathbf{N}_{e,h} = \left(D_e \nabla C_h^l - \frac{t_+}{F} \mathbf{j}_{e,h} + \frac{D_e C_h^l k_{T,e}}{T_h^l} \nabla T_h^l \right) \quad (15)$$

$$\mathbf{N}_{s,h} = \left(D_s \nabla C_h^l + \frac{D_s C_h^l k_{T,s}}{T_h^l} \nabla T_h^l \right) \quad (16)$$

$$\mathbf{j}_{e,h} = \left(\kappa \frac{(1 - t_+) R T_h^l}{F C_h^l} \nabla C_h^l \right) - (\kappa \nabla P_h^l) + \left(\kappa \frac{R \ln C_h^l}{F} \nabla T_h^l \right) \quad (17)$$

$$\mathbf{j}_{s,h} = (\sigma \nabla P_h^l) \quad (18)$$

and

$$\mu = \begin{cases} \mu_0 + RT\ln c, & \mathbf{x} \in \Omega_e \\ \mu_{Li} - FU_0, & \mathbf{x} \in \Omega_s \end{cases} \quad \delta_r = \begin{cases} -1, & \mathbf{x} \in \Omega_e \\ 1, & \mathbf{x} \in \Omega_s \end{cases}.$$

The function $\mathcal{T} = i_{se}(\eta_s + \Pi)$ emerges from the weak formulation of the heat transfer equations. Note that most of the terms on the interface cancel out and $\mathcal{T}$ is what is left.

We solve the nonlinear algebraic system with the Newton method. Since the Jacobi matrix is not symmetric we use BiCGSTAB with ILUT pre-conditioner for solving the linearized problem. For a complete derivation of the weak formulations as well as some of the derivatives for the Newton method, see [12].

Numerical experiments

In this section we present numerical simulations in different settings. We show simulations of the charging and discharging of a battery. We vary the strength of the applied current to show its effect on the temperature. We also vary the thermo-conductivity of the boundary. The total area of the particles in the anode and in the cathode is chosen in such a way that the maximum number of lithium ions that can be stored in each electrode is the same. The parameters that we use in our numerical experiments are given in Table 1. We run our simulations for time $t = 2500$s.

Table 1: Values of the parameters used for the simulations

	$D\left[\frac{\text{cm}^2}{\text{s}}\right]$	t_+	$\kappa\left[\frac{\text{S}}{\text{cm}}\right]$	λ	$\left[\frac{\text{W}}{\text{cm}\cdot\text{K}}\right]$	$\rho\left[\frac{\text{kg}}{\text{cm}^3}\right]$	$c_p\left[\frac{\text{J}}{\text{kg}\cdot\text{K}}\right]$	$c_{max}\left[\frac{\text{mo}}{\text{cm}}\right]$
Electrolyte	7.5×10^{-7}	0.363	0.002	1	0.01	0.001	2000	
Cathode	1.0×10^{-9}	0	0.038	1	0.01	0.0036	7000	0.023671
Anode	3.9×10^{-10}	0	1.0	1	0.01	0.0029	7000	0.024681

Table 2: Values of the parameters for the interface conditions

α_a	α_c	$k_{anode}\left[\frac{\text{A}}{\text{cm}^2}\right]$	$k_{\text{cathode}}\left[\frac{\text{A}}{\text{cm}^2}\right]$	Π_{anode}	$\Pi_{cathode}$
0.5	0.5	0.002	0.2	-0.28	-0.38

Finally for U_0 we use (due to Fuller et al. [5])

$$U_0 = -0.132 + 1.41e^{-3.52\mathrm{soc}}, \mathbf{x} \in \Omega_a$$

$$U_0 = 4.06279 + 0.0677504\tanh(-21.8502\mathrm{soc} + 12.8268) - 0.045e^{-71.69\mathrm{soc}^8} - 0.105734\left(\frac{1}{(00167-\mathrm{soc})^{0.379571}} - 1.576\right) + 0.01e^{-200(\mathrm{soc}-0.19)}, \mathbf{x} \in \Omega_c$$

where soc = $c_s/c_{s,max}$. We have $\Omega = [0{,}5 \times 10^{-3}] \times [0{,}5 \times 10^{-3}]$ and the units are in centimeters.

Table 3: Initial values for the charging process

	$c\left[\frac{\mathrm{mol}}{\mathrm{cm}^3}\right]$	$\Phi[\mathrm{V}]$	$T[\mathrm{K}]$
Electrolyte	0.001	0	300
Cathode	0.0213	4	300
Anode	0.0025	0.8596	300

Table 4: Initial values for the discharging process

	$c\left[\frac{\mathrm{mol}}{\mathrm{cm}^3}\right]$	$\Phi[\mathrm{V}]$	$T[\mathrm{K}]$
Electrolyte	0.001	0	300
Cathode	0.0083	4.13	300
Anode	0.016	0.011	300

The boundary conditions for the charging process are

$$\Phi(\mathbf{x}, t) = 0.8596V, \quad \mathbf{x} \in \partial\Omega_{anode,outer} \tag{20}$$

$$\sigma\frac{\partial\Phi}{\partial\mathbf{n}} = i_{appl}, \quad \mathbf{x} \in \partial\Omega_{cathode,outer} \tag{21}$$

$$\lambda\frac{\partial T}{\partial\mathbf{n}} = \alpha \times (T_{ext} - T), \quad \mathbf{x} \in \partial\Omega \tag{22}$$

and for the discharging process are

$$\Phi(\mathbf{x}, t) = 0.011V, \quad \mathbf{x} \in \partial\Omega_{anode,outer} \tag{23}$$

$$\sigma \frac{\partial \Phi}{\partial \mathbf{n}} = i_{appl} \left[\frac{A}{cm^2}\right], \quad \mathbf{x} \in \partial\Omega_{cathode,outer} \tag{24}$$

$$\lambda \frac{\partial T}{\partial \mathbf{n}} = \alpha \times (T_{ext} - T), \quad \mathbf{x} \in \partial\Omega. \tag{25}$$

$T_{ext} = 300K$ is the ambient temperature. We also take $\alpha = \{0$ W/cm^2K, 5×10^{-6} W/cm^2 K, 10^{-5} W/cm^2 K, 3×10^{-5} W/cm^2 K, 10^{-4} W/cm^2 K$\}$ in each case in order to test different levels of thermal insulation for the battery and $i_{appl} = \{5 \times 10^{-4} \mathrm{A/cm^2}, 10^{-3} \mathrm{A/cm^2}\}$ in order to observe what happens when we drive different currents through the battery. The rest of the boundary conditions have been already defined in 1.2. For our problem the values of the temperature are practically uniform at each time step hence we show only its evolution and not its distribution in the battery. This is a consequence of the fast diffusion related to the scale of the problem. It is obvious from the figures that when driving stronger currents the temperature increases faster, as is expected. The behavior of the concentration is not influenced by the temperature in this test case since the temperature gradient is practically zero. Simulation snapshots for the concentration profile at $t = 2000$ seconds are shown for charge and discharge for $\alpha = 0$ in Figure 8a and Figure 8b respectively. There are two scales - one for the concentration in the electrolyte and one for the concentration in the active particles. We observe that, in contrast to the temperature, the concentration has spatial variation on the same time step. This means that it contributes to the heat generation.

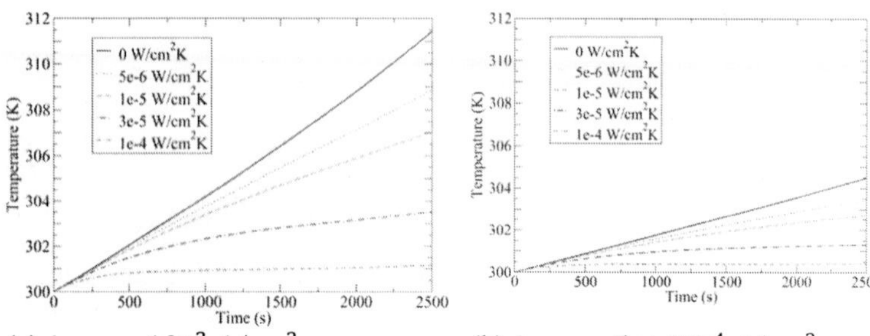

(a) $i_{appl} = 10^{-3}$ A/cm^2 (b) $i_{appl} = 5 \times 10^{-4}$ A/cm^2

Figure 6 Evolution of the temperature in a battery cell during discharge for 2500s for different values of α

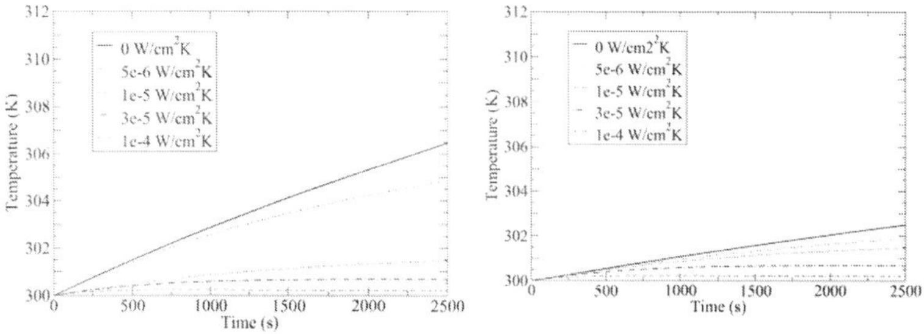

(a) $i_{appl} = 10^{-3}$ A/cm^2 (b) $i_{appl} = 5 \times 10^{-4}$ A/cm^2

Figure 7 Evolution of the temperature in a battery cell during charge for 2500s for different values of α

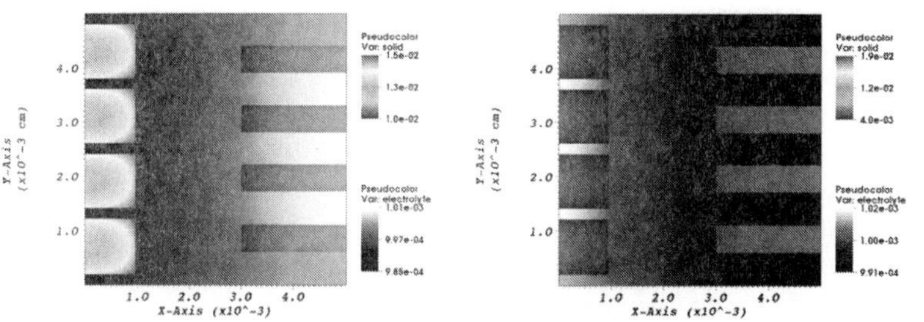

(a) Charging (b) Discharging

Figure 8 Spatial profile of the lithium concentration (mol/cm^2) in a battery cell after 2000s for applied current density 10^{-3} A/cm^2

Conclusion

We have successfully discretized the problem and implemented in C++ code a solver in a two-dimensional setting. The results obtained from the simulations seem physically correct. The node splitting technique that we used for the finite element method allowed us to correctly simulate the discontinuous quantities. A natural extension of our solver would be to move it to a realistic three-dimensional setting. In its current form the geometry does not allow us to have the complicated structures observed in real batteries. If the model is extended to include the degradation processes for the particles, which may include changing their geometry, the mesh should allow for modification.

Acknowledgement The research was funded in part by EU FP7 project PIRG03-GA-2008-230919.

Bibliography

[1] M. Broussely and Ph. Biensan and F. Bonhomme and Ph. Blanchard and S. Herreyre and K. Nechev and R.J. Staniewicz. Main aging mechanisms in Li-ion batteries. *Journal of Power Sources*, 146(1-2):90 - 96, 2005. Selected papers pressented at the 12th International Meeting on Lithium Batteries.

[2] C.W. Wang and A.M. Sastry. Mesoscale modeling of Li-ion polymer cell. *Journal of the Electrochemical Society*, 154(11): A1035-A1047, 2007.

[3] Yufei Chen and J.W. Evans. Heat transfer phenomena in lithium/polymer-electrolyte batteries for electric vehicle application. *Journal of the Electrochemical Society*, 140:7, 1993.

[4] Yufei Chen and J.W. Evans. Three-dimensional thermal modeling of lithium-polymer batteries under galvanostatic discharge and dynamic power profile. *Journal of the Electrochemical Society*, 141:11, 1994.

[5] T. Fuller and M. Doyle and J. Newman. Modeling of galvanostatic charge and discharge of the lithium/polymer/insertion cell. *Journal of the Electrochemical Society*, 140(6):1526-1533.

[6] A. Latz and J. Zausch. Thermodynamic consistent transport theory of Li-ion batteries. *Journal of Power Sources*, 196(6):3296 - 3302, 2011.

[7] Latz, Arnulf and Zausch, Jochen and Iliev, Oleg. Modeling of species and charge transport in Li-ion batteries based on non-equilibrium thermodynamics. *Proceedings of the 7th international conference on Numerical methods and applications* in NMA'10, pages 329--337, Berlin, Heidelberg, 2011. Springer-Verlag.

[8] G. B. Less and J. H. Seo and S. Han and A. M. Sastry and J. Zausch and A. Latz and S. Schmidt and C. Wieser and D. Kehrwald and S. Fell. Micro-Scale Modeling of Li-ion Batteries: Parameterization and Validation. *Journal of The Electrochemical Society*, 159(6):A697-A704, 2012.

[9] John Newman and Karen E. Thomas-Alyea. *Electrochemical Systems*. Wiley-Interscience, 3rd edition, 2004.

[10] Pals, Carolyn R and Newman, John. Thermal Modeling of the Lithium/Polymer Battery .1. Discharge Behavior of a Single-Cell. *Journal of the Electrochemical Society*, 142(10): 3274--3281, 1995.

[11] Popov, P. and Vutov, Y. and Margenov, S. and Iliev, O. Finite volume discretization of equations describing nonlinear diffusion in Li-

ion batteries. *Proceedings of the 7th international conference on Numerical methods and applications* in NMA'10, pages 338--346, Berlin, Heidelberg, 2011. Springer-Verlag.

[12] M. Taralov and V. Taralova and P. Popov and O. Iliev and A. Latz and J. Zausch. Report on Finite Element Simulations of Electrochemical Processes in Li-ion Batteries with Thermic Effects. *Scientific Reports of the ITWM*, 2012.

[13] J. Vetter and P. NovпïSk and M.R. Wagner and C. Veit and K.-C. MпïSller and J.O. Besenhard and M. Winter and M. Wohlfahrt-Mehrens and C. Vogler and A. Hammouche. Ageing mechanisms in lithium-ion batteries. *Journal of Power Sources*, 147(1-2):269 - 281, 2005.

Spatiotemporal Stabilization of Ultrashort Light Pulses Propagating in a Nonlinear Ionized Medium

T. P. Todorov, M. E. Todorova, M. D.Todorov and I. G. Koprinkov

Introduction

The propagation of high-intensity ultrashort light pulses (HULP) in a nonlinear bulk medium is a challenging problem that is not yet fully investigated and understood. The early studies reveal broadening and splitting of the pulse at positive group velocity dispersion (GVD) [1, 2, 3, 4], while the existence of spatiotemporal solitons (STSs) is predicted at negative GVD [5]. That understanding has been changed with the discovery of self-compression (SC) of HULP in various media (atomic and molecular gases and fused silica) at positive GVD [6, 7]. In fact, the SC is part of an entire rearrangement of the pulse that can be summarized as [7]: self-focusing in the transversal direction (space) leading to formation of a light filament; SC in the longitudinal direction (time); increasing of the peak intensity; improvement of the spatio-temporal pulse shape; and stable propagation over a given distance. The simultaneous self-focusing and SC represents complete space-time trapping of the light matter, demonstrated for the first time in [7]. The experimental and theoretical studies have revealed the great potential of the SC [6]-[15]. Efficient generation of few-cycle high-intensity pulses due to the SC has been achieved experimentally [9], whereas the theoretical studies predict generation of even shorter pulses - single-cycle [11] and even sub-cycle high-intensity pulses [12]. Intensive work in the field of propagation equations and physical models [14, 15] has given strong contributions to the understanding of the highly complicated dynamics of HULP.

The SC reverses the natural trend of the pulse to broaden in time, and the problem of stable spatiotemporal propagation of HULP can be considered on that ground. The SC is an effect resulting from the common action of a number of individual processes. The *minimal set* of processes at normal dispersion at which SC can be observed includes GVD, diffraction and cubic nonlinearity within a (3+1)D nonlinear Schrödinger equation [16, 17]. The physical mechanism of SC in that case has been determined as an effective shortening of the pulse duration (full width at half maximum characterization) at a rapid increase of the peak intensity due to self-focusing, which strongly competes with the dispersion broadening at low GVD [16]. However, no signature of pulse stabilization has been found within that simple physics.

The problem of stable propagation of optical pulses and, particularly, generation of STS is not yet fully understood although there is marked progress in that field [18, 19]. One may distinguish two main approaches to the problem: solving the problem under abstract conditions [19, 20] and under realistic physical conditions [21, 22]. In the first case, stable pulse propagation and the existence of STS are predicted within a given range of parameters of the propagation equation. There is no guarantee, however, that such a set of parameters occurs in a real medium. Also, such works are based on the analysis of propagation equations only, assuming no change of the building particles of the medium due to ionization, as for the case of HULP. In the second case, stable pulse propagation has been found assuming that the shape of the initial pulse remains constant and only the parameters of the shape may change. The assumption of constant pulse shape along the whole propagation distance is a strong restriction on pulse evolution which, as experiments show, does not take place in reality. Instead, the shape of the pulse may change strongly, leading to formation of highly complicated structures. In the more simple cases, it may split into two or more well resolved individual pulses.

In this work, we investigate the problem of stable propagation of high-intensity ultrashort laser pulses in a nonlinear bulk medium without any preliminary constraints on pulse evolution. Thus, pulse propagation is ruled solely by the underlying physical processes. Finally, the problem is solved under realistic physical conditions. We find a general stabilization of the pulse shape and the relevant parameters over a given propagation range. To the best of our knowledge, such stabilization is found for the first time within this realistic approach.

Physical model and numerical method of solution

Our study is based on the physical model developed in [14]. The propagation equation is a (3+1)D nonlinear envelope equation (NEE) for the field amplitude *A* (in standard notations [23, 14])

$$\frac{\partial A}{\partial z} = \frac{\mathrm{i}}{2k_0}\hat{T}^{-1}\nabla_\perp^2 A + \mathrm{i}\hat{D}A + \mathrm{i}\frac{\omega_0}{c}n_2\hat{T}|A|^2A - \mathrm{i}\frac{\omega_0}{c}n_4\hat{T}|A|^4A$$

$$-\mathrm{i}\frac{k_0}{2n_0^2\rho_c}\hat{T}^{-1}\rho A - \frac{\sigma}{2}\rho A - \frac{\beta_{MPI}(A)}{2}A, \qquad (1)$$

where the non-instantaneous effects in the nonlinearity are neglected due to the fast electronic response of the atomic medium (gaseous argon) considered here. The physical processes involved in the pulse propagation can be grouped into: *linear processes*, diffraction and dispersion (the first and the second terms, respectively, at the right hand side of the equation); *nonlinear processes in neutrals*, cubic and quintic nonlinearity of neutral particles (third and the fourth terms, respectively); and *processes due to ionization*, ionization modification of the refractive index, collision ionization by inverse bremsstrahlung and multi-photon ionization (fifth, sixth and the seventh terms, respectively). The material parameters in the NEE, particularly those associated with ionization, are field-dependent functions, which also require a realistic self-consistent approach to the problem.

The electron number density ρ is described by the kinetic equation (in standard notations) [14]

$$\frac{\partial \rho}{\partial t} = W(I)(\rho_n - \rho) + \frac{\sigma(\omega_0)}{I_p}|A|^2\rho - f(\rho) \qquad (2)$$

where σ is the inverse *bremsstrahlung* cross-section, and $f(\rho) = \alpha\rho^2$ is the plasma recombination term. The ionization rate $W(I)$ is described by the simple multiphoton formula [14]

$$W = \sigma_k I^k, \qquad (3)$$

where σ_k is the k-photon ionization coefficient, I is the intensity and k is the number of photons to directly ionize the atoms of the propagation medium from their ground state. If necessary, the ionization rate can be refined using, *e.g.*, Perelomov-Popov-Terent'ev theory [24].

In this work, the physical model of [14] is further developed, including the influence of ionization on GVD predicted in [25]. The total GVD of the medium $\beta^{(2)}$ can thus be presented as a sum of GVD of neutrals $\beta_0^{(2)}$ and GVD of plasma $\beta_i^{(2)}$, $\beta^{(2)} = \beta_0^{(2)} + \beta_i^{(2)}$. The ionization contribution to the GVD is given by [25]

$$\beta_i^{(2)} = -\frac{e^2\lambda^3\rho}{2\pi^2 m_e c^4\left(1-\frac{e^2\lambda^2\rho}{\pi m_e c^2}\right)^{\frac{3}{2}}}, \qquad (4)$$

where λ is the wavelength, e, m_e and c are the charge and mass of an electron and the velocity of light, respectively. The ionization contribution to the GVD accomplishes the physical model because the negative contribution of ionization to the GVD is the only process that acts directly against the positive GVD of neutrals. In this way, to each strong physical process in the model corresponds at least one other strong physical process acting in the opposite direction. The latter, in our opinion, is important for a more complete understanding of pulse behavior, including the problem of stable pulse propagation. For that reason, the above model will be put into *a minimal sufficient model*.

The main difficulties in the solution of the problem come from the propagation Eq.(1). It is a (3+1)-dimensional highly nonlinear equation. At first glance, it is of cubic-quintic ($\chi^{(3)} - \chi^{(5)}$) type but the actual nonlinearity is much higher due to the multiphoton ionization term. To ionize the argon atom, having 15.76eV ionization potential, simultaneous absorption of eleven photons from the electromagnetic field of 800nm central wavelength (considered here) is required. Direct (non-resonant) eleven-photon absorption is ruled by $\chi^{(21)}$-nonlinearity that stays behind the multi-photon ionization coefficient $\beta_{(\mathrm{MPI})}$. Thus, the leading nonlinearity in Eq.(1) is the 21st, assuming some perturbation of the field-matter interaction.

The pulse propagation rests on self-consistent numerical solutions of Eqs. (1)-(4) at the following pulse and medium parameters: initial pulse of Gaussian shape in space and in time, having 0.5mJ pulse energy, 150fs pulse duration (full width at half maximum), 300μm transversal width (beam diameter), and 800nm central wave length, and gaseous argon at 18atm pressure as the propagation medium.

The numerical method of solving the problem can be briefly described by the following. We split the original Eq. (1) twice. First, we split it into two z-evolutionary equations by physical processes (see, *e.g.*, [26]). Next, following [27], we apply a coordinate splitting to the resultant equations by using the Crank-Nicolson difference scheme direct in complex arithmetic combining it with inner iterations with respect to the nonlinear terms. We use 512 and 1024 grid nodes in the transversal direction and in time, respectively, and the longitudinal step along the propagation direction z is 5×10^{-7} dimensionless units. We have found that such a small value of propagation step is crucial in order to achieve both independence of the numerical results in the spatial and temporal steps and the stability and fast convergence of the inner iterations with respect to the evolutionary z-step. We have met just one other work in this field where such independence is declared [19]. On the other hand, the small size of the grid dramatically increases the computation time. At the specified conditions, a single numerical run takes about one week on our computer. To make the problem tractable on a standard PC, we have chosen relatively high gas pressure (18atm) in comparison to the typical values of argon gas pressure for most of the simulations and/or experiments. The stronger self-focusing at that high gas pressure ensures a more rapid passage from the initial stage of propagation of the pulse to the stage of stable pulse propagation. It substantially reduces the computation time. Such a high pressure case, however, leads to a more rapid growth of peak intensity (due to higher nonlinearity of the medium) and, consequently, to a rapid growth of free electrons once the pulse intensity reaches the level of substantial ionization. Rapid growth of free electrons may destabilize the pulse, as they strongly and non-instantaneously affect pulse behavior.

Results and discussion

The influence of ionization on the material parameters, more particularly the GVD, as well as on the whole pulse propagation dynamics of HULP, plays a substantial role in our model. Recently, filamentation without ionization has been proposed as a result of a dynamic balance between the focusing ($\chi^{(3)}, \chi^{(7)}$ ...) and defocusing ($\chi^{(5)}, \chi^{(9)}$ ...) nonlinearities [13]. Although the possibility of formation of stable light filaments without ionization is well illustrated and probably takes place at long distance propagation, the whole phenomenology that accompanies filamentation, including strong SC of the pulse and, eventually, stable propagation, is not demonstrated within such a mechanism. The evolution of the electron number density along the propagation distance, and the respective change of the GVD at the specified pulse and medium parameters, are shown in Figures 1 and 2 respectively. Ionization at the initial stage of propagation is negligible, Figure 1, due to the relatively low peak intensity of the pulse. The pulse intensity grows with the propagation due to self-focusing of the pulse, and rapid growth of free electrons (in agreement with the high nonlinearity of the ionization) takes place at around $1 \times 10^{13}\mathrm{W/cm^2}$ for the case of argon atoms.

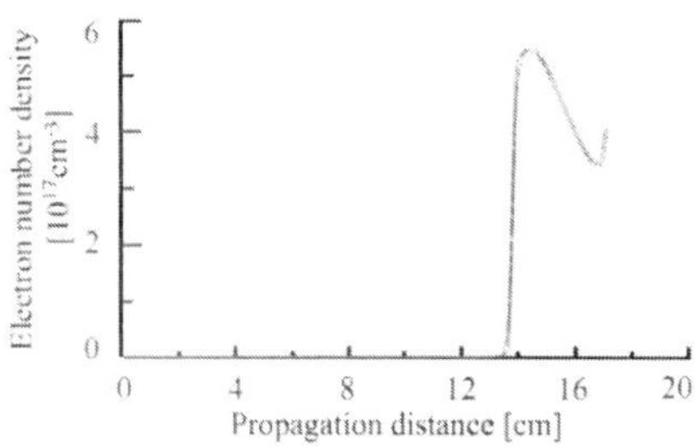

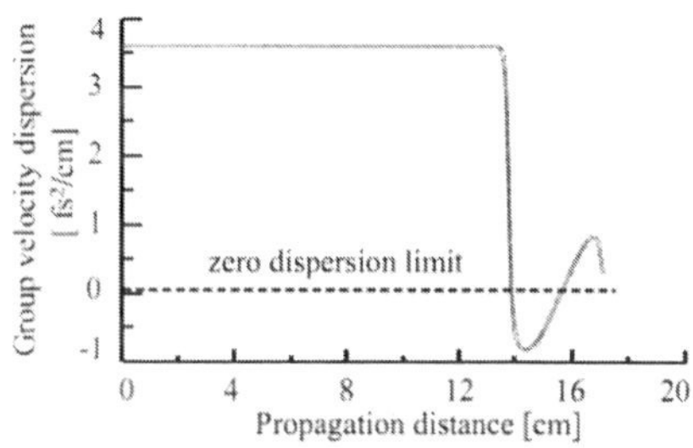

Figure 1: Evolution of the electron number density versus propagation distance (left panel). Figure 2: Evolution of the total GVD of the medium (neutrals plus plasma contribution) versus propagation distance (right panel).

Dissipation of the pulse energy due to ionization leads to stabilization of the peak intensity around $4.5 \times 10^{13}\mathrm{W/cm^2}$ for about 3cm of propagation distance, between 13.7cm and 16.7cm from the beginning of the propagation. Stabilization of the peak intensity from “above” due to ionization plays a substantial role in the entire stabilization of the pulse. The maximal value of the electron number density reaches $5.3 \times 10^{17}\mathrm{cm^{-3}}$, Figure 1. It is substantially higher in comparison to other simulations [14], which can be attributed to the much higher pressure (number density of neutrals) of the gaseous medium considered here.

Ionization contribution to the GVD results in a rapid drop of the total GVD, Figure 2, once the ionization rate becomes substantial - around and above $3 \times 10^{13}\mathrm{W/cm^2}$ of peak intensity. At a high enough density of electron numbers the total GVD may become negative. Such an effect has been predicted in principle at a given level of ionization and has been called ionization induced inversion (I^3) of the GVD [25]. Here, the I^3 created by the pulse itself is proved for the first time under realistic propagation conditions. It can be considered as a new self-action effect, because the ionization created by the pulse acts on the pulse behavior by means of ionization contribution to the GVD. The GVD of the medium changes from the initial value of $+3.6\text{fs}^2/\text{cm}$ to $-0.8\text{fs}^2/\text{cm}$ (the minimal GVD found here) as a result of the ionization. The inverted GVD is of the same order of magnitude as the initial one. The possibility to change and even to invert the GVD by ionization offers, in principle, a new way to control pulse propagation. However, as our calculations show (putting as zero all orders of dispersion of neutrals in the NEE and neglecting the ionization contribution to the GVD), neither the magnitude of the initial positive GVD of the medium nor the magnitude of the inverted negative GVD is sufficient to change substantially the pulse propagation dynamics over the (laboratory scale) distances studied here. The results obtained when all dispersion contributions (including that of ionization) are totally excluded differ insubstantially from those presented here, where all specified quantities are taken into account. This is because the propagation distances are substantially shorter than the dispersion length over which the GVD alone may provide a substantial effect. The latter, however, does not make the problem of stable propagation any less important, because the pulse may strongly change its shape and parameters over such short distances due to the action of nonlinear processes.

Increasing the peak intensity by self-focusing due to $\chi^{(3)}$-nonlinearity results in a growing contribution of $\chi^{(5)}$ (defocusing) nonlinearity. The two processes (in fact, the contribution of all orders of nonlinearity of opposite sign [28]) tend to balance each other. At a given stage of pulse evolution, a balance between self-focusing due to $\chi^{(3)}$-nonlinearity (and higher orders of self-focusing nonlinear terms) from one side, and defocusing due to the common action of diffraction (the most universal defocusing effect but not dominating here) $\chi^{(5)}$-nonlinearity (and higher orders of defocusing nonlinearities), and ionization, takes place. It leads to formation of stable light filaments. The

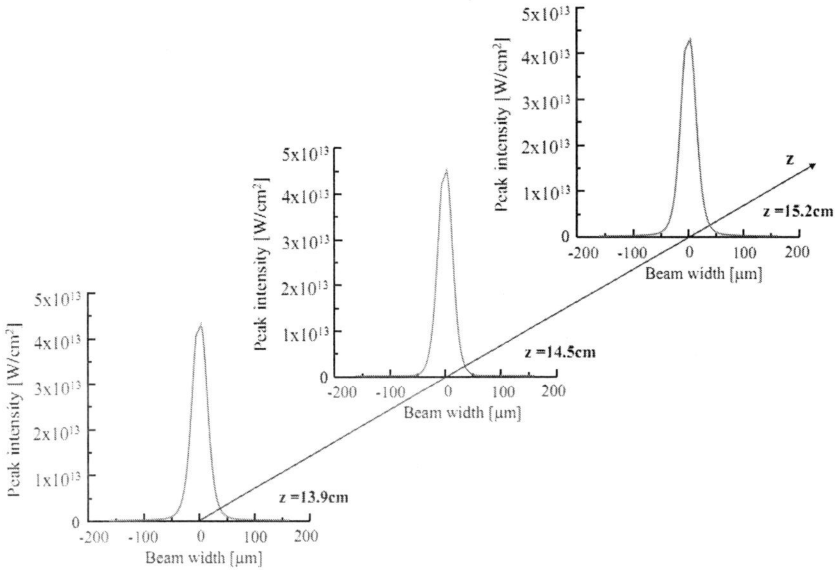

Figure 3: Stabilization of the spatial profile and beam diameter of the pulse between 13.9cm and 15.2cm of the propagation distance.

contribution of higher orders of nonlinearities, together with ionization, as a possible mechanism of SC and spatiotemporal stabilization of the pulses, has been proposed in [7]. In our case, the stabilization of the transversal beam profile, as a whole, and the beam diameter within the specified propagation range is shown in Figure 3. The same results also illustrate the stabilization of the peak intensity within that propagation range. The peak intensity is clamped around $4.5 \times 10^{13}\mathrm{W/cm^2}$ due to ionization losses and has less than 3% variation in magnitude between 13.7cm and 16.7cm. The stabilization of the pulse in the transversal direction (beam diameter) as well as with respect to the peak intensity helps to stabilize the pulse in time. The evolution of the pulse in time is shown in Figure 4. As can be seen, the temporal profile of the pulse remains practically unchanged between 13.9cm and 15.2cm of the propagation distance. The profile of the pulse in time has well expressed

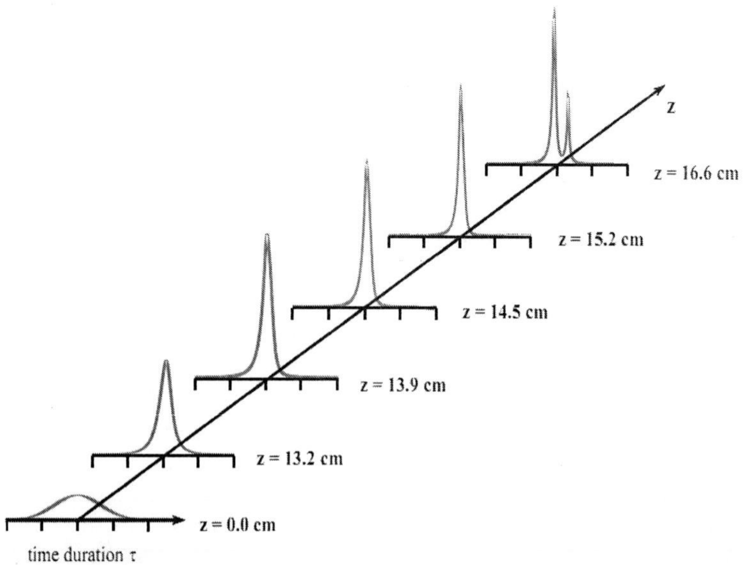

Figure 4: Evolution of the temporal profile of the pulse with the propagation and stabilization between 13.9 cm and 15.2 cm of the propagation distance.

asymmetry - the trailing edge is substantially shorter than the leading edge, which is an indication of self-steepening. The time duration also shows a stabilization signature, although it does not remain constant - the SC of the pulse still continues within the specified range (the pulse duration decreases from 29fs at the beginning to 18fs at the end of the range) but much more slowly than outside that range. As a whole, the pulse undergoes more than 8 times time compression - from 150fs of the initial pulse to 18fs of the shortest single pulse. At longer distances, a second pulse starts appearing at the trailing edge of the main pulse - see the time profile at z = 16.6cm in Figure 4. Such deterioration of the pulse is irreversible and, for that reason, the results at longer distances are not shown. The spatiotemporal evolution of the pulse found here is in agreement with the experimental observations [7]. In order to get entire representation of the spatiotemporal structure of the pulse, the latter is shown in Figure 5 at the same propagation positions as those of the temporal profiles in Figure 4. As can be seen, we have well expressed clean single pulses within the range of stabilization, with no sub-pulses or other strong field structures superimposed on the pulse.

One may distinguish three main cases in the spatiotemporal dynamics of high-intensity ultrashort pulses: (*i*) formation of stable light filament due to a balance of self-focusing and self-defocusing effects, *i.e.* transversal stabilization of the pulse only; (*ii*) formation of stable light filament accompanied by SC of the pulse; and (*iii*) formation of stable light filament with SC of the pulse and stable propagation of the self-compressed pulse over a given distance. It is the last case, indeed, which is the subject of the present studies. The formation of stable light filament is a necessary condition for the entire stabilization of the pulse. However, it plays only a secondary role in the general stabilization of the pulse because filamentation may take place without [29, 30] or with [6]-[11] SC and stabilization [7] of the pulse in the time domain. The main point in the entire pulse stabilization problem is stabilization in the time domain. Stabilization of the pulse in time is the most complicated part of the problem because it depends strongly on the stabilization of the pulse with respect to the other two parameters. Based on above result, we may conclude that a general stabilization of the pulse takes place over a propagation distance of length between 1.5cm and 3cm for the different pulse parameters. Such a conclusion is reinforced if we compare the behavior of the pulse predicted by the NLSE [16, 17]. In that case, the pulse immediately starts splitting after the peak intensity reaches maximum and none of the pulse parameters show stabilization. It means that the above-cubic nonlinearities and ionization play substantial roles, not only in pulse compression but also in stabilization of the pulse parameters. Recently, cubic-quintic STS stabilized by ionization over a 50μm propagation distance was reported [21], but neither the temporal profile nor the stable evolution of the pulse in time was presented. Instead, only the stability of the spatial profile was shown.

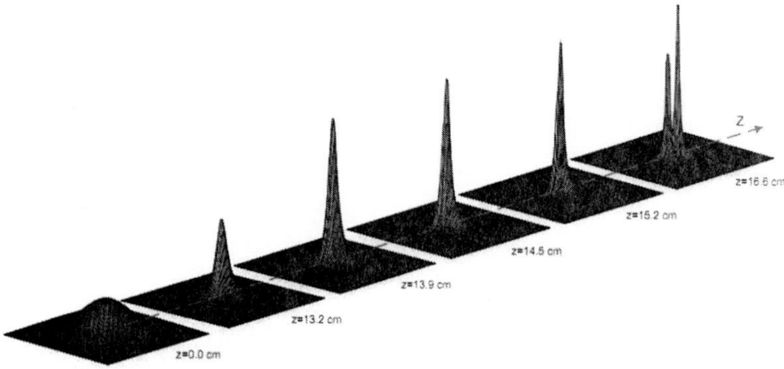

Figure 5: Evolution of the spatiotemporal structure of the pulse with the propagation.

In the typical conditions of HULP, the pulse develops in a strongly nonlinear regime throughout the propagation distance, *i.e.*, $L_{\mathrm{NL}} = L_{\mathrm{DF}} = L_{\mathrm{DS}}$, where L_{NL}, L_{DF}, and L_{DS} are nonlinear, diffraction and dispersion lengths, whose values determined on the base of the pulse parameters in the region of stabilization are 0.16cm, 0.43cm, and 663cm respectively. It is an indication that the linear processes, dispersion and diffraction, play only a secondary role in the SC of HULP and the formation of stable pulses. Instead, a number of nonlinear processes start playing the main roles in that case. Thus, the strong difference between the characteristic lengths in our case is not a big problem for pulse stabilization, because the main balance mechanism results from the interplay between strong nonlinear processes. Although the distance of pulse stabilization is shorter than the dispersion length, it substantially exceeds the nonlinear length L_{NL}, determined by the cubic nonlinearity, which is only 0.16cm. Stabilization of the pulse over laboratory scale distances L, in our case $L \ll L_{\mathrm{DS}}$, must be referred to a balance between nonlinear processes. It seems unjustified to use the linear characteristic lengths L_{DF}, and L_{DS} as stability criteria for HULP, if much more intense nonlinear processes may cause strong modification of the pulse over much shorter distances than L_{DF} and L_{DS}.

Conclusions

To conclude: we have studied numerically the spatiotemporal dynamics of high-intensity ultrashort light pulses, solving self-consistently the propagation and the material equations under realistic physical conditions.

At proper pulse and material parameters, self-compression and stable propagation of the compressed pulse are found. The stabilization of the pulse results mainly from a balance between competitive nonlinear optical processes. We have found, for the first time in a real propagation regime, inversion of the group velocity dispersion due to ionization from a positive to a negative value.

Acknowledgement. These results were presented at the 8th Annual Meeting of the Bulgarian Section of SIAM, 18-19 December 2013.

Bibliography

[1] P. Chernev and V. Petrov, *Opt. Lett.* **17** 172 (1992).
[2] J. E. Rothenberg, *Opt. Lett.* **17** 583 (1992).
[3] J. K. Ranka, R. W. Schirmer, and A. L. Gaeta, *Phys. Rev. Lett.* **77** 3783 (1996).
[4] A. A. Zozulya, S. A. Diddams, A. G. Van Engen, and T. S. Clement, *Phys. Rev. Lett.* **82** 1430 (1999).
[5] Y. Silberberg, *Opt. Lett.* **15** 1282 (1990).
[6] I. G. Koprinkov, A. Suda, P. Wang, and K. Midorikawa, *Jap. J. Appl. Phys.* **38** L978 (1999).
[7] I. G. Koprinkov, A. Suda, P. Wang, and K. Midorikawa, *Phys. Rev. Lett.* **84** 3847 (2000).
[8] S. Tzortzakis, B. Lamouroux, A. Chiron, S. D. Moustaizis, D. Anglos, M. Franco, B. Prade, and A. Mysyrowicz, *Opt. Commun.* **197** 131 (2001).
[9] C. P. Hauri, W. Kornelis, F. W. Helbing, A. Heinrich, A. Couairon, A. Mysyrowicz, J. Biegert, and U. Keller, *Appl. Phys. B* **79**, 673 (2004).
[10] A. Couairon, M. Francko, A. Mysyrowicz, J. Biegert, and U. Keller, *Opt. Lett.* **30** 2657 (2005).
[11] S. Skupin, G. Stibenz, L. Berge, F. Lederer, T. Sokollik, M. Schnürer, N. Zhavoronkov, and G. Steinmeyer, *Phys. Rev. E* **74** 056604 (2006).
[12] M. B. Gaarde and A. Couairon, *Phys. Rev. Lett.*, **103**, 043901 (2009).
[13] P. Béjot, J. Kasparian, S. Henin, V. Loriot, T. Vieillard, E. Hertz, O. Faucher, B. Lavorel, and J.-P. Wolf, *Phys. Rev. Lett.* **104** 103903 (2010).
[14] L. Berge, S. Skupin, R. Nuter, J. Casparian, and J.-P. Wolf, *Rep. Prog. Phys.* **70** 1633 (2007).
[15] A. Couairon, E. Brambilla, T. Gorti, D. Majus, O. de J. Ramires-Gongora, and M. Kolesik, *Eur. Phys. J. Special Topics* **199** 5 (2007).

[16] I. G. Koprinkov, M. D. Todorov, M. E. Todorova, and T. P. Todorov, *J. Phys. B: At. Mol. Opt. Phys.* **40** F231 (2007).
[17] M. D. Todorov, M. E. Todorova, T. P. Todorov, and I. G. Koprinkov, *Opt. Commun.* **281** 2549 (2008).
[18] B. A. Malomed, D. Mihalache, F. Wise, and L. Torner, *J. Opt. B, Quantum Semiclass. Opt.* **7** R53 (2005).
[19] N. Akhmediev, J. M. Soto-Crespo, and P. Grelu, *Chaos* **17** 037112 (2007).
[20] D. Mihalache and D. Mazilu, *Rom. Rep. Phys.* **60** 749 (2008).
[21] C. P. Jisha, V. C. Kuriakose, and K. Porsezian, *Phys. Lett. A* **352** 496 (2006).
[22] M. A. Porras and F. J. Redondo, *Opt. Soc. Am. B* **30** 603 (2013).
[23] T. Brabec and F. Krausz, *Phys. Rev. Lett.* **78** 3282 (1997).
[24] A. M. Perelomov, V. S. Popov, and M. V. Terent'ev, *Sov. Phys. JETP* **23** 924 (1966).
[25] I. G. Koprinkov, *Applied Physics B - Lasers and Optics* **79** 359 (2004).
[26] G. I. Marchuk, "Mathematical Models in Environmental Problems (Studies in Mathematics and its Applications)", North Holland, The Netherlands, 1986.
[27] M. D. Todorov and C. I. Christov, *Mathematics and Computers in Simulation* **80** 46 (2009).
[28] M. Nurhuda, A. Suda, and K. Midorikawa, *New J. Phys.* **10** 053006 (2008).
[29] A. Braun, G. Korn, X. Liu, D. Du, J. Squier, and G. Mourou, *Opt. Lett.* **20**, 73 (1995).
[30] E. T. J. Nibbering, P. F. Curley, G. Grillon, B. S. Prade, M. A. Franc, F. Salin, and A. Mysyrowicz, *Opt. Lett.* **21**, 62 (1996).

CHAPTER FIVE

ALGORITHMS IN INDUSTRIAL MATHEMATICS

THE GEOMETRY OF PYTHAGOREAN QUADRUPLES AND RATIONAL DECOMPOSITION OF PSEUDO-ROTATIONS

DANAIL BREZOV,CLEMENTINA MLADENOVA, IVAÏLO MLADENOV

Preliminaries

The positive integer solutions of the Diophantine equation $x_1^2 + x_2^2 = x_3^2$, referred to as *Pythagorean triples*, or at least some of them, have been known since the time of ancient Babylon and even earlier [11]. Perhaps the most famous method for generating *primitive triples*, i.e. those formed by relatively prime x_k, is the one due to Euclid, based on the rational parameterization of the unit circle

$$x_1 = m^2 - n^2, \qquad x_2 = 2mn, \qquad x_3 = m^2 + n^2 \tag{1}$$

where m, n are relatively prime integers of different parity. The ratio $\tau = n/m$ provides the Euler trigonometric substitution for $\mathbb{S}^1$. On the other hand, interpreting Pythagorean triples as integer points on the future light cone allows for mapping one such point onto another by means of integer-valued Lorentz transformations, i.e. with the aid of the modular group $\mathrm{PSL}_2(\mathbb{Z}) \cong \mathrm{SO}^+_{2,1}(\mathbb{Z})$. This observation was first used by Barning [3], who managed to find a minimal set of matrices in the form

$$L = \begin{pmatrix} 1 & -2 & 2 \\ 2 & -1 & 2 \\ 2 & -2 & 3 \end{pmatrix}, \qquad U = \begin{pmatrix} 1 & 2 & 2 \\ 2 & 1 & 2 \\ 2 & 2 & 3 \end{pmatrix},$$

$$R = \begin{pmatrix} -1 & 2 & 2 \\ -2 & 1 & 2 \\ -2 & 2 & 3 \end{pmatrix} \tag{2}$$

known as the *Hall matrices* [9], which generate all primitive Pythagorean triples without repetition acting on the *Egyptian vector* $v = (3, 4, 5)^t$. As a consequence, we have a ternary branching tree of Pythagorean triples [2, 4], which represents the orbit of v with respect to the subgroup of $\mathrm{O}_{2,1}(\mathbb{Z})$ generated by the matrices (2). As has been shown in [16] with an alternative choice of generators in the modular group $\mathrm{PSL}_2(\mathbb{Z})$, this construction is equivalent to the well-known Euclid's method. One major advantage of the group-theoretic approach is that it allows for natural generalizations. For example, in the case of quadruples $x_1^2 + x_2^2 + x_3^2 = x_4^2$ the set of integer points on the future light cone in Minkowski space $\mathbb{R}^{3,1}$ is preserved by the action of the group $\mathrm{SO}^+_{3,1}(\mathbb{Z})$, and the latter may be used to generate solutions starting from a certain "base point". This idea was used in [12], where it has been shown that all primitive Pythagorean quadruples can be obtained from the orbits of the elements

$$v_1 = (1,0,0,1)^t, \qquad v_2 = (0,1,0,1)^t, \qquad v_3 = (0,0,1,1)^t$$

with respect to the subgroup of $\mathrm{SO}^+_{3,1}(\mathbb{Z})$ generated by the six matrices

$$H_1 = \begin{pmatrix} -1 & -2 & 0 & 2 \\ 2 & 1 & 0 & -2 \\ 0 & 0 & 1 & 0 \\ -2 & -2 & 0 & 3 \end{pmatrix}, \quad H_2 = \begin{pmatrix} 1 & 0 & -2 & 2 \\ 0 & 1 & 0 & 0 \\ 2 & 0 & -1 & 2 \\ 2 & 0 & -2 & 3 \end{pmatrix},$$

$$H_3 = \begin{pmatrix} 1 & 0 & 0 & 0 \\ 0 & -1 & -2 & 2 \\ 0 & 2 & 1 & -2 \\ 0 & -2 & -2 & 3 \end{pmatrix} \tag{3}$$

$$H_4 = \begin{pmatrix} 1 & 0 & 0 & 0 \\ 0 & -1 & 2 & 2 \\ 0 & -2 & 1 & 2 \\ 0 & -2 & 2 & 3 \end{pmatrix}, \qquad H_5 = \begin{pmatrix} 1 & 0 & 2 & -2 \\ 0 & 1 & 0 & 0 \\ -2 & 0 & -1 & 2 \\ -2 & 0 & -2 & 3 \end{pmatrix},$$

$$H_6 = \begin{pmatrix} -1 & 2 & 0 & 2 \\ -2 & 1 & 0 & 2 \\ 0 & 0 & 1 & 0 \\ -2 & 2 & 0 & 3 \end{pmatrix}.$$

Here we suggest a slight modification of this method. Namely, we exploit the idea of vector parameterization for the special orthogonal and pseudo-orthogonal groups (of dimension three and six), briefly described in the next section. This construction allows for simplicity of notation as well as more efficient computations, especially in the case of powers of the same matrix which occur regularly in the procedure. At the end we also consider pairs of Gaussian integers at equal distance from the origin. This second type of Pythagorean quadruples can be realized by means of a similar construction involving $\mathrm{O}(2,2)$-like transformations. The non-trivial solutions here may be identified with convex quadrangles with integer sides, having two opposite right angles. Some of them also have an integer diagonal and thus correspond to pairs of Pythagorean triangles with glued hypotenuses [7]. An interesting physical interpretation of this setting would be an infinite plane crystal with quadratic lattice formed by charges of equal magnitude. If we introduce a reference frame centered at a vertex, the test charge at the origin would experience Coulomb forces of the same intensity from charges described by the coordinates $x_1 + \mathrm{i}x_2$ and $x_3 + \mathrm{i}x_4$, if they satisfy the condition $x_1^2 + x_2^2 = x_3^2 + x_4^2$.

Quaternions and Vector-Parameters

We choose a basis in $\mathfrak{su}(2)$ in the form

$$i = \begin{pmatrix} \mathrm{i} & 0 \\ 0 & -i \end{pmatrix}, \quad j = \begin{pmatrix} 0 & 1 \\ -1 & 0 \end{pmatrix}, \quad k = \begin{pmatrix} 0 & \mathrm{i} \\ \mathrm{i} & 0 \end{pmatrix}$$

and introduce the set of unit quaternions as

$$\zeta = \zeta_0 + \zeta_1 i + \zeta_2 j + \zeta_3 k, \qquad |\zeta|^2 = 1, \qquad \zeta_\mu \in \mathbb{R}$$

with norm given by

$$|\,\zeta\,|^2 = \frac{1}{2}\mathrm{tr}(\zeta\bar{\zeta}) = det(\zeta) = \textstyle\sum_{\mu=0}^{3} \zeta_\mu^2$$

where $\bar{\zeta} = \zeta_0 - \zeta_1 i - \zeta_2 j - \zeta_3 k$ stands for the *conjugate quaternion.* Next, we associate with each vector $x \in \mathbb{R}^3$ a skew-hermitian matrix by the rule

$$x \to \Psi = x_1 i + x_2 j + x_3 k$$

where x_i are the Cartesian coordinates of x in the default basis, and let SU(2) act in its Lie algebra via the adjoint representation $\mathrm{Ad}_\zeta\colon \Psi \to \zeta\,\Psi\,\bar{\zeta}$, which can be viewed as a norm-preserving automorphism of $\mathbb{R}^3$. It is not difficult to obtain also the orthogonal matrix transforming the corresponding Cartesian coordinates of three-dimensional vectors

$$\mathcal{R}(\zeta) = (\zeta_0^2 - \boldsymbol{\zeta}^2)\mathcal{I} + 2\boldsymbol{\zeta} \otimes \boldsymbol{\zeta}^t + 2\zeta_0\boldsymbol{\zeta}^\times \tag{4}$$

where $\boldsymbol{\zeta} \in \mathbb{R}^3$ stands for the *imaginary* or *vector* part of the quaternion $\zeta = (\zeta_0, \boldsymbol{\zeta})$ and ζ_0 is referred to as its *real* or *scalar* part. The famous Rodrigues' rotation formula follows directly with the substitution

$$\zeta_0 = \cos\frac{\varphi}{2}, \qquad \boldsymbol{\zeta} = \sin\frac{\varphi}{2}\, n, \qquad (n, n) = 1.$$

Alternatively, we may project $\zeta \to c = \frac{\zeta}{\zeta_0} = \tan\left(\frac{\varphi}{2}\right) n$ and express the matrix entries of (4) as rational functions of the *vector-parameter c*

$$\mathcal{R}(c) = \frac{(1-c^2)\,\mathcal{I} + 2\,c \otimes c^t + 2\,c^\times}{1+c^2}. \tag{5}$$

Quaternion multiplication then gives the composition law for the vector-parameters of two successive rotations $\mathcal{R}(\langle c_2, c_1\rangle) = \mathcal{R}(c_2)\,\mathcal{R}(c_1)$ in the form

$$\langle c_2, c_1\rangle = \frac{c_2 + c_1 + c_2 \times c_1}{1 - (c_2, c_1)}. \tag{6}$$

The latter constitutes a representation since it is associative and satisfies

$$\langle\, c, 0\, \rangle = \langle\, 0, c\, \rangle = c, \qquad \langle c, -c\rangle = 0.$$

Some of the advantages of the vector parameterization involve exact rational expressions for the rotation matrix entries, less calculations for the compositions of group elements and correct description of the orthogonal group's topology $\mathrm{SO}(3) \cong \mathbb{RP}^3$.

Equivalently, in $\mathfrak{sl}_2(\mathbb{R})$ one has the *split quaternion* basis [1, 5]

$$\tilde{\imath} = \begin{pmatrix} 0 & 1 \\ 1 & 0 \end{pmatrix}, \quad \tilde{\jmath} = \begin{pmatrix} 1 & 0 \\ 0 & -1 \end{pmatrix}, \quad \tilde{\mathrm{k}} = \begin{pmatrix} 0 & 1 \\ -1 & 0 \end{pmatrix}$$

which can be mapped to $\mathfrak{su}(1,1)$ via the isometry $z \to \mathrm{i}\,\frac{z-\mathrm{i}}{z+\mathrm{i}}$ as

$$\tilde{\imath}\,' = \begin{pmatrix} 0 & 1 \\ 1 & 0 \end{pmatrix}, \quad \tilde{\mathrm{j}}\,' = \begin{pmatrix} 0 & \mathrm{i} \\ -\mathrm{i} & 0 \end{pmatrix}, \quad \tilde{k}\,' = \begin{pmatrix} \mathrm{i} & 0 \\ 0 & -\mathrm{i} \end{pmatrix}.$$

Expansion in the above bases allows for an explicit isometry $\mathbb{R}^{2,1} \to \mathfrak{sl}_2(\mathbb{R})$: $x \to \Psi = x_1\tilde{\imath} + x_2\tilde{\jmath} + x_3\tilde{k},\ x \cdot x = -\det\Psi$ and the projection onto $\mathrm{SO}^+(2,1)$ is given by the adjoint action of the group of unit split quaternions $\mathrm{SL}_2(\mathbb{R}) \cong \mathrm{SU}(1,1)$ in its Lie algebra $Ad_{\zeta}\colon \Psi \to \zeta\, \Psi\, \bar{\zeta}$, which is a norm-preserving automorphism. Using the familiar notation $\zeta = (\zeta_0, \boldsymbol{\zeta}), \bar{\zeta} = (\zeta_0, -\boldsymbol{\zeta}),\ \boldsymbol{\zeta} \in \mathbb{R}^{2,1}$ we see that the Cartesian coordinates of x are transformed by the pseudo-orthogonal matrix

$$\Lambda(\zeta) = (\zeta_0^2 + \boldsymbol{\zeta}^2)\jmath - 2\,\boldsymbol{\zeta} \otimes \eta\, \boldsymbol{\zeta} + 2\,\zeta_0\, \boldsymbol{\zeta}^{\curlyvee} \tag{7}$$

where $\eta = diag(1,1,-1)$ is the flat metric in $\mathbb{R}^{2,1}$, $(\boldsymbol{\zeta} \otimes \eta\, \boldsymbol{\zeta})^i_j = \eta_{jk}\, \zeta^i \zeta^k$ (summation over repeated indices is assumed) and $\boldsymbol{\zeta}^{\curlyvee} = \eta\, \boldsymbol{\zeta}^{\times}$, so that we also denote $\boldsymbol{\zeta} \curlyvee \boldsymbol{\xi} = \boldsymbol{\zeta}^{\curlyvee}\, \boldsymbol{\xi}$. Furthermore, we may introduce the *hyperbolic vector-parameter* in the usual manner $c = \frac{\boldsymbol{\zeta}}{\zeta_0}$ and write (7) as

$$\Lambda(c) = \frac{(1+c^2)\jmath - 2\, c \otimes \eta\, c + 2\, c^{\curlyvee}}{1-c^2}. \tag{8}$$

The inverse transformation yields

$$\zeta_0^{\pm} = \pm(1-c^2)^{-\frac{1}{2}}, \qquad \boldsymbol{\zeta}^{\pm} = \zeta_0^{\pm} c$$

where the two signs correspond to different sheets of the spin cover.

From the multiplication rule of split quaternions we easily derive the composition law of hyperbolic vector parameters in the form

$$\langle c_2, c_1 \rangle = \frac{c_2 + c_1 + c_2 \curlyvee c_1}{1 + c_2 \cdot c_1}. \tag{9}$$

It is easy to see that this construction constitutes a representation of SO(2,1), which has the obvious advantages we already discussed.

In the $3+1$ dimensional case we exploit a $\mathrm{Mat}(2,\mathbb{C})$ representation of four-vectors, which allows for expressing their pseudo-norm as a determinant

$$x \in \mathbb{R}^{3,1} \quad \to \quad \Psi = \begin{pmatrix} x_4 + x_1 & x_3 - \mathrm{i}x_2 \\ x_3 + \mathrm{i}x_2 & x_4 - x_1 \end{pmatrix}$$

$$x \cdot x \;=\; -\det\Psi = x_1^2 + x_2^2 + x_3^2 - x_4^2.$$

The standard matrix realization of the Lorentz group can be obtained from the norm-preserving action of $\mathrm{SL}_2(\mathbb{C})$: $\Psi \;\to\; \zeta\,\Psi\,\zeta^{\dagger}$, where $\zeta^{\dagger} = (\bar{\zeta}_0, -\bar{\boldsymbol{\zeta}})$ stands for the Hermitian conjugate of the unit *biquaternion* $\zeta = (\zeta_0, \boldsymbol{\zeta}) \in \mathrm{SL}_2(\mathbb{C})$. Finally, we may project dividing by ζ_0 and thus obtain the *complex vector parameter* $c = \frac{\boldsymbol{\zeta}}{\zeta_0}$, which allows for writing the corresponding pseudo-orthogonal matrix transforming the Cartesian coordinates of x in a block form as (see [8])

$$\Lambda(c) =$$

$$\lambda\begin{pmatrix} 1-|c|^2+c\otimes\bar{c}^t+\bar{c}\otimes c^t+(c+\bar{c})^\times & c-\bar{c}+\bar{c}\times c \\ (\bar{c}-c+\bar{c}\times c)^t & 1+|c|^2 \end{pmatrix}$$

where we denote $\lambda=|1+c^2|^{-1}$ and omit the factor $\mathcal{I}$ in the upper-left block in order to keep notations simple.

Finally, we investigate the case SO(2,2), which is more or less similar [10]. Considering the spin cover $\mathrm{U}(1,1)\otimes\mathrm{U}(1,1)\to\mathrm{SO}(2,2)$ we obtain

$$\Lambda(c,\tilde{c})=\tilde{\lambda}\begin{pmatrix} 1+c\cdot\tilde{c}-c\otimes\eta\,\tilde{c}-\tilde{c}\otimes\eta\,c+(c+\tilde{c})^\curlyvee & c-\tilde{c}-c\curlyvee\tilde{c} \\ (c-\tilde{c}+c\curlyvee\tilde{c})^t & 1-c\cdot\tilde{c} \end{pmatrix} \tag{10}$$

with $c=\frac{\zeta}{\zeta_0}$, $\tilde{c}=\frac{\tilde{\zeta}}{\tilde{\zeta}_0}$ and $\tilde{\lambda}=1/\sqrt{(1-c^2)(1-\tilde{c}^2)}$. One may also determine the vector-parameters from the matrix $\widetilde{\Lambda}=\Lambda-\tilde{\eta}\,\Lambda^t\,\tilde{\eta}$, where $\tilde{\eta}=diag(1,1,-1,-1)$, as

$$c=\frac{1}{\mathrm{tr}\Lambda}\begin{pmatrix}\widetilde{\Lambda}_{14}-\widetilde{\Lambda}_{32}\\ \widetilde{\Lambda}_{13}+\widetilde{\Lambda}_{24}\\ \widetilde{\Lambda}_{21}+\widetilde{\Lambda}_{34}\end{pmatrix},\qquad \tilde{c}=-\frac{1}{\mathrm{tr}\Lambda}\begin{pmatrix}\widetilde{\Lambda}_{14}+\widetilde{\Lambda}_{32}\\ \widetilde{\Lambda}_{24}-\widetilde{\Lambda}_{13}\\ \widetilde{\Lambda}_{34}-\widetilde{\Lambda}_{21}\end{pmatrix}. \tag{11}$$

Pythagorean Spinors

As pointed out in [12, 13, 18], the relation between Pythagorean triples and Gaussian integers $\mathbb{Z}[\mathrm{i}]=\{z=m+in\in\mathbb{C}\,;\,m,n\in\mathbb{Z}\}$, given by the Euclid's parameterization $x_1=\Re(z^2)$, $x_2=\Im(z^2)$, $x_3=|z|^2$, providing also a link to rational points on $\mathbb{S}^1$ via Euler's trigonometric substitution

$$\frac{x_2}{x_3}=\sin\varphi=\frac{2\tau}{1+\tau^2},\quad \frac{x_1}{x_3}=\cos\varphi=\frac{1-\tau^2}{1+\tau^2},\quad \tau=\tan\frac{\varphi}{2}=\frac{x_2}{x_1+x_3},$$

is actually a spin correspondence. This becomes more apparent if we map integer null vectors x to the subset of singular matrices in $\mathfrak{sl}_2(\mathbb{Z})$ as

$$x \to \Psi = \begin{pmatrix} x_2 & x_1 + x_3 \\ x_1 - x_3 & -x_2 \end{pmatrix}, \quad |\Psi|^2 = -\det\Psi = x_1^2 + x_2^2 - x_3^2 = 0.$$

This property remains invariant under the adjoint action of unimodular transformations $\zeta \in \mathrm{SL}_2^{\pm}(\mathbb{R})$, since $\det\zeta\Psi\,\zeta^{-1} = \det\Psi$. Thus, we may obtain the subgroup of $\mathrm{O}_{2,1}(\mathbb{Z})$ transformations (2). Euclid's parameterization, on the other hand, provides the alternative representation

$$\Psi = 2\begin{pmatrix} mn & m^2 \\ -n^2 & -mn \end{pmatrix} = 2\begin{pmatrix} m \\ -n \end{pmatrix} \otimes (n, m) = 2\,\psi^* \otimes \psi$$

where $\psi = (n, m)$ is referred to as the *Pythagorean spinor* generating the triple x, and $\psi^* = (m, -n)^t$ is its conjugate (in this case $\psi^* = \psi^{\perp}$). Then, the adjoint action of $\mathrm{PSL}_2(\mathbb{Z})$ on Ψ is reduced to ordinary unimodular transformations on the spinor ψ^t. In particular, the ternary tree given by the Hall matrices [9] can be obtained from the orbits of the base spinor $\psi^t = \begin{pmatrix} 2 \\ 1 \end{pmatrix}$ corresponding to the triple $v = (3\,, \quad 4\,, \quad 5)^t$ under the action of the subgroup of unimodular matrices generated by

$$\mathcal{S}_L = \begin{pmatrix} 2 & -1 \\ 1 & 0 \end{pmatrix}, \quad \mathcal{S}_U = \begin{pmatrix} 2 & 1 \\ 1 & 0 \end{pmatrix}, \quad \mathcal{S}_R = \begin{pmatrix} 1 & 2 \\ 0 & 1 \end{pmatrix} \qquad (12)$$

as shown in [16]. Acting with only one of the three we obtain the families of Plato, Fermat and Pythagoras, respectively. The first one generates the sequence of harmonics (overtones) $\tau_k = \frac{k+1}{k}$, the last one - the sequence of even natural numbers $\tau_k = 2k$ (octaves) and the one in between is the somewhat more complicated $\tau_k = 2, \frac{5}{2}, \frac{12}{5}, \frac{29}{12}, \frac{70}{29}, \frac{169}{70}, \ldots \to 1 + \sqrt{2}$. Combining these families one may easily obtain other useful integer or rational numerical sequences, such as the Fibonacci numbers [9].

The Two Types of Quadruples

In the case of quadruples $x_1^2 + x_2^2 + x_3^2 - x_4^2 = 0$ we use a similar construction

$$x \to \Psi = \begin{pmatrix} x_4 + x_1 & x_3 - \mathrm{i}x_2 \\ x_3 + \mathrm{i}x_2 & x_4 - x_1 \end{pmatrix}$$

and, since the action of $\mathrm{PSL}_2(\mathbb{Z}[\mathrm{i}])\colon \Psi \to \zeta\Psi\,\zeta^{-1}$ preserves the determinant, it yields a Lorentz transformation for the null vector x, parameterized by $\alpha, \beta \in \mathbb{Z}[\mathrm{i}]$, such that $|\alpha|^2 + |\beta|^2 \in 2\mathbb{Z}$ (see [17] for comparison)

$$x^t = \frac{1}{2}\begin{pmatrix} |\alpha|^2 - |\beta|^2, & 2\,\Im(\alpha\bar{\beta}), & 2\,\Re(\alpha\bar{\beta}), & |\alpha|^2 + |\beta|^2 \end{pmatrix} \quad (13)$$

which can be written also as

$$\Psi = \begin{pmatrix} |\alpha|^2 & \bar{\alpha}\beta \\ \alpha\bar{\beta} & |\beta\,|^2 \end{pmatrix} = \begin{pmatrix} \bar{\alpha} \\ \bar{\beta} \end{pmatrix} \otimes \begin{pmatrix} \alpha, & \beta \end{pmatrix} = \psi^* \otimes \psi, \qquad \psi^* = \psi^\dagger.$$

Working with the parameter $\tau = \beta/\alpha \in \mathbb{CP}^1$ (or its conjugate) we relate Pythagorean quadruples to rational points on the unit sphere and (13) becomes the well-known stereographic projection. Next, we let the group $\mathrm{PSL}_2(\mathbb{Z}[\mathrm{i}])$ act in the space of spinors and thus generate all solutions from the orbits of the points v_k given in $\psi = (\,\alpha\,,\beta\,)$ parameterization as

$$v_1 \leftrightarrow (\,1 + \mathrm{i}\,, 0\,), \qquad v_2 \leftrightarrow (\,\mathrm{i}\,, 1\,), \qquad v_1 \leftrightarrow (\,1\,, 1\,)$$

and the $\mathrm{PSL}_2(\mathbb{Z}[\mathrm{i}])$ transformations corresponding to the matrices (??) are

$$h_1 = \begin{pmatrix} 1 & -2\,\mathrm{i} \\ 0 & 1 \end{pmatrix}, \quad h_2 = \begin{pmatrix} 2 & -1 \\ 1 & 0 \end{pmatrix}, \quad h_3 = \begin{pmatrix} 1 - \mathrm{i} & -1 \\ -1 & 1 + \mathrm{i} \end{pmatrix}$$

$$h_4 = \begin{pmatrix} 1+\mathrm{i} & 1 \\ 1 & 1-\mathrm{i} \end{pmatrix}, \qquad h_5 = \begin{pmatrix} 0 & 1 \\ -1 & 2 \end{pmatrix}, \qquad h_6 = \begin{pmatrix} 1 & 2\,\mathrm{i} \\ 0 & 1 \end{pmatrix}.$$

As for the second type $x_1^2 + x_2^2 - x_3^2 - x_4^2 = 0$, it is convenient to represent

$$x \to \Psi = \begin{pmatrix} x_4 + \mathrm{i}x_3 & x_1 + \mathrm{i}x_2 \\ x_1 - \mathrm{i}x_2 & x_4 - \mathrm{i}x_3 \end{pmatrix}$$

and use a pair of complex parameters $\alpha, \beta \in \mathbb{C}$ (not necessarily in $\mathbb{Z}[\mathrm{i}]$)

$$x^t = \begin{pmatrix} \Re(\alpha^2 - \beta^2), & \Im(\alpha^2 - \beta^2), & 2\,\Im(\bar{\alpha}\beta), & |\alpha|^2 - |\beta|^2 \end{pmatrix} \tag{14}$$

which yields with the notation $\xi = \alpha + \beta$ and $\eta = \alpha - \beta$

$$\Psi = \begin{pmatrix} \alpha^2 - \beta^2 & (\alpha + \beta)(\bar{\alpha} - \bar{\beta}) \\ (\bar{\alpha} + \bar{\beta})(\alpha - \beta) & \bar{\alpha}^2 - \bar{\beta}^2 \end{pmatrix}$$

$$= \begin{pmatrix} \xi \\ \bar{\xi} \end{pmatrix} \otimes \begin{pmatrix} \eta, & \bar{\eta} \end{pmatrix}. \tag{15}$$

Since ξ and η are independent, they are transformed by $\Gamma \otimes \Gamma$, where Γ is the subgroup generated by the U(1,1) images of the matrices in (12)

$$\tilde{h}_1 = \begin{pmatrix} 1-\mathrm{i} & \mathrm{i} \\ -\mathrm{i} & 1+\mathrm{i} \end{pmatrix}, \quad \tilde{h}_2 = \begin{pmatrix} 1 & 1+\mathrm{i} \\ 1-\mathrm{i} & 1 \end{pmatrix}, \quad \tilde{h}_3 = \begin{pmatrix} 1+\mathrm{i} & 1 \\ 1 & 1-\mathrm{i} \end{pmatrix} \tag{16}$$

together with their inverses. Note that the results $x_1 x_2 x_3 = 0 \bmod 60$ and $x_1 x_2 x_3 x_4 = 0 \bmod 12$ for the standard triples and quadruples, obtained by considering all possible solutions in $\mathbb{Z}_3$, $\mathbb{Z}_4$ and for the former $\mathbb{Z}_5$, have no analogue in this case. It can be shown by a standard descent procedure [12] that each such quadruple can be shrunk via the action of the subgroup

(16) to a null vector in the box $|x_i| \leq 1$. Then it is straightforward to see that all solutions reside in the orbits of the points

$$\tilde{v}_1 = (1,0,0,1)^t, \qquad \tilde{v}_2 = (0,1,0,1)^t$$

$$\tilde{v}_3 = (1,1,1,1)^t, \qquad \tilde{v}_4 = (1,1,-1,1)^t$$

which are given in α, β -parameterization as

$$\tilde{v}_1 \leftrightarrow (1,0), \qquad \tilde{v}_2 \leftrightarrow \frac{1}{\sqrt{2}}(1+i,0)$$

$$\tilde{v}_3 \leftrightarrow (5/4+i,3/4+i), \qquad \tilde{v}_4 \leftrightarrow (-5/4-i,3/4+i)$$

respectively. Using the correspondence

$$\tilde{h}_k\,\xi \otimes \eta \leftrightarrow \widetilde{H}_k, \qquad \xi \otimes \tilde{h}_k\,\eta \leftrightarrow \widetilde{H}_{k+3}, \qquad k = 1,2,3 \tag{17}$$

we obtain the $\mathbb{Z}^{2,2}$ analogue of the Hall matrices (2) and (3) in the form

$$\widetilde{H}_1 = \begin{pmatrix} 1 & 1 & 1 & 0 \\ -1 & 1 & 0 & 1 \\ 1 & 0 & 1 & -1 \\ 0 & 1 & 1 & 1 \end{pmatrix}, \quad \widetilde{H}_2 = \begin{pmatrix} 1 & 0 & 1 & 1 \\ 0 & 1 & -1 & 1 \\ 1 & -1 & 1 & 0 \\ 1 & 1 & 0 & 1 \end{pmatrix},$$

$$\widetilde{H}_3 = \begin{pmatrix} 1 & -1 & 0 & 1 \\ 1 & 1 & -1 & 0 \\ 0 & -1 & 1 & 1 \\ 1 & 0 & -1 & 1 \end{pmatrix}$$

$$\widetilde{H}_4 = \begin{pmatrix} 1 & 1 & -1 & 0 \\ -1 & 1 & 0 & 1 \\ -1 & 0 & 1 & 1 \\ 0 & 1 & -1 & 1 \end{pmatrix}, \quad \widetilde{H}_5 = \begin{pmatrix} 1 & 0 & -1 & 1 \\ 0 & 1 & 1 & 1 \\ -1 & 1 & 1 & 0 \\ 1 & 1 & 0 & 1 \end{pmatrix},$$

$$\widetilde{H}_6 = \begin{pmatrix} 1 & -1 & 0 & 1 \\ 1 & 1 & 1 & 0 \\ 0 & 1 & 1 & -1 \\ 1 & 0 & 1 & 1 \end{pmatrix}.$$

The orbits of the points $\tilde{v}_k$ with respect to the subgroup spanned by $\widetilde{H}_k, \widetilde{H}_k^{-1}$ contain all primitive null vectors in $\mathbb{Z}^{2,2}/\{0\}$, including solutions of the type $x_1 = x_3$, $x_2 = x_4$, as well as triples, permutations of the same coordinates (generated by the symmetries $\mathcal{S}_{1243}$, $\mathcal{S}_{2134}$ and $\mathcal{S}_{3412}$) and sign inversions. Note that the matrices $\widetilde{H}_k$ are not all in SO(2,2). They belong to a larger class of transformations, defined by the property $\Lambda\tilde{\eta} \pm \tilde{\eta}\Lambda = 0$, which also keeps the isotropic pseudo-cone invariant and may still be obtained from (10) if we omit the pre-factor $\tilde{\lambda}$.

Vector-Parameter Algorithm

Instead of the Hall matrices (2), we use (with the expense of allowing negative solutions) L, R and $\mathring{U} = -U \in \mathrm{SO}(2,1)$, so the vector parameterization is straight forward

$$c_L = (\,0\,,1\,,1\,)^t, \qquad c_{\mathring{U}} = (\,-1\,,1\,,0\,)^t, \qquad c_R = (\,-1\,,0\,,-1\,)^t,$$

and we may take advantage of the composition (9) as an alternative to the matrix multiplication. Afterwards, we obtain the compound transformation matrix explicitly with the aid of (8), thus greatly simplifying calculations, especially for words of the type Λ^n with vector-parameter $c^{\langle n\rangle} = \langle\, c, c^{\langle n-1\rangle}\,\rangle = \langle\, c^{\langle n-1\rangle}, c\,\rangle$, $c^{\langle 0\rangle} = 0$ that can be written as $c^{\langle n\rangle} = \frac{a_n}{b_n}\,c$, where a_k and b_k are determined by the recurrence relations

$$a_{k+1} = a_k + b_k, \qquad b_{k+1} = a_k\,c^2 \pm b_k, \qquad a_1 = b_1 = 1. \tag{18}$$

The sign is positive for triples and quadruples of the second kind, and negative for standard quadruples. In the latter case, the vector-parameters derived from (3) are

$$c_1 = (0, \mathrm{i}, 1)^t, \qquad c_2 = (-\mathrm{i}, -1, 0)^t, \qquad c_3 = (1, 0, \mathrm{i})^t$$

with $= c_{7-k} = -c_k$ and finally, in the $\mathbb{Z}^{2,2}$ case, denoting $c_F = (1,1,0)^t$, one has from (10) and (17)

$$\widetilde{H}_1 \colon \Lambda(c_R + c_F, 0), \quad \widetilde{H}_2 \colon \Lambda(c_F, 0), \quad \widetilde{H}_3 \colon \Lambda(-c_R, 0)$$

$$\widetilde{H}_4 \colon \Lambda(0, -c_L), \quad \widetilde{H}_5 \colon \Lambda(0, -c_F), \quad \widetilde{H}_6 \colon \Lambda(0, c_L - c_F)$$

where the coefficient of proportionality is $\tilde{\lambda}^{-1}$ defined in (10).

The Two Types of Quintuples

For the first type of quintuples $x_1^2 + x_2^2 + x_3^2 + x_4^2 = x_5^2$ the spin cover $\mathrm{U}(2) \times \mathrm{U}(2) \rightarrow \mathrm{SO}(4)$ allows for parameterizing by a pair of quaternions $\zeta_{1,2}$ with complex coordinates α_i, β_i (or two vectors $\boldsymbol{\alpha}, \boldsymbol{\beta} \in \mathbb{C}^2$). Denoting

$$(\boldsymbol{\alpha}, \boldsymbol{\beta})^{\pm} = \alpha_1\beta_2 \pm \beta_1\alpha_2, \qquad \langle \boldsymbol{\alpha}, \boldsymbol{\beta} \rangle^{\pm} = \alpha_1\alpha_2 \pm \beta_1\beta_2$$

$$|\boldsymbol{\alpha\beta}|^{\pm} = \left(\textstyle\prod_{i=1}^{2} (|\alpha_i|^2 \pm |\beta_i|^2)\right)^{1/2}$$

we can express from the corresponding orthogonal matrix (ignoring signs)

$$x^t = (\Re\,(\boldsymbol{\alpha}, \boldsymbol{\beta})^+, \quad \Im\,\langle \boldsymbol{\alpha}, \boldsymbol{\beta} \rangle^+, \quad \Re\,\langle \boldsymbol{\alpha}, \boldsymbol{\beta} \rangle^-, \quad \Im\,(\boldsymbol{\alpha}, \boldsymbol{\beta})^-, \quad |\boldsymbol{\alpha\beta}|^+)$$

which can also be written in terms of $\tau_k = \beta_k/\alpha_k$, $k = 1,2$. The above gives a quintuple if $|\zeta_1\zeta_2| = |\boldsymbol{\alpha\beta}|^+ \in \mathbb{Z}$, e.g. $|\zeta_1| = |\zeta_2|$ or $\zeta_{1,2}$, both correspond to quintuples themselves. The projective action of $\mathrm{Sp}(1,1) \cong \mathrm{Spin}(4,1)$ on the space of spinors, this time realized as block quaternion matrices, propagates the solutions in the usual way.

For the second kind $x_1^2 + x_2^2 + x_3^2 = x_4^2 + x_5^2$ we may use either SO(2,2) or SO(3,1) parameterization. Choosing the former, we write

$$x^t = (\Re\,\langle \boldsymbol{\alpha}, \boldsymbol{\beta}\,\rangle^+,\quad \Im\,\langle \boldsymbol{\alpha}, \boldsymbol{\beta}\,\rangle^-,\quad \Im\,(\boldsymbol{\alpha}, \boldsymbol{\beta}\,)^-,\quad \Re\,(\boldsymbol{\alpha}, \boldsymbol{\beta}\,)^+,\quad |\boldsymbol{\alpha\beta}\,|^-)$$

which can be expressed also in terms of the complex parameters $\tau_k = \beta_k/\alpha_k$. Just as in the previous case, the product of the norms of the two split-quaternions needs to be an integer. To ensure this property one may choose parameters which satisfy it and then obtain the rest of the solutions via the action of a subgroup of $\mathrm{Sp}\,(4,\mathbb{R}) \cong \mathrm{Spin}(3,2)$ which preserves it.

Vector Decomposition in $\mathbb{Q}^3$ and $\mathbb{Q}^{2,1}$

First, we note that unit vectors in $\mathbb{Q}^3$ are related to primitive quadruples

$$x_1^2 + x_2^2 + x_3^2 = x_4^2, \quad x_4 \neq 0 \quad \Leftrightarrow \quad n \in \mathbb{Q}_0^3, \quad n_k = \frac{x_k}{x_4}, \quad k = 1,2,3$$

where $\mathbb{Q}_0^3$ denotes the set of unit vectors in $\mathbb{R}^3$ with rational coordinates. Suppose we are given three rational unit vectors (quadruples) $c_k \in \mathbb{Q}_0^3$ to determine three axes of rotation. Furthermore, the unit vector along the compound rotation's axis is denoted by n, and τ is its scalar parameter $\tau = \tan\frac{\varphi}{2}$, where φ is the angle of rotation[3]. We use the notations

$$g_{ij} = (\hat{c}_i, \hat{c}_j), \qquad r_{ij} = (\hat{c}_i, \mathcal{R}(c)\,\hat{c}_j), \qquad \omega = (\hat{c}_1, \hat{c}_2 \times \hat{c}_3)$$

where $c = \tau n$ is the compound vector-parameter and $\mathcal{R}(c)$ is the matrix transformation associated with it. As shown in [6], the necessary and sufficient condition for the decomposition $\mathcal{R}(c) = \mathcal{R}(c_3)\,\mathcal{R}(c_2)\,\mathcal{R}(c_1)$ (where $c_k = \tau_k \hat{c}_k$) to exist over $\mathbb{RP}^1 \cong \mathbb{R} \cup \infty$ is given by the formula

$$\Delta = \begin{vmatrix} 1 & g_{12} & r_{31} \\ g_{21} & 1 & g_{23} \\ r_{31} & g_{32} & 1 \end{vmatrix} \geq 0. \tag{19}$$

[3]Note that it also has to be Pythagorean in order to have $\tau \in \mathbb{Q}$.

In the rational case considered here we demand $\sqrt{\Delta} \in \mathbb{Q}$ instead, as long as $r_{ij} \neq g_{ij}$ for $i > j$. The solutions for the *scalar parameters* $\tau_k = \tan\frac{\varphi_k}{2}$ are given in the generic setting ($\varphi \neq 0, \pi$ and $\hat{c}_3 \neq \pm\mathcal{R}(c)\,\hat{c}_1$) by

$$\tau_2^{\pm} = \frac{-\omega \pm \sqrt{\Delta}}{r_{31}+g_{31}-2g_{12}g_{23}} \tag{20}$$

and respectively (for most of the results in this section we refer to [6])

$$\tau_1^{\pm} = \frac{(g_{32}-r_{32})\tau_2^{\pm}}{((g_{32}+r_{32})\upsilon_1-(g_{31}+r_{31})\upsilon_2)\tau\tau_2^{\pm}+r_{31}-g_{31}}$$

$$\tau_3^{\pm} = \frac{(g_{21}-r_{21})\tau_2^{\pm}}{((g_{21}+r_{21})\upsilon_3-(g_{31}+r_{31})\upsilon_2)\tau\tau_2^{\pm}+r_{31}-g_{31}} \tag{21}$$

where we make use of the notation

$$\upsilon_k = (\hat{c}_k, n), \quad \tilde{\upsilon}_1 = (\hat{c}_2 \times \hat{c}_3, n), \quad \tilde{\upsilon}_2 = (\hat{c}_3 \times \hat{c}_1, n), \quad \tilde{\upsilon}_3 = (\hat{c}_1 \times \hat{c}_2, n).$$

In the symmetric case of a half-turn ($\tau = \infty$) l'Hôpital's rule yields

$$\tau_1^{\pm} = \frac{(g_{23}-\upsilon_2\upsilon_3)\tau_2^{\pm}}{(\upsilon_1\tilde{\upsilon}_1+\upsilon_2\tilde{\upsilon}_2)\tau_2^{\pm}+\upsilon_1\upsilon_3-g_{13}}$$

$$\tau_3^{\pm} = \frac{(g_{12}-\upsilon_1\upsilon_2)\tau_2^{\pm}}{(\upsilon_2\tilde{\upsilon}_2+\upsilon_3\tilde{\upsilon}_3)\tau_2^{\pm}+\upsilon_1\upsilon_3-g_{13}}.$$

Also, decomposing the identity ($\varphi \equiv 0$) in the case $\omega \neq 0$, one obtains

$$\tau_1 = \frac{\omega}{g_{12}g_{13}-g_{23}}, \quad \tau_2 = \frac{\omega}{g_{12}g_{23}-g_{13}}, \quad \tau_3 = \frac{\omega}{g_{13}g_{23}-g_{12}} \tag{22}$$

and for $\omega = 0$ we have one more (degenerate) solution if $\hat{c}_1 = \pm\hat{c}_3$ - it appears in the form of two mutually inverse matrices, i.e. $\tau_2 = 0$, $\tau_1 \mp \tau_3 = 0$. More generally, such infinite families of solutions emerge for a generic transformation when the *gimbals lock* condition $\hat{c}_3 = \pm\mathcal{R}(c)\,\hat{c}_1$ is satisfied and they are explicitly given by the expressions

$$\tau_2 = \frac{\tilde{\tilde{v}}_3}{g_{12}v_2 - v_1}, \qquad \tau'_1 = \frac{\tau_1 \pm \tau_3}{1 \mp \tau_1\tau_3} = \frac{\tilde{\tilde{v}}_3}{g_{12}v_1 - v_2}. \tag{23}$$

Note that in this case $\Delta = -(r_{21} - g_{21})^2$, so (19) is actually equivalent to

$$r_{21} = g_{21}. \tag{24}$$

Next, we attempt to find a situation in which the condition $\sqrt{\Delta} \in \mathbb{Q}$ is guaranteed. For example, in the Davenport setting $\hat{c}_2 \perp \hat{c}_{1,3}$ that yields $\Delta = 1 - r_{31}^2 \geq 0$ for arbitrary c it is sufficient to demand

$$r_{31} = \frac{(2v_1v_3 - g_{13})\tau^2 - 2\tilde{v}_2\tau + g_{13}}{1+\tau^2} \in \left\{\frac{m^2-n^2}{m^2+n^2}, \frac{2mn}{m^2+n^2}\right\}$$

for some integers $(m,n) \neq (0,0)$, i.e. the acute angle between the directions of $\hat{c}_3$ and $\mathcal{R}(c)\,\hat{c}_1$ to be present in some Pythagorean triangle.

Another solution is obtained when $r_{31} = 2g_{12}g_{23} - g_{13}$ ($\tau_2 = \infty$), e.g. if $\hat{c}_2$ is normal to either $\hat{c}_1$ or $\hat{c}_3$ and $\mathcal{R}(c)$ inverts the projection of $\hat{c}_1$ on $\hat{c}_3$.

One more setting is $r_{31} = g_{31}$, which guarantees that $\Delta = \omega^2$ as a Gram determinant, and for the middle parameter one has $\tau_2^{\pm} = 0, \frac{\omega}{g_{12}g_{23} - g_{13}}$, while τ_1 and τ_3 are determined as before. We note that when one encounters a vanishing parameter, i.e., $r_{ij} = g_{ij}$ for some $i > j$, the remaining two may also be obtained in another way based on two-axes decomposition. For example, in the case $r_{21} = g_{21}$ we may decompose into a pair of rotations about the first two axes and the rational solution is given by

$$\tau_1 = \frac{\tilde{\tilde{v}}_3}{g_{12}v_1 - v_2}, \qquad \tau_2 = \frac{\tilde{v}_3}{g_{12}v_2 - v_1}. \tag{25}$$

Similar expressions can be derived in the cases $r_{31} = g_{31}$ and $r_{32} = g_{32}$. Finally, in the gimbal lock setting $\hat{c}_3 = \pm\mathcal{R}(c)\,\hat{c}_1$ the condition (24) is equivalent to (19), as already discussed. Thus (23) is justified via (25) and the parameters τ_1 and τ_3 take all rational values that satisfy it.

In [6] we also argue that one may consider a system of axes, attached to the rotating object, which we denote by $\{\hat{c}'_k\}$, while $\{\hat{c}_k\}$ stands for the static one. Then, the decompositions in the two systems are related via

$$\mathcal{R}(c) = \mathcal{R}(c'_3)\mathcal{R}(c'_2)\mathcal{R}(c'_1) = \mathcal{R}(c_1)\mathcal{R}(c_2)\mathcal{R}(c_3)$$

and in the case of two axes we have in particular

$$\mathcal{R}(c) = \mathcal{R}(c'_2)\mathcal{R}(c'_1) = \mathcal{R}(c_1)\mathcal{R}(c_2).$$

Using a set of linear relations, it is not hard to see how the gimbals lock condition in $\{\hat{c}'_k\}$ corresponds to Euler decomposition $\{\hat{c}_k\}$ and vice versa.

The Hyperbolic Case

In the hyperbolic setting we consider vectors with positive, negative or vanishing pseudo-norms[4]. Restricted to $\mathbb{Q}^{2,1}$ these three types are related to Pythagorean quadruples and triples. Namely, vectors of positive norm can be rescaled to unit length as $x_1^2 + x_2^2 - x_3^2 = x_4^2 \Rightarrow n \in \mathbb{Q}_0^{2,1}$, $n_k = x_k/x_4$, while in the time-like case we have $x_1^2 + x_2^2 - x_3^2 = -x_4^2$ and $n \cdot n = -1$, i.e. a quadruple of the first kind with x_3 and x_4 exchanged. Finally, a null integer vector $x_1^2 + x_2^2 - x_3^2 = 0$ yields a Pythagorean triple. In order to distinguish between different types of vectors, we introduce the ε-factors $\varepsilon = n \cdot n$ and $\varepsilon_k = \hat{c}_k \cdot \hat{c}_k$. We also denote

$$r_{ij} = \hat{c}_i \cdot \Lambda(c)\, \hat{c}_j, \qquad g_{ij} = \hat{c}_i \cdot \hat{c}_j, \qquad \omega = \hat{c}_1 \cdot \hat{c}_2 \curlyvee \hat{c}_3$$

and obtain the necessary condition for decomposability

$$\Delta = -\begin{vmatrix} \varepsilon_1 & g_{12} & r_{31} \\ g_{21} & \varepsilon_2 & g_{23} \\ r_{31} & g_{32} & \varepsilon_3 \end{vmatrix} \geq 0 \qquad (26)$$

that is also sufficient unless some of the relations $\hat{c}_3 = \pm\Lambda(c)\, \hat{c}_1$ and

[4]Usually referred to as *space-like*, *time-like* and *isotropic* or *null* vectors, respectively.

$$\varepsilon_i = g_{ij} = 0, \qquad |i - j| = 1 \tag{27}$$

holds. The latter impose additional conditions, such as $\omega \neq 0$ or $r_{ij} = g_{ij}$ for certain j and $i > j$. Some examples are provided in the next paragraph.

In the rational case we demand $\sqrt{\Delta} \in \mathbb{Q}$ instead of (26) and in the generic setting ($\tau \neq 0, \infty$, (27) does not hold and $\hat{\mathrm{c}}_3 \neq \pm\Lambda(c)\,\hat{\mathrm{c}}_1$) the solutions are

$$\tau_2^{\pm} = \frac{\omega \pm \sqrt{\Delta}}{\varepsilon_2(r_{31}+g_{31}) - 2g_{12}g_{23}} \tag{28}$$

for the middle parameter, while for the other two we obtain

$$\tau_1^{\pm} = \frac{(r_{32}-g_{32})\tau_2^{\pm}}{((g_{32}+r_{32})v_1-(g_{31}+r_{31})v_2)\tau\tau_2^{\pm}+g_{31}-r_{31}}$$

$$\tag{29}$$

$$\tau_3^{\pm} = \frac{(r_{21}-g_{21})\tau_2^{\pm}}{((g_{21}+r_{21})v_3-(g_{31}+r_{31})v_2)\tau\tau_2^{\pm}+g_{31}-r_{31}}.$$

In the limit $\tau \to \infty$, corresponding to either a half turn if $\varepsilon = -1$ or a non-orthochronous Lorentz transformation for $\varepsilon = +1$, we obtain

$$\tau_1^{\pm} = \frac{(\varepsilon g_{23}-v_2v_3)\tau_2^{\pm}}{(v_1\tilde{\tilde{\mathrm{v}}}^1+v_2\tilde{\mathrm{v}}^2)\tau_2^{\pm}+v_1v_3-\varepsilon g_{13}}$$

$$\tau_3^{\pm} = \frac{(\varepsilon g_{12}-v_1v_2)\tau_2^{\pm}}{(v_2\tilde{\tilde{\mathrm{v}}}^2+v_3\tilde{\mathrm{v}}^3)\tau_2^{\pm}+v_1v_3-\varepsilon g_{13}}.$$

Similarly, for the identity transformation ($\tau \equiv 0$) we have

$$\tau_1 = \frac{\omega}{\varepsilon_1 g_{23}-g_{12}g_{13}}, \quad \tau_2 = \frac{\omega}{\varepsilon_2 g_{13}-g_{12}g_{23}}, \quad \tau_3 = \frac{\omega}{\varepsilon_3 g_{12}-g_{13}g_{23}}$$

in the regular case $\omega \neq 0$, and for $\omega = 0$ the only possibility is $\hat{\mathrm{c}}_3 = \pm\hat{\mathrm{c}}_1$, i.e. decomposition into a pair of two mutually inverse transformations.

Decomposing into a pair of pseudo-rotations about $\hat{c}_1$ and $\hat{c}_2$, on the other hand, demands the condition $r_{21} = g_{21}$ and then the solution is

$$\tau_1 = \frac{\tilde{\tilde{v}}^3}{\varepsilon_1 v_2 - g_{12} v_1}, \qquad \tau_2 = \frac{\tilde{v}^3}{\varepsilon_2 v_1 - g_{12} v_2}. \tag{30}$$

The same relation needs to hold also in the gimbal lock setting

$$\hat{c}_3 = \pm\Lambda(c)\,\hat{c}_1 \tag{31}$$

in which the degenerate solutions are given by the expressions

$$\tau_2 = \frac{\tilde{v}^3}{\varepsilon_2 v_1 - g_{12} v_2}, \qquad \frac{\tau_1 \pm \tau_3}{1 \pm \varepsilon_1 \tau_1 \tau_3} = \frac{\tilde{v}^3}{\varepsilon_1 v_2 - g_{12} v_1}. \tag{32}$$

Just as in the Euclidean case, here (24) guarantees the above decomposition, but this time it is not implied by (26), since $\Delta = \varepsilon_1 (r_{21} - g_{21})^2$ and if c_1 is space-like or null, unlike (24) and (32), it is automatically satisfied.

Orthogonality Conditions and Lattice Cubes

The orthogonality condition between two integer vectors $x, y \in \mathbb{Z}^3$ with equal lengths is given by a null complex vector $z = x + \mathrm{i}y \in \mathbb{Z}^3[\mathrm{i}]$, $z^2 = 0$, i.e., a Wick-rotated complex Pythagorean equation $z_1^2 + z_2^2 = (\mathrm{i}z_3)^2$, the real part of which yields $|x| = |y| = L$ and the imaginary one determines that $x \perp y$. The solutions correspond to lattice squares in $\mathbb{Z}^3$. In particular, if $L \in \mathbb{Z}$, that is if both x and y represent quadruples, it also gives rise to a lattice cube[5] [15] with a third edge determined by $x \times y$ and volume $vol = L^3$. Multiplying all coordinates by L^{-1} rescales to unit cubes in $\mathbb{Q}^3$, which are in one-to-one correspondence with $\mathrm{SO}(3, \mathbb{Q})$ matrices[6], so they can easily be obtained from the spin representation. We see that, unlike in the generic case, directions here cannot be chosen arbitrarily, since they need to correspond to Pythagorean quadruples and, in the orthogonal setting in particular (Bryan angles), to rows (columns) of some

[5] Actually, this determines eight such cubes considering the possible reflections.

[6] In [14] we find an efficient way to generate $\mathrm{O}(n, \mathbb{Q})$ matrices (n-dimensional cubes).

$\mathrm{SO}(3,\mathbb{Q})$ matrix. Then, in order to have $\sqrt{\Delta}\in\mathbb{Q}$ satisfied, we demand that r_{31} is a cosine of an angle in a Pythagorean triangle, e.g.

$$r_{31} = 2\tau\frac{v_1 v_3 \tau - v_2}{1+\tau^2} = \frac{2mn}{m^2+n^2}$$

for some integer $(m,n)\in\mathbb{Z}^2/\{0\}$. One obvious solution is $\tau = m/n$ and $v_1 v_3 = n/m$ in the case $v_2 = 0$ (n lies in the plane determined by $\hat{\mathrm{c}}_{1,3}$).

Similarly, in the case of Euler angles ($g_{12} = q_{23} = 0,\ g_{13} = 1$) one has

$$r_{31} = \frac{2v_1^2+\tau^{-2}-1}{1+\tau^{-2}} = \frac{m^2-n^2}{m^2+n^2}$$

which certainly holds for $\tau = n/m$ and $v_1 = 0$. Thus, we conclude that in the Euler setting for each $\mathrm{SO}(3,\mathbb{Q})$ transformation there exists a rational decomposition as long $n\perp c_1$. For example, one may choose $\hat{\mathrm{c}}_1$, $\hat{\mathrm{c}}_2$ and n to be the three rows of some $\mathrm{SO}(3,\mathbb{Q})$ matrix and let $\tau\in\mathbb{Q}$ be arbitrary.

In the hyperbolic case one obtains an orthogonal pair of unit vectors from the real and imaginary part of a complex null vector $z = x + \mathrm{i}y$, or alternatively from the solutions of the complex Pythagorean equation $z_1^2 + z_2^2 = z_3^2$. The hyperbolic cross product $\hat{\mathrm{E}}$ can be used to construct a third integer vector $x\hat{\mathrm{E}}y$ that is normal (in the Lorentz metric) to both x and y. In the case $|x|,|y|\in\mathbb{Z}$ this relates Pythagorean quadruples of the second kind directly to hyperbolic lattice pseudo-cubes and therefore to orthonormal frames in three-dimensional Minkowski space. We note that such frames can be retrieved from the rows or columns of a rational SO(2,1) matrix, the first two corresponding to space-like vectors, while the third corresponds to a time-like one and thus to a quadruple of the first kind. Moreover, from (26) and (27) we have the configurations[7] (for $i,j = 1,3$)

$$\varepsilon_i = g_{i2} = 0,\quad \omega = \pm g_{31} \neq 0\quad (\varepsilon_2 = 1)\quad \Rightarrow\quad \tau_2 = \frac{\omega\pm r_{31}}{g_{31}+r_{31}} \tag{33}$$

---

[7]Note that (8) naturally imposes the restrictions $\varepsilon = 1 \Rightarrow |\tau| \neq 1$ and $\varepsilon = 0 \Rightarrow |\tau| \neq \infty$.

$$\varepsilon_2 = g_{2i} = 0, \quad \omega = \pm g_{2j} \neq 0 \quad (\varepsilon_i = 1) \quad \Rightarrow \quad \tau_2 = \frac{g_{31}-r_{31}}{2\,\omega}$$

that guarantee the existence of a rational decomposition for arbitrary rational pseudo-rotation away from the gimbals lock setting (31) - a construction with no analogue in the Euclidean case, as we show above. On the other hand, the Davenport condition $g_{12} = g_{23} = 0$, if we have additionally $\varepsilon_1\varepsilon_3 = 1$ and $r_{31} = \pm 1$, provides a degenerate solution with

$$\Delta = \varepsilon_2(r_{31}^2 - \varepsilon_1\varepsilon_3) = 0.$$

The latter is an exact (rational) square also when $\hat{c}_2$ is time-like and $r_{31} = p/q$, where p and q are, respectively, one of the legs and the hypotenuse of a Pythagorean triangle, or when $\hat{c}_2$ is space-like, $\varepsilon_1\varepsilon_3 = -1$ and p, q are the two legs of such a triangle. More generally, if we have $g_{12} = \pm g_{23}$ and $\varepsilon_1 = \varepsilon_3 = \pm 1$, then all four expressions

$$\Delta = \varepsilon_2(r_{31}^2 - 1) \pm 2g_{12}^2(r_{31} \pm 1)$$

vanish for either $r_{31} = 1$ or $r_{31} = -1$ and that also yields degenerate solutions.

For a detailed discussion on the real case we refer to [6], while [5] provides some useful relations to hyperbolic geometry and physics. A recommended reading for the vector parameterization technique would be [8] and for Pythagorean triples and quadruples - [2, 13] and [17], respectively.

Bibliography

[1] Ahlfors L., *Möbius Transformations in Several Dimensions*, School of Mathematics, University of Minessota 1981.

[2] Alperin R., *The Modular Tree of Pythagoras*, Am. Math. Monthly **112** (2005) 807-816.

[3] Barning F., On Pythagorean and Quasi-Pythagorean Triangles and a Generation Process with the help of Unimodular Matrices (in Dutch), Math. Centrum Amsterdam Afd. Zuivere Wisk, ZW-001, 1963.

[4] Berggren B., *Pytagoreiska trianglar* (in Swedish), Tidskrift för elementär matematik, fysik och kemi **17** (1934) 129-139.

[5] Brezov D., Mladenova C. and Mladenov I., *Vector Parameters in Classical Hyperbolic Geometry*, J. Geom. Symmetry Phys. **30** (2013) 21-50.

[6] Brezov D., Mladenova C. and Mladenov I., *Some New Results on Three-Dimensional Rotations and Pseudo-Rotations*, AIP Conf. Proc. **1561** (2013) 275-288.

[7] Eckert E., *Groups of Primitive Pythagorean Triangles*, Math. Magazine **51** (1984) 22-27.

[8] Fedorov F., *The Lorentz Group* (in Russian), Science, Moskow 1979.

[9] Hall A., *Genealogy of Pythagorean Triads*, Math. Gazette **54** (1970) 377-379.

[10] Helgason S., Differential Geometry, Lie Groups and Symmetric Spaces, Academic Press, San Diego 1978.

[11] Ivanov I. and Avramova I., *Varna Necropolis*: *The Dawn of European Civilization*, Agató Publishers, Sofia 2000.

[12] Jadczyk A. and Kocik J., *Pythagoreans Quadruples on the Future Light Cone*, http://arkadiusz-jadczyk.eu/docs/parabolic1.pdf (2013).

[13] Kocik J., Clifford Algebras and Euclid's Parameterization of Pythagorean Triples, Adv. Appl. Clifford Algebras **17** (2007) 71-93.

[14] Liebeck H. and Osborne A., *The Generation of All Rational Orthogonal Matrices*, Am. Math. Monthly **98** (1991) 131-133.

[15] Parris R., *Lattice Cubes*, College Math. Journal **42** (2011) 118-125.

[16] Saunders R. and Randall T., *The Family Tree of the Pythagorean Triplets Revisited*, Math. Gazette **78** (1994) 190-193.

[17] Spira R., *The Diophantine equation* $x^2 + y^2 + z^2 = m^2$, Am. Math. Monthly **69** (1962) 360-365.

[18] Trautman A., *Pythagorean Spinors and Penrose Twistors*, The Geometric Universe: Science, Geometry and the Work of Rodger Penrose, Oxford University Press, 1998.

SERIES IN PRABHAKAR FUNCTIONS AND THE GEOMETRY OF THEIR CONVERGENCE

JORDANKA PANEVA-KONOVSKA

Introduction

In 1971 Prabhakar introduced and studied a 3-index generalization of the Mittag-Leffler function.

Let $E^{\gamma}_{\alpha,\beta}$ denote the Prabhakar generalization (see [18]) of the Mittag-Leffler (M-L) functions E_{α} and $E_{\alpha,\beta}$, defined in the whole complex plane $\mathbb{C}$ by the power series:

$$E^{\gamma}_{\alpha,\beta}(z) = \sum_{k=0}^{\infty} \frac{(\gamma)_k}{\Gamma(\alpha k+\beta)} \frac{z^k}{k!}, \qquad \alpha,\ \beta,\ \gamma \in \mathbb{C},\ \ Re(\alpha) > 0, \qquad (1)$$

where $(\gamma)_k$ is the Pochhammer symbol ([1], Section 2.1.1)

$$(\gamma)_0 = 1, \quad (\gamma)_k = \gamma(\gamma+1)\dots(\gamma+k-1).$$

It is clear that for $\gamma = 1$, (1) coincides with the M-L function $E_{\alpha,\beta}$ (see e.g. [7], [6], [11], [3]), while for $\gamma = \beta = 1$ with E_{α} ([1], Vol. 3), i.e.:

$$E^{1}_{\alpha,\beta}(z) = E_{\alpha,\beta}(z), \quad E^{1}_{\alpha,1}(z) = E_{\alpha}(z), \qquad (2)$$

with $\alpha,\ \beta \in \mathbb{C},\ \ Re(\alpha) > 0$.

Consider now Prabhakar's generalization (1) for indices $\beta = n$ with integer $n = 0,1,2,\dots$, i.e.

$$E^{\gamma}_{\alpha,n}(z) = \sum_{k=0}^{\infty} \frac{(\gamma)_k}{\Gamma(\alpha k+n)} \frac{z^k}{k!}, \quad \alpha,\ \gamma \in \mathbb{C},\ \ Re(\alpha) > 0,\ \ n \in \mathbb{N}_0. \quad (3)$$

Depending on γ and n, some coefficients in (3) may be equal to zero, that is, the summation in (3) begins from some $p \geq 0$. So, (3) can be written as follows:

$$E_{\alpha,n}^{\gamma}(z) = z^p \sum_{k=p}^{\infty} \frac{(\gamma)_k}{\Gamma(\alpha k+n)} \frac{z^{k-p}}{k!}, \quad or$$

$$E_{\alpha,n}^{\gamma}(z) = z^p \sum_{k=p}^{M} \frac{(\gamma)_k}{\Gamma(\alpha k+n)} \frac{z^{k-p}}{k!}. \tag{4}$$

More precisely, as it is seen above, if γ is different from zero, then $p = 1$ for $n = 0$, whereas $p = 0$ for each positive integer n. In the case $\gamma = 0$, the following remark can be made:

Remark 1.1 If $\gamma = 0$, then the functions in (3) take the simplest form

1. $E_{\alpha,n}^{0}(z) = 0$ for $n = 0$,

2. $E_{\alpha,n}^{0}(z) = \frac{1}{\Gamma(n)}$ for $n \in \mathbb{N}$.

Furthermore, an asymptotic formula for "large" values of the indices n is valid as follows - a proof can be seen in [13]:

Theorem 1.2 *Let $z, \alpha, \gamma \in \mathbb{C}, n \in \mathbb{N}_0, Re(\alpha) > 0, \gamma \neq 0$. Then there exist entire functions $\theta_{\alpha,n}^{\gamma}$ such that the generalized Mittag-Leffler function (3) has the following asymptotic formula*

$$E_{\alpha,n}^{\gamma}(z) = \frac{(\gamma)_p}{\Gamma(\alpha p+n)} z^p \left(1 + \theta_{\alpha,n}^{\gamma}(z)\right), \tag{5}$$

where $\theta_{\alpha,n}^{\gamma}(z) \to 0$ as $n \to \infty$, with a corresponding p, depending on the index n. Moreover, on the compact subsets of the complex plane $\mathbb{C}$, the convergence is uniform and

$$\theta_{\alpha,n}^{\gamma}(z) = O\left(\frac{1}{n^{Re(\alpha)}}\right) \quad (n \in \mathbb{N}). \tag{6}$$

Remark 1.3 According to the asymptotic formula (5), it follows that there exists a positive integer N_0 such that the functions (3) have no zeros for $n > N_0$, possibly except for the origin.

Remark 1.4 Each function in (3) ($n \in \mathbb{N}$), being an entire function, not identically zero, has no more than a finite number of zeros in the closed and bounded set $|z| \leq R$. Moreover, because of Remark 1.3, no more than a finite number of these functions have some zeros, possibly except for the origin.

Series in Prabhakar's functions

Let $\tilde{E}^{\gamma}_{\alpha,n}$ be the functions given by the following relations:

$$\tilde{E}^{0}_{\alpha,0}(z) = 0, \quad \tilde{E}^{0}_{\alpha,n}(z) = \Gamma(n)\ z^{n}\ E^{0}_{\alpha,n}(z), \quad n \in \mathbb{N},$$

$$\tilde{E}^{\gamma}_{\alpha,n}(z) = \frac{\Gamma(\alpha p+n)}{(\gamma)_p}\ z^{n-p}\ E^{\gamma}_{\alpha,n}(z), \quad n \in \mathbb{N}_0 \quad (\gamma \neq 0), \tag{7}$$

(with the corresponding values of p).

We consider the series in these functions of the form:

$$\sum_{n=0}^{\infty} a_n \tilde{E}^{\gamma}_{\alpha,n}(z), \tag{8}$$

with complex coefficients a_n ($n = 0,1,2,\dots$).

In the process of studying the convergence of such kinds of series we give their regions of convergence in the complex plane, and investigate the behavior of the series on the boundaries of these regions. We determine where the series converge and where they do not, where the convergence is uniform. Finding their disks of convergence, we study the series behavior inside the found disks and "near" their boundaries, as well as on the boundaries, giving Cauchy-Hadamard, Abel, and Fatou type theorems. Such kinds of results are provoked by the fact that the solutions of some fractional order differential and integral equations can be written in terms of series (or series of integrals) of Mittag-Leffler type functions and their generalizations (as for example in works of V. Kiryakova [5], T. Sandev,

Ž. Tomovski and J. Dubbeldam [20], A. Khamzin, R. Nigmatullin and I. Popov [2], and many others).

The asymptotic formula (5) for the Prabhakar functions, earlier established by the author, in the cases of "large" values of indices, is used in the proofs of the convergence theorems for the considered series.

The same type convergence theorems have been previously obtained for series in some other special functions, for example, for series in: Laguerre and Hermite polynomials, by Rusev ([19]) , and, respectively Bessel functions, their Wright's 2-, 3-, and 4-indices generalizations, and also more general multi-index Mittag-Leffler functions (in the sense of Yu. Luchko - V. Kiryakova [4], [6], [3]) (see e.g. [9] - [17]) by the author .

Note that for $\gamma = 0$ the series (8) reduces to the power one and because of that the discussed proofs are only for $\gamma \neq 0$.

Cauchy-Hadamard and Abel type theorems

In the beginning, we state a theorem of Cauchy-Hadamard type and a corollary for the series (8).

In what follows we use the notations $D(0;R)$ and $C(0;R)$ respectively for the open disk with a radius R centered at the origin and with a radius R at its boundary, i.e.

$$D(0;R) = \{z: |z| < R,\ z \in \mathbb{C}\}, \quad C(0;R) = \{z: |z| = R,\ z \in \mathbb{C}\}.$$

Theorem 3.1 (of Cauchy-Hadamard type) *The region of convergence of the series (8) with complex coefficients a_n is the disk $D(0;R)$ with a radius of convergence R, where*

$$R = \left(\limsup_{n\to\infty} \left(\ |a_n| \ \right)^{1/n} \right)^{-1}. \tag{9}$$

More precisely, the series (8) is absolutely convergent on the disk $D(0;R)$ and divergent on the region $|z| > R$. The cases $R = 0$ and $R = \infty$ fall in the general case.

Thus, the considered series (8) converges in a disk, as in the classical theory of the power series. Analogously, inside the disk, the convergence of the discussed series is uniform, i.e. the following corollary, similar to the classical Abel lemma, holds.

Corollary 3.2 *Let the series (8) converge at the point* $z_0 \neq 0$. *Then it is absolutely convergent on the disk* $D(0;|z_0|)$. *Inside the disk* $D(0;R)$, *i.e. on each closed disk* $|z| \leq r < R$ *(R defined by (9)), the convergence is uniform.*

Proof. Indeed, since the considered series converges at the point $z_0 \neq 0$, its radius of convergence is the positive number R, and moreover the point z_0 lies either in the disk $D(0;R)$ or on its boundary - the circle $C(0;R)$. That is why the disk $D(0;|z_0|)$ is either a part of the region of convergence or it coincides with it, whence the absolute convergence follows. To prove uniformity of the convergence inside the disk $D(0;R)$, it is sufficient to show that the series is uniformly convergent on each closed disk $|z| \leq r < R$. To this purpose, choosing a point ζ, $|\zeta| = \rho$, $r < \rho < R$ and considering the series (8), we estimate $|a_n \tilde{E}^{\gamma}_{\alpha,n}(z)|$. First, note that some of the values of $\tilde{E}^{\gamma}_{\alpha,n}(\zeta)$, but only a finite number of them, can be zero. Then there exists a number P such that

$$|a_n \tilde{E}^{\gamma}_{\alpha,n}(z)| = |a_n \tilde{E}^{\gamma}_{\alpha,n}(\zeta)| \frac{|\tilde{E}^{\gamma}_{\alpha,n}(z)|}{|\tilde{E}^{\gamma}_{\alpha,n}(\zeta)|} \leq |a_n \tilde{E}^{\gamma}_{\alpha,n}(\zeta)| \; \frac{|1+\theta^{\gamma}_{\alpha,n}(z)|}{|1+\theta^{\gamma}_{\alpha,n}(\zeta)|}$$

for all $n > P$ and $|z| \leq r$.

Because of (6) and the relations $\lim\limits_{n\to\infty} \frac{1}{n^{Re(\alpha)}} = 0$, $\lim\limits_{n\to\infty} (1 + \theta^{\gamma}_{\alpha,n}(\zeta))^{-1} = 1$, there exist numbers A and B such that $|1 + \theta_n(z)||1 + \theta^{\gamma}_{\alpha,n}(\zeta)|^{-1} \leq AB$ and hence, $|a_n \tilde{E}^{\gamma}_{\alpha,n}(z)| \leq AB|a_n \tilde{E}^{\gamma}_{\alpha,n}(\zeta)|$, for all the values of $n > P$ and $|z| \leq r$. Since the series $\sum_{n=0}^{\infty} a_n \tilde{E}^{\gamma}_{\alpha,n}(\zeta)$ is absolutely convergent and by the Weierstrass comparison criterion, the uniform convergence is proved.

The disk of convergence itself is not necessarily a region of uniform convergence, and on its boundary the series may even be divergent.

Let $z_0 \in \mathbb{C}$, $0 < R < \infty$, $|z_0| = R$ and g_{φ} be an arbitrary angular region with size $2\varphi < \pi$ and with a vertex at the point $z = z_0$, which is symmetric with respect to the straight line defined by the points 0 and z_0, and d_φ be the part of the angular region g_φ, closed between the angle's arms and the arc of the circle with center at the point 0 and touching the arms of the angle.

The following inequality can be verified inside d_φ,

$$|z - z_0|\cos\varphi < 2(|z_0| - |z|). \tag{10}$$

The next theorem refers to the uniform convergence of the series (8) on the set d_φ and the limit of its sum at the point z_0, provided $z \in D(0; R) \cap g_\varphi$.

Theorem 3.3 (of Abel type) *Let $\{a_n\}_{n=0}^{\infty}$ be a sequence of complex numbers, R be the real number defined by (9) and $0 < R < \infty$. If $\tilde{f}(z;\ \alpha, \gamma)$ is the sum of the series (8) on the region $D(0; R)$, i.e.*

$$\tilde{f}(z;\ \alpha, \gamma) = \textstyle\sum_{n=0}^{\infty} a_n \tilde{E}_{\alpha,n}^{\gamma}(z), \quad z \in D(0; R)$$

and this series converges at the point z_0 of the boundary $C(0; R)$, then:

(i) The following relation holds

$$\lim_{z \to z_0} \tilde{f}(z;\ \alpha, \gamma) = \textstyle\sum_{n=0}^{\infty} a_n \tilde{E}_{\alpha,n}^{\gamma}(z_0), \tag{11}$$

provided $z \in D(0; R) \cap g_\varphi$; and

(ii) The series (8) is uniformly convergent on the region d_φ.

The proofs of Theorems 3.1 and 3.3, excepting the uniformity, are given in [14].

Remark 3.4 If the series (8) has a finite and non-zero radius of convergence R, it converges at the point $z_0 \in C(0, R)$ and F is the holomorphic function defined by this series in its region of convergence, then by the Theorem 3.3 it follows that

$$\lim_{z \to z_0, z \in d_\varphi} F(z) = F(z_0),$$

i.e. the restriction of the function F to each set of the kind d_φ is continuous at the point z_0.

Proof. Here we consider the series (8) whose convergence has been proved in [14]. To prove its uniform convergence we use the inequality (10) that is the crucial point of the proof.

So, let $z \in d_\varphi$. Setting

$$S_k(z) = \sum_{n=0}^{k} a_n \tilde{E}_{\alpha,n}^{\gamma}(z),$$

$$S_k(z_0) = \sum_{n=0}^{k} a_n \tilde{E}_{\alpha,n}^{\gamma}(z_0), \quad \lim_{k \to \infty} S_k(z_0) = s, \tag{12}$$

$$\beta_n = S_n(z_0) - s, \quad \beta_n - \beta_{n-1} = a_n \tilde{E}_{\alpha,n}^{\gamma}(z_0),$$

we obtain

$$S_{k+p}(z) - S_k(z) = \sum_{n=0}^{k+p} a_n \tilde{E}_{\alpha,n}^{\gamma}(z) - \sum_{n=0}^{k} a_n \tilde{E}_{\alpha,n}^{\gamma}(z)$$

$$= \sum_{n=k+1}^{k+p} a_n \tilde{E}_{\alpha,n}^{\gamma}(z).$$

According to Remark 3.4, there exists a natural number N_0 such that $\tilde{E}_{\alpha,n}^{\gamma}(z_0) \neq 0$ when $n > N_0$. Let $k > N_0$ and $p > 0$. Then, using the denotation $\gamma_n(z; z_0) = \tilde{E}_{\alpha,n}^{\gamma}(z) / \tilde{E}_{\alpha,n}^{\gamma}(z_0)$, we can write the difference $S_{k+p}(z) - S_k(z)$ as follows:

$$S_{k+p}(z) - S_k(z) = \sum_{n=k+1}^{k+p} a_n \tilde{E}_{\alpha,n}^{\gamma}(z_0) \frac{\tilde{E}_{\alpha,n}^{\gamma}(z)}{\tilde{E}_{\alpha,n}^{\gamma}(z_0)}$$

$$= \sum_{n=k+1}^{k+p} a_n \tilde{E}_{\alpha,n}^{\gamma}(z_0) \gamma_n(z; z_0).$$

Now, by the Abel transformation (see in [8], vol.1, ch.1, p.32, 3.4:7), we obtain consecutively:

$$S_{k+p}(z) - S_k(z) = \sum_{n=k+1}^{k+p} (\beta_n - \beta_{n-1})\gamma_n(z; z_0)$$

$$= \beta_{k+p}\gamma_{k+p}(z) - \beta_k\gamma_{k+1}(z) - \sum_{n=k+1}^{k+p-1} \beta_n(\gamma_{n+1}(z; z_0) - \gamma_n(z; z_0)),$$

and

$$|S_{k+p}(z) - S_k(z)| \leq |S_{k+p}(z_0) - s||\gamma_{k+p}(z)| + |S_k(z_0) - s||\gamma_{k+1}(z)|$$

$$+ \sum_{n=k+1}^{k+p-1} |S_n(z_0) - s| \times \left|\frac{\tilde{E}_{\alpha,n}^{\gamma}(z)}{\tilde{E}_{\alpha,n}^{\gamma}(z_0)} - \frac{\tilde{E}_{\alpha,n+1}^{\gamma}(z)}{\tilde{E}_{\alpha,n+1}^{\gamma}(z_0)}\right|. \quad (13)$$

So, using the last relation, we are going to estimate the module of the difference $S_{k+p}(z) - S_k(z)$. Because of (6) and the equalities $\lim\limits_{n\to\infty} \frac{1}{n^{Re(\alpha)}} = 0$, $\lim\limits_{n\to\infty} (1 + \theta_n(z_0))^{-1} = 1$, there exist numbers $A > 0$ and $N_1 > N_0$ such that $|1 + \theta_n(z)| \leq A/2$ for all the natural values of n and $|1 + \theta_n(\zeta)|^{-1} < 2$ for $n > N_1$, whence

$$|\gamma_n(z, z_0)| \leq A \quad for \quad n > N_1. \quad (14)$$

Further, setting

$$e_n(z, z_0) = \frac{\tilde{E}_{\alpha,n}^{\gamma}(z)}{\tilde{E}_{\alpha,n}^{\gamma}(z_0)} - \frac{\tilde{E}_{\alpha,n+1}^{\gamma}(z)}{\tilde{E}_{\alpha,n+1}^{\gamma}(z_0)}$$

and observing that $e_n(z_0, z_0) = 0$, we apply the Schwartz lemma for $e_n(z, z_0)$. So, we get that there exists a constant C:

$$|e_n(z, z_0)| = \left|\frac{\tilde{E}_{\alpha,n}^{\gamma}(z)}{\tilde{E}_{\alpha,n}^{\gamma}(z_0)} - \frac{\tilde{E}_{\alpha,n+1}^{\gamma}(z)}{\tilde{E}_{\alpha,n+1}^{\gamma}(z_0)}\right| \leq C|z - z_0||z/z_0|^n,$$

whence, and according to (10):

$$\sum_{n=k+1}^{k+p+1} |e_n(z, z_0)| \leq \sum_{n=0}^{\infty} C|z - z_0||z/z_0|^n$$

$$= C|z_0| \times \frac{|z-z_0|}{|z_0|-|z|} < \frac{2C|z_0|}{\cos\varphi}. \tag{15}$$

Let ε be an arbitrary positive number. Taking in view the third of the relations (12), we can confirm that there exists a positive number $N_2 > N_0$ so large that

$$|S_n(z_0) - s| < min\left(\frac{\varepsilon}{3A}, \frac{\varepsilon\cos\varphi}{6C|z_0|}\right) \quad for \quad n > N_2. \tag{16}$$

Now, let $N = N(\varepsilon) = \max(N_1, N_2)$ and $k > N$. Therefore (13)-(16) give

$$\left|S_{k+p}(z) - S_k(z)\right| < \frac{2\varepsilon}{3} + \frac{\varepsilon\cos\varphi}{6C|z_0|}\sum_{n=k+1}^{k+p+1} |e_n(z, z_0)|$$

$$< \frac{2\varepsilon}{3} + \frac{\varepsilon\cos\varphi}{6C|z_0|}\frac{2C|z_0|}{\cos\varphi} = \varepsilon,$$

which completes the proof.

Fatou type theorem

Let $\{a_n\}_{n=0}^{\infty}$ be a sequence of complex numbers with $\limsup_{n\to\infty} (|a_n|)^{1/n} = R^{-1}$, $0 < R < \infty$ and $f(z)$ be the sum of the power series $\sum_{n=0}^{\infty} a_n z^n$ on the open disk $D(0; R)$, i.e.

$$f(z) = \sum_{n=0}^{\infty} a_n z^n, \quad z \in D(0; R). \tag{17}$$

Definition 4.1 *A point $z_0 \in \partial D(0; R)$ is called regular for the function f if there exist a neighbourhood $U(z_0; \rho)$ and a function $f_{z_0}^* \in \mathcal{H}(U(z_0; \rho))$ (the space of complex-valued functions, holomorphic in the set $U(z_0; \rho)$), such that $f_{z_0}^*(z) = f(z)$ for $z \in U(z_0; \rho) \cap D(0; R)$.*

By this definition it follows that the set of regular points of the power series is an open subset of the circle $C(0; R) = \partial D(0; R)$ with respect to the relative topology on $\partial D(0; R)$, i.e. the topology induced by that of $\mathbb{C}$.

In general, there is no relation between the convergence (divergence) of a power series at points on the boundary of its disk of convergence and the regularity (singularity) of its sum of such points. For example, the power series $\sum_{n=0}^{\infty} z^n$ is divergent at each point of the unit circle $C(0;1)$ regardless of the fact that all the points of this circle, except for $z = 1$, are regular for its sum. The series $\sum_{n=1}^{\infty} n^{-2} z^n$ is (absolutely) convergent at each point of the circle $C(0;1)$, but nevertheless one of them, namely $z = 1$, is a singular (i.e. not regular) for its sum.

However, under additional conditions on the sequence $\{a_n\}_{n=0}^{\infty}$, such a relation does exist (for details, see the Fatou theorem in [8], Vol.1, Ch.3, § 7, 7.3, p.357), namely:

If the coefficients of the power series (17) with the unit disk of convergence tend to zero, i.e. $R = 1$ and $\lim\limits_{n\to\infty} a_n = 0$, then the power series converges, even uniformly, on each arc of the unit circle, all the points of which (including the ends of the arc) are regular for the sum f of the series.

Let us point out that under the hypothesis of the above assertion there exists a region $G \supset \sigma$ and a function $f^* \in H(G)$ such that $f^*(z) = f(z)$ for $z \in G \cap D(0;1)$.

This means that the function f^* is an analytical continuation of the function f outside the disk $D(0;1)$. Moreover, as it is not hard to see, the series (17) converges on that open arc $\tilde{\sigma} \subset C(0;1)$ which contains σ and is included in the region G. Then Abel's theorem yields that the sum of the series (17) is $f(z)$ for each $z \in \tilde{\sigma}$. Therefore, we may assume that the power series (17) represents the function f even on the arc $\tilde{\sigma}$.

Propositions referring to the properties discussed above have also been established for series in the Laguerre and Hermite polynomials, as well as in Mittag-Leffler systems (see e.g. [19], resp. [15]). Here we give such a type of theorem for the Prabhakar systems, as follows.

Theorem 4.2 (of Fatou type) *Let $\{a_n\}_{n=0}^{\infty}$ be a sequence of complex numbers satisfying the conditions*

$$\lim_{n\to\infty} a_n = 0, \qquad \limsup_{n\to\infty} (\ |a_n|\ )^{1/n} = 1, \tag{18}$$

and $F(z;\alpha,\gamma)$ *be the sum of the series (8) on the unit disk* $D(0;1)$, *i.e.*

$$F(z;\alpha,\gamma)=\sum_{n=0}^{\infty} a_n\tilde{E}_{\alpha,n}^{\gamma}(z),\quad z\in D(0;1).$$

Let σ be an arbitrary arc of the unit circle $C(0;1)$ with all its points (including the ends) regular to the function F. Then the series (8) converges, even uniformly, on the arc σ.

Proof. Since all the points of the arc σ are regular to the function $F(z;\alpha,\gamma)$ there exists a region $G\supset\sigma$ where the function F can be continued. Denoting $\tilde{G}=G\cup D(0;1)$, we define the function ψ in the region $\tilde{G}$ by the equality

$$\psi(z)=F(z;\alpha,\gamma),\quad z\in D(0;1).$$

More precisely, it means that F has a single-valued analytical continuation in $\tilde{G}$.

Let $\rho>0$ be the distance between the boundary $\partial\tilde{G}$ of the region $\tilde{G}$ and the arc σ ($\partial\tilde{G}$ contains a part of the unit circle $C(0;1)$), and take the points $\zeta_1,\ \zeta_2\notin\sigma,\ |\zeta_1|=|\zeta_2|=1$, such that the distances between each of the points ζ_1,ζ_2 and the respective closer end of the arc σ are equal to $\rho/2$, and $z_1=\zeta_1(1+\rho/2)$, $z_2=\zeta_2(1+\rho/2)$.

Define the auxiliary functions

$$\varphi_n(z)=\psi(z)-\sum_{k=0}^{n} a_k\tilde{E}_{\alpha,k}^{\gamma}(z),\quad \omega_n(z)=\frac{\varphi_n(z)}{\tilde{E}_{\alpha,n+1}^{\gamma}(z)}(z-\zeta_1)(z-\zeta_2).\tag{19}$$

In order to prove that the sequence $\left\{\sum_{k=0}^{n} a_k\tilde{E}_{\alpha,k}^{\gamma}(z)\right\}$ is uniformly convergent on the arc σ, it is sufficient to show that the sequence $\{\omega_n(z)\}_{n=0}^{\infty}$ tends uniformly to zero on the boundary $\partial\Delta$ of the sector $\Delta=Oz_1z_2$ which is a compact set.

To this end, we come back to (6). Just to mention that since $\lim\limits_{n\to\infty}\frac{1}{n^{Re(\alpha)}}=0$, then there exist numbers C and $\widetilde{N}$ such that $|1+\theta_{\alpha,n}^{\gamma}(z)|\leq C/2$ for all the values of $n \in \mathbb{N}_0$ and $1/2 \leq |1+\theta_{\alpha,n}^{\gamma}(z)| \leq 2$ for $n > \widetilde{N}$ on an arbitrary compact subset of $\mathbb{C}$.

Now, taking $\varepsilon > 0$ and setting

$$R = 1+\rho/2, \quad \varepsilon_1 = \frac{\varepsilon\rho^3}{8(8CR^2+\rho)}, \quad M = \max_{z\in[\Delta]}|\psi(z)| \qquad ([\Delta] = \Delta \cup \partial\Delta),$$

we separate the considerations in four cases as follows:

1) First, let $z \in (O,\zeta_1) \cup (O,\zeta_2) \subset D(0;1)$.

In the unit disk, according to (19) and (7), we have:

$$\omega_n(z) = \sum_{k=n+1}^{\infty} a_k z^{k-n-1} \frac{(1+\theta_{\alpha,k}^{\gamma}(z))}{(1+\theta_{\alpha,n+1}^{\gamma}(z))}(z-\zeta_1)(z-\zeta_2).$$

Since $a_n \to 0$, there exists a number $N_1 = N_1(\varepsilon_1) > \widetilde{N}$, such that

$$|\omega_n(z)| \leq \varepsilon_1 \sum_{k=n+1}^{\infty} |z|^{k-n-1} \left|\frac{(1+\theta_{\alpha,k}^{\gamma}(z))}{(1+\theta_{\alpha,n+1}^{\gamma}(z))}\right| |(z-\zeta_1)||(z-\zeta_2)|$$

$$< 2C\varepsilon_1 \sum_{k=n+1}^{\infty} |z|^{k-n-1}(1-|z|) = 2C\varepsilon_1$$

for $n > N_1$, i.e.

$$|\omega_n(z)| < 2C\varepsilon_1. \tag{20}$$

2) $z \in (\zeta_1, z_1) \cup (\zeta_2, z_2)$.

In this case $|z-\zeta_1| = |z|-1$, $|z-\zeta_2| \leq |z|+|\zeta_2| < 2R$, and taking into account (5), (7) and (19) we can write the following inequalities for the absolute value of $\omega_n(z)$

$$\omega_n(z) = \frac{\psi(z)-\sum_{k=0}^{n} a_k z^k (1+\theta_{\alpha,k}^{\gamma}(z))}{z^{n+1}(1+\theta_{\alpha,n+1}^{\gamma}(z))}(z-\zeta_1)(z-\zeta_2),$$

namely

$$|\omega_n(z)| \leq \frac{M+\sum_{k=0}^{n}|a_k||z|^k|(1+\theta_{\alpha,k}^{\gamma}(z))|}{|z|^{n+1}|(1+\theta_{\alpha,n+1}^{\gamma}(z))|}\ 2R(|z|-1)$$

$$< 2R\left(2M+\sum_{k=0}^{N_1} C|a_k|R^k\right)\ \frac{(|z|-1)}{|z|^{n+1}} + 2\varepsilon_1 RC\ \frac{(|z|-1)}{|z|^{n+1}}\sum_{k=N_1+1}^{n}|z|^k\ .$$

Furthermore, having in mind that

$$\frac{(|z|-1)}{|z|^{n+1}} < \frac{(|z|-1)}{|z|^{n+1}-1} = \frac{1}{|z|^{n}+\cdots+1} < \frac{1}{n+1},$$

$$\sum_{k=N_1+1}^{n}|z|^k = \frac{|z|^{n+1}-|z|^{N_1+1}}{(|z|-1)} < \frac{|z|^{n+1}}{(|z|-1)},$$

we conclude that

$$|\omega_n(z)| < \frac{2R}{n+1}\left(2M+\sum_{k=0}^{N_1} C|a_k|R^k\right)\ + 2\varepsilon_1 RC.$$

Then, since $n^{-1}\to 0$, there exists a number $N_2 = N_2(\varepsilon_1) > N_1$ such that

$$\frac{2R}{n+1}\left(2M+\sum_{k=0}^{N_1} C|a_k|R^k\right) < \varepsilon_1$$

for $n > N_2$, i.e.

$$|\omega_n(z)| < (1+2RC)\varepsilon_1. \tag{21}$$

3) z belongs to the arc z_1z_2 (including the ends).

Then $|z-\zeta_1| < 2R$, $|z-\zeta_2| < 2R$ and hence

$$|\omega_n(z)| < \frac{4R^2\left(2M+\sum_{k=0}^{n} C|a_k|R^k\right)}{R^{n+1}} < \frac{4\left(2M+\sum_{k=0}^{N_1} C|a_k|R^k\right)}{R^{n-1}}$$

$$+\frac{4\varepsilon_1 C\left(\sum_{k=N_1+1}^{n} R^k\right)}{R^{n-1}} < \frac{4\left(2M+\sum_{k=0}^{N_1} C|a_k|R^k\right)}{R^{n-1}} + \frac{8\varepsilon_1 CR^2}{\rho}.$$

Since $R^{-n} \to 0$, there exists a number $N_3 = N_3(\varepsilon_1) > N_1$, such that

$$|\omega_n(z)| < \left(\frac{8CR^2}{\rho} + 1\right)\varepsilon_1 \tag{22}$$

for $n > N_3$.

4) $z \in \{O, \zeta_1, \zeta_2\}$.

In this case we have $\omega_n(0) = a_{n+1}\zeta_1\zeta_2$, whence $|\omega_n(0)| = |a_{n+1}| < \varepsilon_1$ for $n > N_1$, and $\omega_n(\zeta_{1,2}) = 0$.

Let $N = \max\{N_1, N_2, N_3\}$ and $n > N$, then having in view the inequalities (20) - (22), we can write on the boundary of the region Δ:

$$|\omega_n(z)| < max\left(2C\varepsilon_1, (2RC+1)\varepsilon_1, \left(\frac{8CR^2}{\rho} + 1\right)\varepsilon_1\right) = \left(\frac{8CR^2}{\rho} + 1\right)\varepsilon_1,$$

which also verifies the uniform convergence of $\omega_n(z)$ on the boundary $\partial\Delta$. Having in view the last estimate, according to the principle of the maximum of the modulus, we can write

$$|\omega_n(z)| < \left(\frac{8CR^2+\rho}{\rho}\right)\varepsilon_1 \tag{23}$$

on the arc σ.

Finally, according to (5), (7) and (19), since $|z| = 1$ on the arc σ,

$$|\omega_n(z)| = \frac{\left|\psi(z) - \sum_{k=0}^{n} a_k \tilde{E}_{\alpha,k}^{\gamma}(z)\right|}{|z^{n+1}|\left|1+\theta_{\alpha,n+1}^{\gamma}(z)\right|}|z-\zeta_1||z-\zeta_2|$$

$$> \frac{\rho^2}{8}\left|\psi(z) - \sum_{k=0}^{n} a_k \tilde{E}_{\alpha,k}^{\gamma}(z)\right|,$$

whence the inequality (23) yields

$$\left|\psi(z) - \sum_{k=0}^{n} a_k \tilde{E}_{\alpha,k}^{\gamma}(z)\right| < \frac{8}{\rho^2}|\omega_n(z)| < \frac{8\varepsilon_1}{\rho^3}(8CR^2 + \rho) = \varepsilon$$

on the arc σ.

Special cases

In particular, as has been discussed in the Introduction, for $\gamma = 1$ the Prabhakar function $E_{\alpha,\beta}^{\gamma}$, defined by (1), coincides with M-L's function $E_{\alpha,\beta}$, i.e. $E_{\alpha,\beta}^{1}(z) = E_{\alpha,\beta}(z)$ (see (2)). So in this case the series (8) takes the form

$$\sum_{n=0}^{\infty} a_n \tilde{E}_{\alpha,n}^{1}(z) = \sum_{n=0}^{\infty} a_n \tilde{E}_{\alpha,n}(z), \tag{24}$$

with complex coefficients a_n $(n = 0,1,2,\ldots)$.

Such a kind of series was studied in detail e.g. in [16] and [15], but note also that all the results obtained concerning them follow as particular cases from the preceding sections.

Conclusion

We emphasize that the results obtained for the series (8) are quite analogous to these for the classical power series (17). As seen, they have one and the same radius of convergence R, and are both absolutely and uniformly convergent on each closed disk $|z| \leq r$ $(r < R)$. Moreover, if each one of them converges at the point z_0 of the boundary of $D(0; R)$, then the theorems of Abel type hold for both series in one and the same angular region. Finally, if $\{a_n\}_{n=0}^{\infty}$ is a sequence of complex numbers satisfying the conditions (18), and all the points (including the ends) of the arc σ of the unit circle $C(0; 1)$ are regular to the sums of both considered series, then the series (8) and (17) converge even uniformly, on the arc σ.

Acknowledgements: This paper is produced within the framework of the Bilateral Res. Project "Mathematical Modeling by Means of Integral Transform Methods, Partial Differential Equations, Special and Generalized Functions" between BAS and SANU (2012-2014).

Bibliography

[1] A. Erdélyi et al. (Ed-s), *Higher Transcendental Functions.* **1 - 3**, 1st Ed., McGraw-Hill, New York-Toronto-London (1953-1955).

[2] A.A. Khamzin, R.R. Nigmatullin, I.I. Popov, *Justification of the empirical laws of the anomalous dielectric relaxation in the framework of the memory function formalism*, Fract. Calc. Appl. Anal., **17** (2014), no. 1, 247-258, doi:10.2478/s13540-014-0165-5.

[3] A.A. Kilbas, A.A. Koroleva, S.V. Rogosin, *Multi-parametric Mittag-Leffler functions and their extension*, Fract. Calc. Appl. Anal. **16** (2013), no. 2, 378-404, DOI:10.2478/s13540-013-0024-9.

[4] V. Kiryakova, *The special functions of fractional calculus as generalized fractional calculus operators of some basic functions*, Computers and Mathematics with Appl. **59** (2010), no. 3, 1128-1141, doi:10.1016/j.camwa.2009.05.014.

[5] V. Kiryakova, Fractional order differential and integral equations with Erdélyi-Kober operators: Explicit solutions by means of the transmutation method, AIP Conf. Proc., **1410** (2011), 247–258, doi:10.1063/1.3664376.

[6] V. Kiryakova, Yu. Luchko, The multi-index Mittag-Leffler functions and their applications for solving fractional order problems in applied analysis, AIP Conf. Proc., **1301** (2010), 597-613, doi:10.1063/1.3526661.

[7] F. Mainardi, *Fractional Calculus and Waves in Linear Viscoelasticity*. Imperial College Press, London (2010).

[8] A. Markushevich, *A Theory of Analytic Functions.* **1, 2** (In Russian), Nauka, Moscow (1967).

[9] J. Paneva-Konovska, Theorems on the convergence of series in generalized Lommel-Wright functions, Fract. Calc. Appl. Anal. **10** (2007), no. 1, 59-74.

[10] J. Paneva-Konovska, Cauchy-Hadamard, *Abel and Tauber type theorems for series in generalized Bessel-Maitland functions*, Compt. Rend. Acad. Bulg. Sci. **61** (2008), no. 1, 9-14.

[11] J. Paneva-Konovska, *Series in Mittag-Leffler functions: Inequalities and convergence theorems*, Fract. Calc. Appl. Anal. **13** (2010), no. 4, 403-414.

[12] J. Paneva-Konovska, *The convergence of series in multi-index Mittag-Leffler functions*. Integral Transforms Spec. Funct. **23** (2012), no. 3, 207-221, DOI:10.1080/10652469.2011.575567.

[13] J. Paneva-Konovska. *Inequalities and Asymptotic Formulae for the Three Parametric Mittag-Leffler Functions*. Mathematica Balkanica, New Ser. **26** (2012), fasc. 1-2, 203-210.

[14] J. Paneva-Konovska, *Convergence of series in three parametric Mittag-Leffler functions*. Mathematica Slovaca. 64(2014), Issue 1,73-84 DOI:10.2478/s12175-013-0188-0

[15] J. Paneva-Konovska, *Fatou type theorems for series in Mittag-Leffler functions.* AIP Conf. Proc. **1497** (2012), 318-325, doi: 10.1063/1.4766800.

[16] J. Paneva-Konovska, *Series in Mittag-Leffler functions: Geometry of convergence*, Adv. Math. Sci. Journal, **1**, no. 2 (2012), 73-79, UDC: 517.58:517.521.

[17] J. Paneva-Konovska. *On the multi-index (3m-parametric) Mittag-Leffler functions, fractional calculus relations and series convergence*, Central European Journal of Physics **11**, no. 10 (2013), 1164-1177, DOI: 10.2478/s11534-013-0263-8.

[18] T.R. Prabhakar, A singular integral equation with a generalized Mittag-Leffler function in the kernel. Yokohama Math. J. **19** (1971), 7-15.

[19] P. Rusev, Classical Orthogonal Polynomials and Their Associated Functions in Complex Domain. Publ. House Bulg. Acad. Sci., Sofia (2005).

[20] T. Sandev, Ž. Tomovski, J. Dubbeldam, *Generalized Langevin equation with a three parameter Mittag-Leffler noise*, Physica A, **390** (2011), issue 21-22, 3627–3636, doi:10.1016/j.physa.2011.05.039.

CHAPTER SIX

NETWORK APPLICATIONS IN INDUSTRY

APPLICATION OF PROBABILITY NEURAL NETWORKS FOR CLASSIFICATION OF EXPLOSIVES WITH BLASTING ACTION

VALERIJ I. DZHUROV, MILENA P. KOSTOVA AND KALOYAN V. DZHUROV

Introduction

The fight against terrorism is a research priority for scientists from around the world. There are many devices created and used for detection of explosives and traces of explosives, on documents, clothes, letters and other objects. These are built by a number of companies in Russia, Great Britain, China, the USA and Israel.

In Russia there are in use devices of the Pilot-M series for detecting explosives in non-airtight spaces and traces of dangerous substances on examined objects. The price of such devices is around 10000 euro. The device of series Pilot M-1 makes it possible to detect trinitrotoluene (TNT), nitroglycerin (NGl), hexogen (XG), pentaeritrittetranitrat and their derivatives. For examination of trace explosives on documents, passports and others, a certificate is offered by the GOVERNMENT OF DEFENCE-2D. The price of the device varies depending on the client's requirements. Detecting and classifying the type of blasting material is realized by the device MO-2M with a price of around 13000 euro. Most portable devices and instruments are combined as they are designed for detecting explosives, chemical materials, toxic chemicals and drugs (SABRE 5000, IONSCAN 400B). The device EVD 3000, which has an ICAO certificate, is used for detection of plastic explosives and other such materials made by high pressure methods, with an approximate price of 44000 euro [3,6].

For detecting and identifying traces of exploding materials on the surface of packages, clothes, hands, guns, etc., a device called Poisk -- XT is used, with a price around 60000 euro. There is also a spray from the same series.

The device Dropex Plus, which is very small, has a similar functionality. The device ExPen 1, 2, 3 is used for detecting blasting materials when testing suspicious packages at customs and checkpoints [8].

For identifying and detecting blasting materials of group A (TNT, TNB and others) and blasting materials of group B (plastic explosives -- SemtexH, RDX, C4 and others) the device EXPRAY is used. There is a modified type of this device -- Mini EXPRAY with the same functionalities (identification of explosives and non-organic compounds containing nitrates) [11,12].

The Israeli company International Technologies Lasers (ITL) has developed a device with remote action for detecting explosives, drugs and different types of forbidden chemical substances. It uses a sensitive laser, whose three components are easily distinguishable: a laser emitter which scans the object, a spectrometer and a computer. A reaction starts in the molecules of the substance which the laser hits, causing emanation with a specific wavelength. The spectrometer then analyzes the results of the emanation, and the computer compares the results with a database of forbidden substances. A similar device is ITMS or VaporTracer-2. The applied laser spectrometry is very sensitive. It gives the opportunity for detection of more than 40 known narcotic substances (cocaine, heroin, methamphetamine) and explosives (such as trotyl, hexogen, PETN, EGDN, dynamite) [9].

The Israeli-American company AR Challenges implements in Europe and other continents an automated complex system Trust Based Security (TBS). The detectors used enable detection of random explosives of 0.5 micrograms in shipping containers, luggage, handbags and clothes. A patented technology Video Synopsis is used for the processing of a huge amount of video data [13,14].

From this brief overview we can draw some conclusions as to drawbacks when using these popular control systems:

1. The hardware used requires significant financial resources and well-trained staff;
2. A UVW frequency range (in THz) is used, which hides some health risks in multiple reuse;

3. Existing systems can be ``misleading" when using fragrances and softeners and can also give false information for substances placed in colored glass bottles.

We offer here an approach to building a system for classification of discovered traces of explosives, which in some ways compensates for the above disadvantages. This method of approach offers an alternative working principle, by analyzing the spectral picture of a coherent signal reflected by samples of explosive materials.

A classification system for traces of explosives with a blasting action

Building a system for classifying traces of blasting materials suggests knowledge of their composition, action and the specific features of those materials and compounds.

Synopsis of the major explosives

Explosives are chemicals or compounds capable, under the force of external impulse (a hit, friction, heating or other), of turning into a self-spreading chemical transformation in the form of a blast, with release of heat and the formation of gaseous products. The process is related to a spurt of spreading, which for modern substances can reach up to 9000 [m/s]. In this way, the range of explosives most used by terrorists can be limited by the blasting materials, and in separate cases to the use of some brands of gunpowder and easily accessible pyrotechnic mixtures [15].

A general classification of blasting materials can be:

- initiating (fulminated mercury, nitroglycerin);
- blasting (TNT hexogen, nitropenta, plastic materials, C4, P4, D-5A);
- propelling (gunpowder);
- pyrotechnic mixtures.

The main pyrotechnic characteristics of the blasting materials in most common use are given in Table 1

Table 1. Main parameters of Blasting explosive substances

Parameters	Trotyl	Nitroglycerin	Hexogen	TNT	Ammonite
Density $[g/sm^2]$	1,6	1,6	1,7	1,7	1,5
Flashpoint [deg]	290	290	230	220	220
Heat of the blast [deg]	1000	6552	1200	5964	4350
Spreading speed [m/s]	7000	7600	8200	8240	4800

Experimental part

Laboratory tests in the University Of Rousse's chemical laboratory were conducted under the following conditions:

Temperature of the environment – 21 [deg];
Humidity – 80%;
Altitude – 50 [m];
Power of the probing signal – up to 80 [mW];
Distance between source of the signal and the examined sample – 0,5 [m];
Distance between reflected signal and the receiving aperture – 0,6 [m];
Azimuth angle – 90 [deg];
Site corner – 30 [deg].

A 3D scenario of the experimental set-up is presented in *Fig. 1*

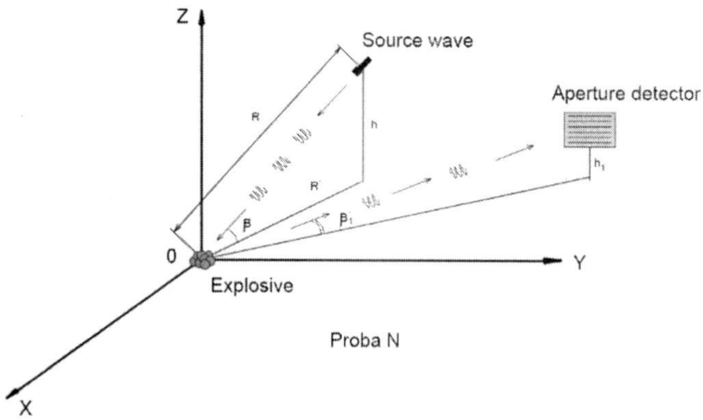

Fig.1. 3D scenario of the experimental set-up

The following samples (patterns) have been used:

> Trinitrotoluol (Sample 2), Hexogen (Sample 3), Nitropenta (Sample 4), Plastit (Sample 1). The mass of the samples, for safety's sake, is not more than 6 [mg].

Every sample is irradiated consecutively with coherent probing signals with wavelength $\lambda = 630$ [nm], $\lambda = 530$[nm], $\lambda = 440$[nm] respectively. The time of each irradiation is 10 [sec].

The reflected signals are received in the form of colored images upon a RGB aperture with 18 Mpxl (5184 x 3456 pixels). Each image is divided into nine ``areas" (facets) on axis Ox and Oy, in which a filter for resulting color is applied (filter of type Gaussian blur). For every area are reported the average intensities of the pixels' brightness for the corresponding waves (colors). The intensity level for each color ranges from 0 to 255. The average brightness intensities of the pixels for the corresponding waves (colors) of the four patterns are shown in *Fig. 2* and presented in tabular form in Table 1÷4.

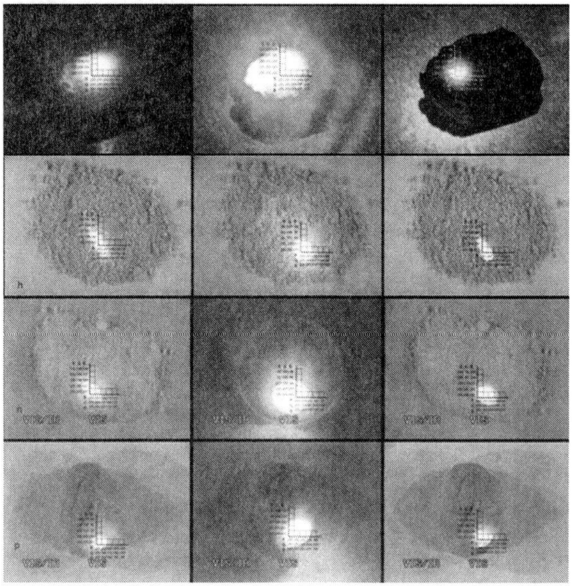

Fig.2. Reflected coherent signals from trinitrotoluol (T), hexogen (h), nitropenta (n) and plastit (p)

Table 2. Average brightness intensity for corresponding mask of Sample1

Sample 1 (plastit-5D-1A)									
Average brightness intensity for corresponding mask at $\lambda = 650$ nm (red), $\lambda = 530$ nm (green), λ=430 nm (blue)									
№	R/R	R/G	R/B	G/R	G/G	G/B	B/R	B/G	B/B
1.	251	251	251	255	255	255	253	254	200
2.	251	251	251	255	255	255	254	255	200
3.	251	248	251	255	255	255	254	254	200
4.	248	188	216	255	255	255	225	241	200
5.	244	152	163	255	255	250	211	234	160
6.	245	139	97	255	255	250	171	190	150
7.	245	140	93	255	254	250	170	173	130

8.	246	141	93	154	236	240	167	158	100
9.	236	139	95	49	227	200	159	142	100
10	236	139	95	51	228	150	153	132	80
11	251	251	251	255	255	255	253	254	160
12	252	200	210	255	255	255	215	241	150
13	252	200	210	255	255	255	191	224	130
14	252	155	61	240	252	255	174	172	110
15	252	147	77	237	255	255	168	154	120
16	252	141	92	139	250	255	163	142	120
17	236	130	92	89	254	250	154	120	110
18	236	130	92	60	236	180	154	119	110
19	191	119	84	65	186	160	152	108	100
20	183	121	81	67	163	130	149	107	80

Table 3. Average brightness intensity for corresponding mask of Samples 2, 3, 4

Average brightness intensity for corresponding mask at λ = 650 nm (red), λ = 530 nm (green), λ=430 nm (blue)									
Sample 2 (trinitrotoluon)									
№	R/R	R/G	R/B	G/R	G/G	G/B	B/R	B/G	B/B
1.	255	249	255	255	255	220	255	255	253
2.	255	253	255	255	255	220	255	254	252
3.	255	252	255	255	255	220	255	249	252
4.	255	242	254	255	255	220	255	137	252
5.	255	211	249	255	255	210	253	137	252
6.	252	119	180	255	255	210	243	104	250
7.	252	119	180	245	253	210	128	17	180
8.	246	70	120	224	248	210	100	9	150
9.	233	26	50	127	235	180	58	2	100

10	218	14	28	74	233	180	38	2	100
11	255	249	255	255	255	210	255	255	250
12	255	252	255	255	255	210	255	229	250
13	255	252	255	255	255	210	255	185	250
14	255	237	253	255	255	210	255	128	250
15	255	199	242	253	253	210	247	63	240
16	254	168	225	224	247	200	208	50	240
17	246	96	165	210	247	200	95	10	150
18	234	59	105	192	242	150	57	3	110
19	209	21	40	127	234	150	31	2	60
20	194	10	17	79	233	130	22	2	60
Sample 3 (Hexogen)									
21.	253	253	253	255	255	220	251	251	251
22.	247	242	247	255	255	220	252	251	252
23.	250	217	250	253	254	210	247	186	253
24.	250	151	113	220	251	200	249	227	252
25.	198	145	101	57	252	160	170	117	233
26.	188	141	88	60	249	150	249	227	252
27.	198	145	101	65	240	100	150	114	158
28.	165	138	88	93	212	100	149	144	148
29.	198	145	101	96	206	100	158	128	148
30	149	130	81	97	207	90	161	131	146
31	253	253	253	255	255	230	251	251	251
32	247	241	248	255	255	220	249	227	252
33	252	199	243	255	255	190	251	243	251
34	250	152	116	247	254	180	240	160	248
35	251	197	115	86	251	160	215	116	251
36	188	140	89	74	250	150	200	106	251
37	207	144	97	82	248	130	170	109	244

38	165	138	89	94	250	80	175	113	243
39	176	144	90	99	229	100	172	121	234
40	148	129	82	98	220	90	165	115	224
Sample 4 (Nitropenta)									
41	255	255	255	255	255	180	252	250	252
42	254	254	254	255	255	180	255	255	255
43	250	244	250	254	254	180	253	253	253
44	250	165	221	103	247	180	255	255	255
45	250	152	165	103	247	160	253	251	250
46	234	149	154	35	249	150	249	211	250
47	219	144	153	46	241	130	249	169	250
48	180	140	141	52	210	80	253	91	254
49	171	139	135	56	190	80	246	75	254
50	159	138	134	64	172	70	235	70	254
51	255	255	255	255	255	180	252	250	252
52	252	252	252	255	255	180	254	251	254
53	251	241	251	255	255	180	253	241	253
54	250	154	150	254	254	170	253	150	254
55	250	146	164	227	250	150	254	118	254
56	233	146	156	193	249	130	249	80	249
57	232	149	157	52	250	110	254	85	254
58	180	140	141	59	245	100	253	87	254
59	172	148	139	63	238	90	254	76	254
60	159	137	133	68	226	120	233	71	254

Key to symbols presented in the tables:
R / R -- pixel brightness level at probing coherent signal with $\lambda = 630$[nm] and reflected signal with $\lambda = 630$ [nm];
R / G -- pixel brightness level at probing coherent signal with $\lambda = 630$ [nm] and reflected signal with $\lambda = 530$ [nm];
R / B -- pixel brightness level at probing coherent signal with $\lambda = 630$ [nm] and reflected signal with $\lambda = 440$ [nm];

G / R -- pixel brightness level at probing coherent signal with
$\lambda = 530$ [nm] and reflected signal with $\lambda = 630$ [nm];
G / G -- pixel brightness level at probing coherent signal with
$\lambda = 530$ [nm] and reflected signal with $\lambda = 530$[nm];
G / B -- pixel brightness level at probing coherent signal with
$\lambda = 530$[nm] and reflected signal with $\lambda = 440$[nm];
B / R -- pixel brightness level at probing coherent signal with
$\lambda = 440$[nm] and reflected signal with $\lambda = 630$[nm];
B / G -- pixel brightness level at probing coherent signal with
$\lambda = 440$[nm] and reflected signal with $\lambda = 530$[nm];
B / B -- pixel brightness level at probing coherent signal with
$\lambda = 440$[nm] and reflected signal with $\lambda = 440$[nm].

Mathematical, algorithmic and software procedures for information and examination of classification signs for traces from substances with a blasting action

For trace classification of substances with a blasting action there are defined nine signs and two classes -- Class 1, non-plastic exploding substances (trinitrotoluol, hexogen and nitropenta); and Class 2 plastic exploding substances (plastit 5D- 1A). The signs are determined by the average brightness intensity of the spectral picture in the corresponding wave range, consequent on irradiation of the pattern using coherent signals with wavelengths respectively of $\lambda = 630$ [nm], $\lambda = 530$[nm], $\lambda = 440$[nm]. They are conditionally marked with: R/R ; R/G; R/B; G/R; G/G; G/B; B/R; B/G; B/B .

Interval marks are used to evaluate the information of the inputted signs. These marks with probability γ contain P. 100% from all elements of the general combination studied. Practical borders of dispersing (PBD) are used, marked respectively with U_D (bottom border) U_G (top border).

They are evaluated by the formulae [2]:

$$U_D = \overline{X} - k \cdot s, U_G = \overline{X} + k \cdot s \quad (1)$$

The value of the coefficient κ depends on the trust probability γ, the quantity P $(0 < P < 1)$ and the volume of the sample, except n which is taken from tables, and $s$which is the mark of the mean-squared diversion.

Some criteria for information are introduced:

The sign is informative if:

- The intersection of the intervals, determined by the practical borders for the corresponding classes, is an empty set;
- The intersection of the intervals is not an empty set, but there is no interval which is not a subset of another and the corresponding probabilities $P\ (U_{D2} < X < x_0)$ and $P\ (x_0 < X < U_{G1})$which we will provisionally call "probability for wrong classification" (P_{GK}), are smaller than 25%. With $x_0$is marked the abscissa of the intersection $\mathrm{f}(x_1; a_1; b_1)$ and $\mathrm{f}(x_2, a_{2,} b_2)$.

The determination of $P(U_{D2} < X < x_0)$ and $P(x_0 < X < U_{G1})$ is based on one of the main tasks from probability theory for determining a random quantity to fall within a given range. It is accepted that the values of the informative signs are random quantities which are distributed in normal law (Gauss's law) [2].

It is known that in a finite closed interval with length 6σ[X]and environment matching the average value $E[X]$, fall practically all possible values of the distributed in normal law random quality X [2]. More preciselyin this interval fall 99,73% of the X values, but in practive this is accepted as 100%. Therefore, if the practical borders of disperse are evaluated, based on the test data we can determine the approximate value for σ[X].

$$\sigma[X] \approx \frac{1}{6}(U_G - U_D) \tag{2}$$

When a given random quantity X is distributed in normal law it has the density distribution:

$$f(x; a; b) = \frac{1}{\sqrt{2\pi} b} e^{-\frac{(x-a)^2}{2b^2}} \tag{3}$$

where a =$E[X]$ and $b = \sigma(X)$ are the distribution parameters.

One of the main tasks in each distribution is the determination of the probability of a random quantity falling within a given interval (x_1, x_2). For the random quantity X distributed in normal law this probability is:

$$P(x_1 < X < x_2) = \int_{x_1}^{x_2} f(x)dx = \Phi(t_1) - \Phi(t_2), \tag{4}$$

$$\text{where } t_1 = \frac{x_1 - E[X]}{\sigma[X]}, t_2 = \frac{x_2 - E[X]}{\sigma[X]};$$

$$\Phi(t) = \frac{1}{\sqrt{2\pi}} \int_{-\infty}^{t} e^{-\frac{t^2}{2}} dt \tag{5}$$

The value of Φ(t) is taken from tables [2].

Distribution densities for class 1 and class 2 are given with the formulae (6) and (7) respectively:

$$f(X_1, a_1; b_1) = \frac{1}{\sqrt{2\pi} b_1} . e^{-\frac{(x-a_1)^2}{2b_1^2}} \tag{6}$$

$$f(X_2, a_2, b_2) = \frac{1}{\sqrt{2\pi} b_2} . e^{-\frac{(x-a_2)^2}{2b_2^2}} . \tag{7}$$

For class 1 it is accepted that in this class U_G is larger. The probability for the classification being wrong can be determined for every class using the formula (4) as

$$\text{For class } 1 \Rightarrow t_1 = \frac{x_0 - E[X_1]}{\sigma[X_1]}; t_2 = \frac{U_{G1} - E[X_1]}{\sigma[X_1]} \tag{8}$$

$$\text{For class } 2 \Rightarrow t_1 = \frac{U_{D2} - E[X_2]}{\sigma[X_2]}; t_2 = \frac{x_0 - E[X_2]}{\sigma[X_2]},$$

where x_0 is the abscissa of the intersection point of f(X_1; $a_1; b_1$) and f(X_2, a_2, b_2).

The value of x_0 is found by resolving the equation:

$$\frac{1}{\sqrt{2\pi}b_1}e^{-\frac{(x-a_1)^2}{2b_1^2}} = \frac{1}{\sqrt{2\pi}b_2}e^{-\frac{(x-a_2)^2}{2b_2^2}} \tag{9}$$

which is equivalent of

$$\frac{(x-a_2)^2}{2b_2^2} - \frac{(x-a_1)^2}{2b_1^2} = \ln\frac{b_1}{b_2} \tag{10}$$

The solution is sought in interval x∈ (U_{D2}, U_{G1}).

The approach for the selection of informative signs can be described using the algorithm shown on *Fig. 3*. The following symbols are used: P_R – sign K_i – i – th class, U_{Di_i}, U_{Gi}, bottom and top practical border of class i, x_0- abscissa of the intersection point for the graphics of distribution density of the sign and the two classes.

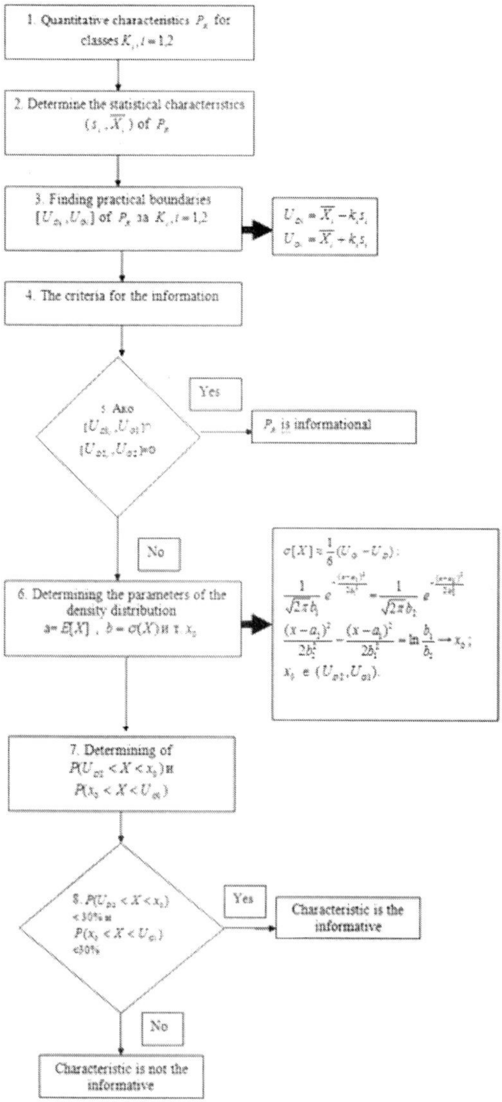

Fig.3. A flowchart of an algorithm for determining information of input indications

Results of the statistical processing of the examined test patterns

The practical borders of disperse for all inputted signs for the two classes are given in Table 4÷ Table 12 and presented graphically in *Fig.*4 ÷*Fig.6*

Table 4. Practical borders for sign R/R

Class	n	$\overline{x}$	S	k	Practical borders
Class1 (plastit-5D-1A)	20	240,5	19,30	3,615	170,73÷310,27
Class2 (Trinitrotoluol, Hexogen, Nitropenta)	60	218,7	51,05	3,06	63,48÷375,91

Table 5. Practical borders for sign R / G

Class	n	$\overline{x}$	S	k	Practical borders
Class1 (plastit-5D-1A)	20	169,1	47,36	3,615	-36,36÷373
Class2 (Trinitrotoluol, Hexogen, Nitropenta)	60	168,17	66,84	3,06	-2,11÷340

Table 6. Practical borders for sign R / B

Class	n	$\overline{x}$	S	k	Practical borders
Class1 (plastit-5D-1A)	20	142,75	72	3,615	-117,5÷403
Class2 (Trinitrotoluol, Hexogen, Nitropenta)	60	169,05	73,66	3,06	-56,39 ÷394,45

Table 7. Practical borders for sign G / R

Class	n	$\overline{x}$	s	k	Practical borders
Class1 (plastit-5D-1A)	20	185,05	88,08	3,615	-133,36÷503
Class2 (Trinitrotoluol, Hexogen, Nitropenta)	60	170,08	86,44	3,06	-94,43÷434,6

Table 8. Practical borders for sign G / G

Class	n	$\overline{x}$	S	k	Practical borders
Class1 (plastit-5D-1A)	20	241,8	25,14	3,615	150,91÷337
Class2 (Trinitrotoluol, Hexogen, Nitropenta)	60	239,93	33,33	3,06	137,94÷341,39

Table 9. Practical borders for sign G / B

Class	n	$\overline{x}$	S	k	Practical borders
Class1 (plastit-5D-1A)	20	233,3	41,4	3,615	83,64÷382,96
Class2 (Trinitrotoluol, Hexogen, Nitropenta)	60	164	43,38	3,06	-32÷297

Table 10. Practical borders for sign B / R

Class	n	$\overline{x}$	S	k	Practical borders
Class1 (plastit-5D-1A)	20	189,5	39,22	3,615	47,72÷331, 3
Class2 (Trinitrotoluol, Hexogen, Nitropenta)	60	210,88	66,83	3,06	10,39÷411,37

Table 11. Practical borders for sign B / G

Class	n	$\overline{x}$	S	k	Practical borders
Class1 (plastit-5D-1A)	20	183,7	55,47	3,615	-16,82÷384,22
Class2 (Trinitrotoluol, Hexogen, Nitropenta)	60	130,23	88,57	3,06	-140,79÷401,24

Table 12. Practical borders for sign B /B

Class	n	$\overline{x}$	S	k	Practical borders
Class1 (plastit-5D-1A)	20	135,5	40,5	3,615	-10 ÷ 282
Class2 (Trinitrotoluol, Hexogen, Nitropenta)	60	223,3	56	3,615	55÷336

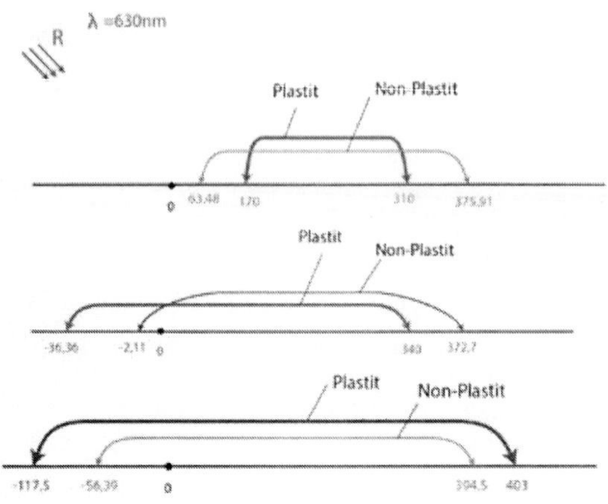

Fig.4. Practical borders for probing signal with wavelength $\lambda = 630\ nm$

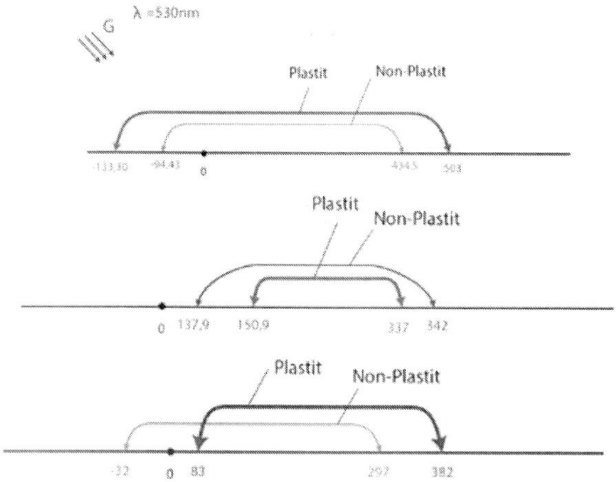

Fig.5. Practical borders for probing signal with wavelength $\lambda = 530\ nm$

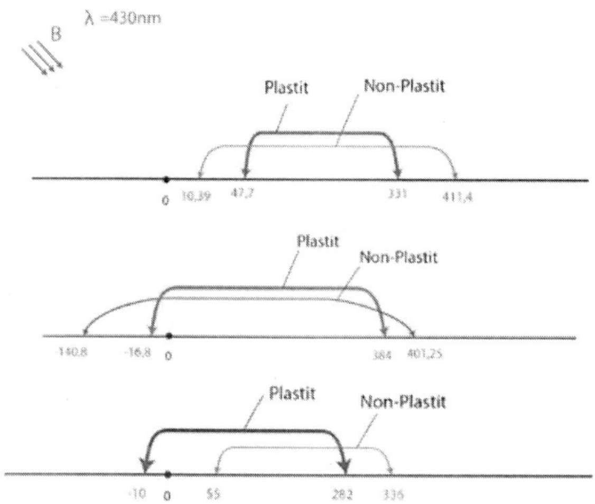

Fig.6. Practical borders for probing signal with wavelength $\lambda = 430\ nm$

A software application in MATLAB environment has been developed for determining the probability percentage for ``wrong classification" (Table13, Table 14). It is applied for signs G / B and B / B, which meet the requirement that none of the intervals of the practical borders are to be a subset of the other. The graphics of the density distribution for sign G / B (a) and sign B / B (b), are presented in *Fig. 7 (a) and (b)* respectively.

Table 13. Parameters of distribution density, X_0,P_{GK} for sign G / B

Class	Density parameters for sign G /B		Value for X_0	P_{GK}%
Class1 -- Plastic	a_1 =233.3	b_1 =49,83	199.9024	22
Class2 -- non-plastic	a_2 =164	b_2 =46	199.9024	24

Table 14. Parameters of distribution density,$X_0, P^{GK}_{\max}$ for sign B / B

Class	Density parameters for sign B /B		Value for X_0	P_{GK}%
Class1 -- Plastic	a_1 =135,5	b_1 =48,66	180.61	18
Class2-- non-plastic	a_2 =223.3	b_2 =56	180.61	22

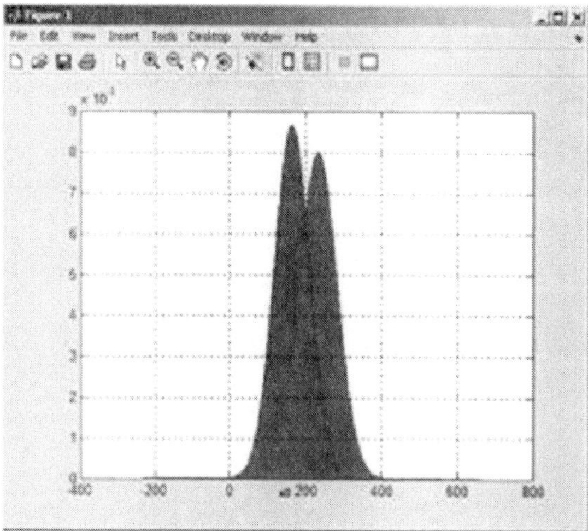

(a)

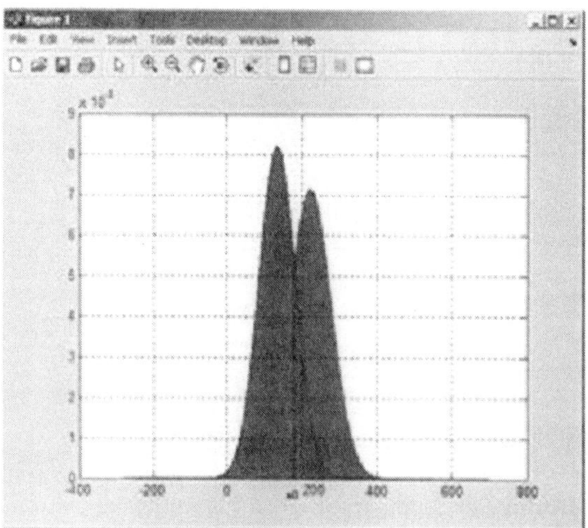

(b)

Fig.7. Graphics of the functions for density distribution for sign G / B (a) and sign B / B (b)

```
x1=-300:0.02:800;
x2=-300:0.02:800;
m1=233.3;
s1=49.83;
m2=164;
s2=46;

x0 =

  199.9024

sinq plo6t pl1

pl1 =

   0.7823

4ervena plo6t pl2

pl2 =

   0.2414
```

a)

```
x1=-300:0.02:800;
x2=-300:0.02:800;
m1=223.3;
s1=55.83;
m2=135.5;
s2=48.66;

x0 =

  180.6127

sinq plo6t pl1

pl1 =

   0.8231

4ervena plo6t pl2

pl2 =

   0.2223
```

b)

Fig.8. Input data and received results for signs G / B (a) and B / B (b) with software processing in MATLAB environment

The received statistical characteristics are used for creating a classifier based on Probability neural network.

Synthesis of classifier with Probability neural network

There is a large amount of developed and well-studied algorithms for classification based on artificial neural networks [1,4]. The main disadvantages of the neural networks are that in most cases they represent "black boxes" as regards to their way of working, and they require significant time for training.

Probability neural classifiers (PNCs) are of type one-way multilayer perceptrons (MLP-Multilayer Perceptrons). Typically they use a strategy similar to the statistical and probability Bayes's strategy for minimizing error, which gives an opportunity to provide a strict mathematical rationale for the way they work. In classification tasks the probability neural network evaluates the probabilities of a given observation belonging to each of the classes and compares them to the most likely class chosen [7,10].

The most frequent architecture of a PNC is built on four layers -- input, radial, summing and output *(Fig.9)*.

The input layer (IL) contains as many neurons as the number of informative signs.

The number of elements of the radial layer, also called Example Layer (EL), is equal to the number of elements of the training sample (the training vectors), grouped in K groups, where K is the number of classes. Every neuron from the radial layer contains Gaussian activation function. The Summation Layer (SL) is built of as many elements as the number of classes. In the output layer (the Decision Layer) a decision is taken about the class to which the input vector belongs.

In the summation layer every neuron performs an aggregation of the received probabilistic densities for the corresponding class in the previous layer,

$$S_k(X) = \sum_{i=1}^{M_k} w_{ki} P_{ki}(X), k \in \{1,2,\dots,K\}, \qquad (11)$$

where $M_k -$ number of neurons from class K, w_{ki} - weight coefficients. The probabilistic sense of (11) is the posterior probability

$$S_k(X) = P_r(K_i \backslash X) = \frac{P_{ki}(X).P_r(K_i)}{P(X)}, \text{where} \quad (12)$$

$P_r(K_i)$ is the posterior probability for class i, and $P(X)$ is the complete probability for the pattern X.

In the output layer (DL) is chosen the class to which the input vector belongs, based on the rule: "The winner takes it all"

$$C(k) = \underset{1 \le k \le K}{\operatorname{argmax}}(S_k). \quad (13)$$

The mechanism of PNC operation is as follows:

To each element of the input layer is given the input vector $X = (x_1, x_2, \dots, x_n)$. Every neuron from layer IL gives the input data to each of the EL elements. In this layer the output of the i-th neuron from k-th class is evaluated by Gaussian function, which has the following look:

$$P(i/k) = P_{ki}(X) = \frac{1}{(2\pi\sigma^2)^{\frac{n}{2}}} \exp\left(-\frac{\|X - X_{ki}\|^2}{2\sigma_{k,i}^2}\right), \quad (14)$$

where $X_{k,i}$ is the “center” of the core, and $\sigma_{k,i}$ is the smoothing parameter [2]. We can accept $\sigma_{k,i}$ equals the parameter σ, which is experimentally determined using methodology determined by the expert. For the current task σ gets the minimum value at which the two classes give the optimal recognition accuracy [1].

$P(i/k) = P_{ki}(X)$ shows the probability of the j-th pattern belonging to the K-th class.

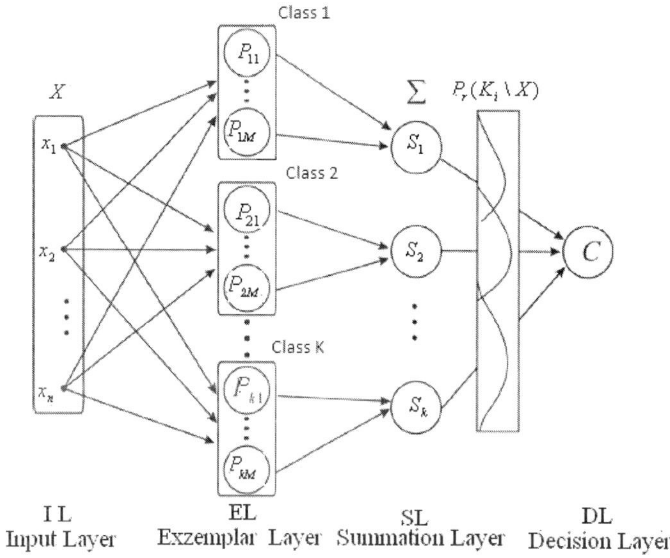

Fig.9. Architecture of PNN

The main advantage of the probability neural network is that it is relatively quick to train for. Training time is reduced to the time needed for giving the training data to the input. This gives the opportunity to carry out a large volume of sample tests. Studies show that the classifiers based on PNN neural networks are effective at highly noised data [7].

As a disadvantage can be highlighted the need for a large amount of memory. The network must store all training data. When the task is not related to a large amount of data this fact does not influence the work speed of the network.

The synthesis of the classifier based on the probability neural network can be described with the following algorithm:

1. Determining the number of input vectors depending on the informative signs;
2. Determining the average values of the informative signs for each of the classes;
3. Determining the value of smoothing parameter σ.

The probability classifier (PNC) for the specific task has 2 inputs, which correspond to the two informative signs and 2 outputs determined by the number of classes. The radial layer contains 4 neurons, at 2 for each class. The aggregation layer consists of two neurons, which is the same as the number of classes. The architecture of PNC with 2 inputs and 2 outputs is presented on *Fig.10.*

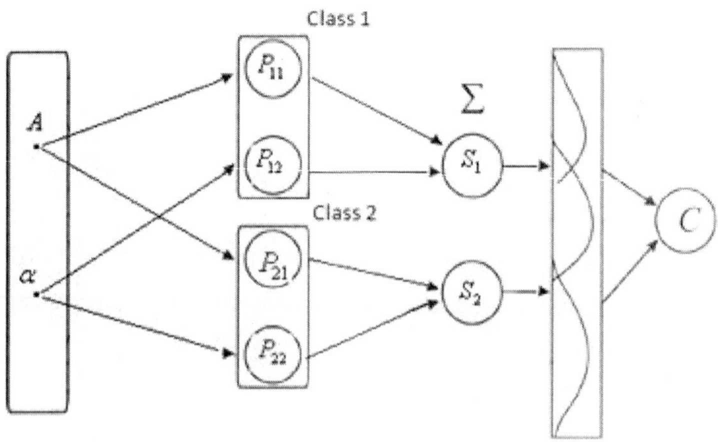

Fig.10. Architecture of PNC with 2 inputs and 2 outputs

The classifier is trained using training vectors with coordinates which are the average values of the corresponding signs $(\frac{\overline{G}}{B}, \frac{\overline{B}}{B})$ for each class - $\frac{\overline{G}}{B_p} =$ 183,7; $\frac{\overline{G}}{B_n} = 130{,}23$; $\frac{\overline{B}}{B_p} = 135{,}5$; $\frac{\overline{B}}{B_n} = 223{,}3$.

The rule for making a decision as to which class to put the input vector into is given with the formula: $C(k) = \underset{1 \le k \le K}{\operatorname{argmax}}(S_k)$,

which means that it is put into that class for which the posterior probability for correct classification is highest [7]. It is established experimentally that the optimal value of σ is 0,4.

To determine the accuracy of the classifier we made 20 tests with patterns of every class. The error for the first class is 15% (from 20 attempts 3 give the wrong result). For the second class the error is 10% (from 20 attempts 2 give the wrong result).

The work window of PNC in MATLAB environment is given in *Fig.11.*

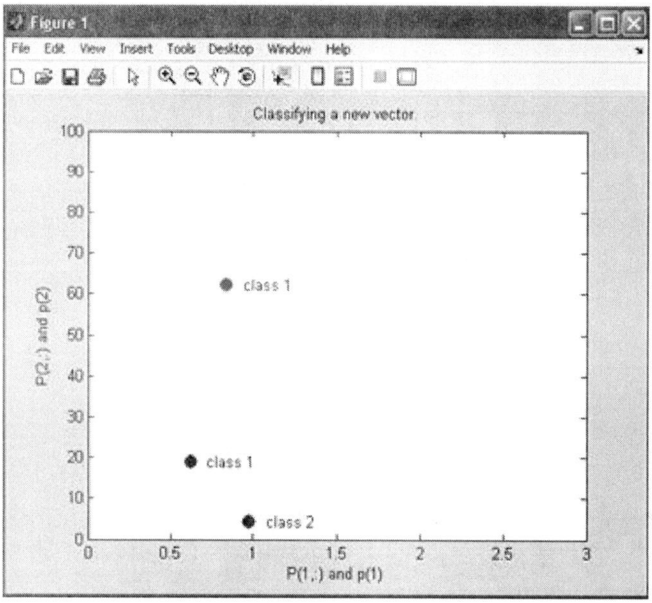

Fig.11. Work window of PNC in MATLAB environment

Conclusions

- Information of the inputted signs is established at probing signals with wavelengths respectively $\lambda = 530$ nm and $\lambda = 440$ nm and reflected by the examined object with signal of wavelength λ, which equals or is smaller than the wavelength of the probing signal.

- The PNC classifying accuracy is higher for Class 2 (10%) compared to Class 1 (15%).

- The frequency range used has fewer risks for the health of the serving staff and of the objects which are subject to control, compared to systems using X-ray waves in the terahertz range.

- The approach offered is suitable for control of access to authorized companies which work with explosives.

- The proposed approach could be used for other explosives (gunpowder), and also for other plastic substances (C4, RDX, Semtex).

- The studies related to the approach offered do not hold any risks for the research team.

Bibliography

[1] Баранов, В.Г., В. В. Кондратьев, В. Р. Милов, Ю. Х. Зарипова. Нейросетевые алгоритмы распознавания образов, Журнал "Нейрокомпьютеры: Разработка, применение", 2007, 11.
[2] Гмурман, В. Е. Теория вероятностей и математическая статистика, 9-е изд.,стер., М. Высшая школа, 2003, СТАТИСТИКА.
[3] Дойков, Н. Взрив. Взривни вещества и самоделни взривни устройства, използвани за терористични актове и други престъпления. С., 2002, стр. 299-307.
[4] Органезов, А. И. Приложение нейронных сетях в задачах распознавания образов, дисерт.доктор физикоматем. Наук-Тбилиси, 2006.
[5] Станкевич, Л. А., Н. Д. Хоа. Классификация объектов с использованием нормализующего фильтра и нечеткой нейросети. Журнал "Нейрокомпьютеры: разработка, применение", 2009, 3.
[6] Cassese, Antonio. Terrorism is Also Disrupting Some Crucial Legal Categories of International Law. European Journal of International Law, Vol. 12, No. 5, 2001, pp. 993-1001.
[7] Damynov, Ch., V.Nachev. Probabilistic Neural Classifiers, "Fundamental Sciences and Applications", 2006, Bulgaria, Vol.13(4).
[8] Greenwood, Christopher. International Law and the Conduct of Military Operations. Newport & Naval War College Press, March 2001, pp. 179-198.
[9] Marphy, Sean. Terrorizmus und das Konzept des bewaffneten Angriffs in Arikel 51 der UNO-Charta. Harvard International Law Journal, Band. 3, Winter 2002, S. 41-51. (Translated from the English by Jana Fedotova).

[10] Wasserman, Ph.D. Neural computing, ANZA Research , New York, USA, 1992, Robert, Adam. Counter terrorism, Armed Force and the Laws of War. Survival, Vol. 44, No. 1, Spring 2002, pp. 7-32.
[11] Wynn C. M., S. Palmacci, R. R. Kunz, and M. Aernecke Noncontact optical detection of explosive particles via photodissociation followed by laser-induced fluorescence Optics Express, Vol. 19, Issue 19, pp. 18671-18677 (2011).
[12] http://www.state.gov/www/global/arms/bureau_ac/wmeat98/
[13] http://www.state.gov/documents/organization/18726.pdf
[14] http://dx.doi.org/10.1364/OE.19.018671
[15] http://www. Explosive Detector/Makro Detector-makrodetector.com

Comparative Study on the Efficiency of Hybrid Learning Procedures Used for Training of Fuzzy-Neural Networks

Yancho Todorov
and Margarita Terziyska

Introduction

Neural networks and fuzzy logic have been successfully used for the identification and control of nonlinear plants for many years. In addition, there have been developed many structures combining these two techniques, such as ANFIS [6], DENFIS[7], NEFCON [9] and others. The fusion of fuzzy logic with neural networks makes it possible to combine the learning and computational ability of neural networks with the human-like IF-THEN thinking and reasoning of a fuzzy system. This can be compared with the human brain [1] – a neural network concentrates on the structure of human brain, i.e. on the hardware, whereas a fuzzy logic system concentrates on the software.

Many recent developments show that recurrent neural networks (RNNs) and recurrent fuzzy-neural networks (RFNNs) are a better way of describing complicated dynamic systems than feed-forward NNs/FNNs, because they can handle time-varying inputs or outputs through their own natural temporal operation. RFNNs have an internal feedback loop that allows them to capture the dynamic response of a system with external feedback through delays.

The error back-propagation (EBP) algorithm is the most commonly used training approach for neural networks/neuro-fuzzy structures. The method belongs to the group of first-order gradient algorithms, but it has some disadvantages, such as a slow convergence rate and becoming stuck in local minima. During recent years, many training improvements of the EBP algorithm have been developed, such as momentum [11], Quickprop [18], Resilient [13] and others.

Other group algorithms for neural networks/fuzzy-neural networks training are so-called second-order methods like the Newton [20] and Levenberg-Marcguard [19] approaches. Typical for these algorithms is that they provide a fast convergence. A paper on Fast and Efficient Second-Order Method for Training Radial Basis Function Networks is presented in [21]. Basic requirements for training algorithms and a brief commentary on first- and second-order gradient algorithms can be found in [15].

Biologically-inspired algorithms are also used as learning algorithms, for global optimization. Along with the established genetic algorithms (GA) [12], there is Ant Colony Optimization [2], [14], as well as Artificial Bee Colony [3] and Particle Swarm Optimization [5], [8]. Typical of this group of algorithms is that they do not require calculation of derivatives; but they are significantly slower compared to the well-known gradient approaches.

In this paper we demonstrate the development of a hybrid EBP algorithm for training a Takagi-Sugeno recurrent fuzzy-neural network with a global feedback. The proposed algorithm represents a combination of Gauss-Newton and Levenberg-Marcquardt approaches used to adjust the fuzzy rule-consequent parameters, while the premises are adjusted by the Gradient Descent approach. The efficiency of the proposed hybrid algorithm is studied through prediction by the proposed Fuzzy-Neural Network of two common Chaotic Time Series: Rossler and Mackey-Glass.

Recurrent Takagi-Sugeno Fuzzy-Neural Network

The fuzzy model proposed by Takagi and Sugeno (TS) [16] is described by fuzzy IF-THEN rules which represent the local input-output relations of a nonlinear system. The main feature of a Takagi-Sugeno fuzzy-neural model is that it expresses the local dynamics of each fuzzy implication (rule) by a linear system model. The overall fuzzy model of the system is achieved by fuzzy "blending" of the linear system models. Thus, in discrete time by using the NARX representation model (Nonlinear AutoregRessive model with eXogenous inputs), can be derived:

$$y(k) = f_y(x(k)) \tag{1}$$

where the unknown nonlinear function f_y can be approximated by Takagi-Sugeno type fuzzy rules:

$$R^{(i)}: if x_1 is \tilde{A}_1{}^{(i)} and.....x_p is \tilde{A}_p{}^{(i)} then f_y^{(i)}(k) \tag{2}$$

$$f_y^{(i)}(k) = a_1^{(i)} y_m(k-1) + a_2^{(i)} y_m(k-2)+\ldots+a_{ny}^{(i)} y_m(k-n_y) + \\ +b_1^{(i)} u(k) + b_2^{(i)} u(k-1)+\ldots+b_{nu}^{(i)} u(k-n_u) + b_0^{(i)} \tag{3}$$

where *(i)=1,2,N* denotes the number of fuzzy rules *R(i)*. A_i is an activated fuzzy set defined in the universe of discourse of the input x_i and the crisp coefficients a_1, a_2,...a_{ny}, b_1, b_2,...b_{nu} are the coefficients in the Sugeno function f_y. The input vector *x* contains notional regression of the input/output history dependence. Fig. 1 shows a schematic diagram of the proposed recurrent TS fuzzy-neural network. Identification of the proposed recurrent network requires the two main groups of unknown parameters to be determined: the number of membership functions, their shape and the parameters of the function f_y in the consequent part of the rules. For this purpose, in this work a simplified fuzzy-neural approach is applied [17].

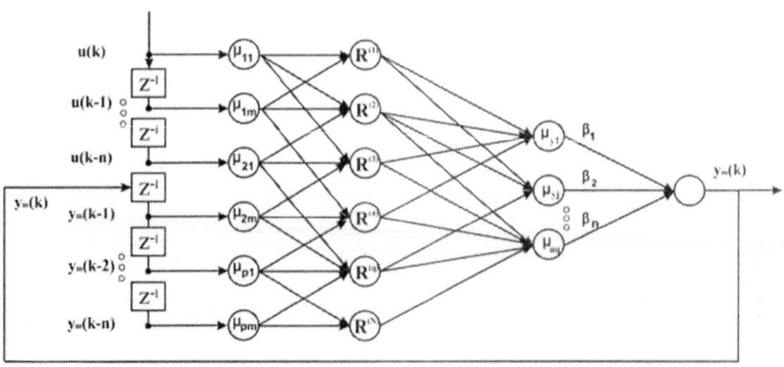

Figure 1: Schematic diagram of the proposed recurrent fuzzy-neural network.

Classical learning algorithm for the designed recurrent fuzzy-neural network

We implement a two-step gradient learning procedure based on minimization of an instant error measurement function $\Xi(k)=\varepsilon^2/2$, where $\varepsilon(k)=y(k)-\hat{y}(k)$, between the process output and the model output. During the learning process, two groups of parameters in the fuzzy-neural architecture premise and consequent parameters are under adaptation. The consequent parameters are the coefficients a_1, a_2,...a_{ny}, b_1, b_2,...b_{nu} in the Sugeno function, and they are calculated by making a step defined by the following chain rule:

$$\frac{\partial \Xi}{\partial \beta_{ij}} = \frac{\partial \Xi}{\partial y_M} \frac{\partial y_M}{\partial f_y} \frac{\partial f_y}{\partial \beta_{ij}} \tag{4}$$

where f_y is the Sugeno function, y_M is the model output and β_{ij} is an adjustable i^{-th} coefficient (a_i or b_i) of the j^{-th} activated fuzzy rule. Thus, the recurrent equations for calculation of the rule-consequent parameters can be expressed as:

$$\beta_{ij}(k+1) = \beta_{ij}(k) + \eta(y - y_M)\bar{\mu}_y{}^{(j)}(k)x_i(k), \tag{5}$$

$$\beta_{0j}(k+1) = \beta_{0j}(k) + \eta(y - y_M)\bar{\mu}_y{}^{(j)}(k) \tag{6}$$

where η is the learning rate and $\bar{\mu}_y$ is the normalized value of the membership function degree defined by the fuzzy implication realized by means of the product composition:

$$\mu_y^{(i)} = \mu_{1j}^{(i)} * \mu_{2j}^{(i)} * \ldots\ldots * \mu_{pj}^{(i)} \tag{7}$$

and

$$\bar{\mu}_y^{(i)} = \mu_y{}^{(i)} / \textstyle\sum_{i=1}^{q} \mu_y{}^{(i)}. \tag{8}$$

The premise parameters are a_{ij} (the centre c_{ij} and the deviation σ_{ij}) of an input Gaussian fuzzy set defined as:

$$\mu_{ij}{}^{(i)}(x_i) = \exp(-(x_i - c_i)/2\sigma_i)^2. \tag{9}$$

During the learning process, the error is being propagated through the links composed of the corresponding membership degrees from the last to the first network layer. Hence, again using the chain rule, the gradient step is defined as:

$$\frac{\partial \Xi}{\partial \alpha_{ij}} = \frac{\partial \Xi}{\partial y_M}\frac{\partial y_M}{\partial \mu_{ij}}\frac{\partial \mu_{ij}}{\partial \alpha_{ij}}. \tag{10}$$

The final recurrent equations for the premise parameters are expressed as:

$$c_{ij}(k+1) = c_{ij}(k) + \eta(y - y_M)\bar{\mu}_y{}^{(j)}(k)[f_y^{(i)} - \hat{y}(k)]\frac{[x_i(k) - c_{ij}(k)]}{c_{ij}^2(k)} \tag{11}$$

$$\sigma_{ij}(k+1) = \sigma_{ij}(k) + \eta(y - \hat{y})\bar{\mu}_y{}^{(j)}(k)[f_y^{(i)} - \hat{y}(k)]\frac{[x_i(k) - \sigma_{ij}(k)]^3}{\sigma_{ij}^2(k)}. \tag{12}$$

Newtonian approaches for learning the rule-consequent parameters

Gauss-Newton approach

To improve the efficiency of the proposed recurrent fuzzy-neural network, a Gauss-Newton method for adjusting the rule-consequent parameters is applied. Since the Newton method requires computation of the second-order derivative of the defined error cost term, taking into account (5 and 6) it can be rewritten:

$$\Delta\beta = -[\nabla^2 \Xi(\beta)]^{-1}\nabla\Xi(\beta). \tag{13}$$

The Hessian and the Gradient of $\Xi(\beta)$ are expressed as:

$$\nabla\Xi(\beta) = J^T(\beta)e(k), \nabla^2\Xi(\beta) = J^T(\beta)J(\beta) + \sum_{j=1}^{N} e_j(k)\nabla^2 e_j(k) \tag{14}$$

where the dimension of the *Jacobian* matrix is ($\mathrm{N}x\mathrm{N}_p$); where N is the number of training samples and N_p is the number of adjustable parameters in the network. Using the Newton approach the second term in (14) is assumed to be equal to zero. Therefore the update rule, according to (13), becomes:

$$\Delta\beta = -[J^T(\beta)J(\beta)]^{-1}J^T(\beta)e(k) \tag{15}$$

where the *Jacobian* according to adjustable parameters is calculated as:

$$J^T(\beta_{ij}) = \bar{\mu}_y{}^{(j)} x_i, J(\beta_{ij}) = \left[\bar{\mu}_y{}^{(j)} x_i\right]^T \tag{16}$$

$$J^T(\beta_{0j}) = \bar{\mu}_y{}^{(j)}, J(\beta_{0j}) = \left[\bar{\mu}_y{}^{(j)}\right]^T. \tag{17}$$

Finally, the recurrent equations for the rule-consequent parameters are derived as follows:

$$\begin{aligned}\beta_{ij}(k+1) = \beta_{ij}(k) + \eta\left[\bar{\mu}_y{}^{(j)} x_i(k)\left(\bar{\mu}_y{}^{(j)} x_i(k)\right)^T\right]^{-1}(y - y_M)\bar{\mu}_y{}^{(j)} \\ +\zeta(\beta_{ij}(k) - \beta_{ij}(k-1))\end{aligned} \tag{18}$$

$$\begin{aligned}\beta_{0j}(k+1) = \beta_{0j}(k) + \eta\left[\bar{\mu}_y{}^{(j)}\left(\bar{\mu}_y{}^{(j)}\right)^T\right]^{-1}(y - y_M)\bar{\mu}_y{}^{(j)}(k) \\ +\zeta(\beta_{0j}(k) - \beta_{0j}(k-1))\end{aligned} \tag{19}$$

where the second term represents an introduced momentum ζ in relation to a previous increment of the adjusted parameter.

Levenberg-Marquardt approach

The Levenberg-Marquardt algorithm also uses the approximated Hessian and the information in the gradient, taking into account some regularization factors. The algorithm iterates using the following general equation:

$$\Delta\beta = -[J^T(\beta)J(\beta) + \lambda I]^{-1} J^T(\beta) e(k) \tag{20}$$

where H is the Hessian as computed in (14), I is the identity matrix and λ is the Levenberg-Marquardt parameter, which adjusts the direction of movement to extremes. When the λ parameter is small, the method represents a quadratic approximation, and when it is large the Hessian is negligible and the LM method works similarly to the Gradient Descent algorithm. At first iterations, LM works as a gradient method and as it gets near the optimal point it gradually switches to a Newton-based approach.

When the LM parameter gets smaller, LM finds a locally linear solution, precisely and quickly. After each iteration of the search, Hessian is checked to be positive definite (convex optimization). If Hessian is not positive definite, λ is increased until this happens. To investigate the positive definiteness of Hessian, Cholesky factorization has to be used [10]. To find the minimum using LM it is necessary to calculate the Gradient and the Hessian of the cost function. Using a TS fuzzy-neural network model of the system, it is a straightforward task.

$$\beta_{ij}(k+1) = \beta_{ij}(k) + \eta\left[\bar{\mu}_y{}^{(j)}x_i(k)\left(\bar{\mu}_y{}^{(j)}x_i(k)\right)^T + \lambda I\right]^{-1}(y - y_M)\bar{\mu}_y{}^{(j)} + \\ +\zeta(\beta_{ij}(k) - \beta_{ij}(k-1)) \tag{21}$$

$$\beta_{0j}(k+1) = \beta_{0j}(k) + \eta\left[\bar{\mu}_y{}^{(j)}(\bar{\mu}_y{}^{(j)})^T + \lambda I\right]^{-1}(y - y_M)\bar{\mu}_y{}^{(j)}(k) + . \\ +\zeta(\beta_{0j}(k) - \beta_{0j}(k-1)) \tag{22}$$

Chaotic Time-Series prediction by the proposed recurrent Fuzzy-Neural Network

Chaos is a common dynamic phenomenon in various fields [22] and different definitions exist as series representations. Chaotic time series are inherently nonlinear, sensitive to initial conditions and difficult to predict. In the mathematical sense, a chaotic process is one where there is positive feedback of some kind. Under some circumstances such processes can create time series that appear to be completely random - the corollary of this is that some seemingly random series are in fact chaotic, and thus to a certain extent predictable. Chaotic systems are never completely predictable; because of feedback the simulation and the real series will always rapidly diverge. This effect is caused by small differences between the initial real state and the simulation growing geometrically as the simulation is advanced in time. Chaotic time series commonly occur in physics, biology, meteorology, engineering and finance.

Experimental results in Chaotic Time Series Prediction

The prediction of chaotic time series based on measurement is a practical technique for studying the characteristics of complicated dynamics and for evaluation of the accuracy of different types of nonlinear models such as RNNвЂ™s. In this study, two chaotic time series, Mackey-Glass [?] and Rossler [4], are used to assess the performance prediction of the proposed recurrent TS network, with chosen fixed momentum of ζ=0.098.

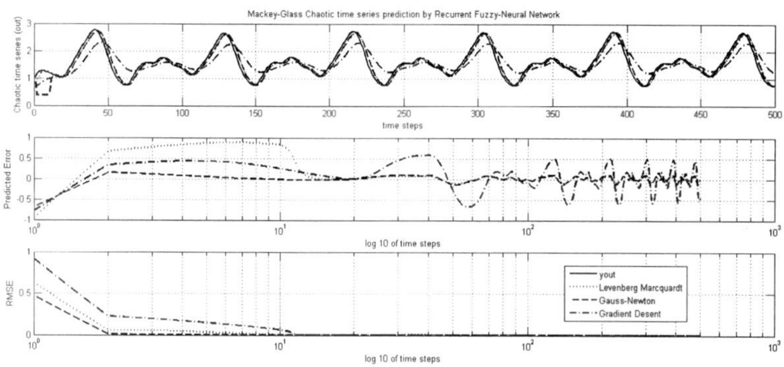

Figure 2: Prediction of Mackey-Glass Chaotic Time Series.

Fig. 2 demonstrates the model's performance in prediction of the Mackey-Glass chaotic time series, with the following parameters: a=0.3; b=0.1; C=10; initial conditions x_0=0.1 and τ= 17s. As can be seen, the classical Gradient Descent approach predicts the time series with a great error compared to the proposed hybrid learning using the Gauss-Newton and Levenberg-Marquardt approaches. In order to illustrate the fluctuations in the prediction error and the Root Mean Squared Error, they are given on a logarithmic scale, which again proves the positive effect of the designed approach.

Fig. 3 shows the results obtained in prediction of the Rossler chaotic series, with the following parameters: a=0.2; b=0.4; c=5.7; initial conditions x_0=0.1; y_0=0.1; z_0=0.1. The results obtained again show that by using the adopted hybrid approach the model performs well on prediction of minimum error, and there is a fast transient response of the predicted error and a RMSE (illustrated on a logarithmic scale) approaching zero.

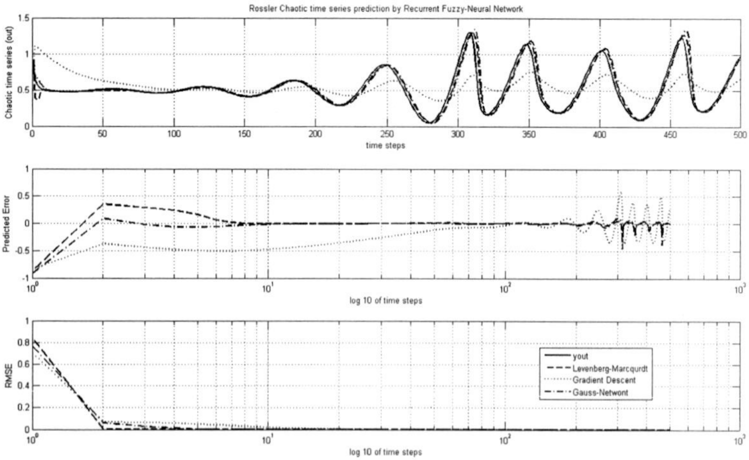

Figure 3: Prediction of Rossler Chaotic Time Series.

Conclusions

This paper presents a hybrid learning approach for a Recurrent Fuzzy-Neural Network with a global feedback. The proposed algorithm relies on the Gradient Descent approach for adjusting the rule premise and the Gauss-Newton and Levenberg-Marquardt approaches for scheduling the rule-consequent parameters. The experimental simulations of two common Chaotic Time Series (Mackey-Glass and Rossler) and the model's performance in prediction show the potential of the adopted approach.

The modeling error is smaller and its transient response is faster with values closer to zero, compared to the classical Gradient learning methodology. A potential extension of the proposed approach is that the model can be coupled with an Optimization procedure into a Model Predictive Control scheme.

Bibliography

[1] C. R. Alavala. New Age. International Pvt Ltd Publishers, 2008.

[2] Blum, C. Ant colony optimization: Introduction and recent trends. *Physics of Life*, 2(4):353-373, 2005.

[3] Bullinaria, J. A. and Alyahya, K. Artificial Bee Colony Training of Neural Networks. In: G. Terrazas, F. E. B. Otero & A. D. Masegosa (Eds), Nature Inspired Cooperative Strategies for Optimization (NICSO, :191-201, 2013.

[4] Diaconescu, E. The use of NARX Neural Networks to predict Chaotic Time Series. *WSEAS Trans. on Computer Research*, 3(3):182-191, 2008.

[5] Esmaili, A. and M. Shahbazian and B. Moslemi. Nonlinear Process Identification Using Fuzzy Wavelet Neural Network Based on Particle Swarm Optimization Algorithm. *J. Basic. Appl. Sci. Res.*, 3(5), 2013.

[6] Jang, J. S. R. ANFIS: Adaptive Network based Fuzzy Inference Systems. *IEEE Transactions, System, Man & Cybernetics*, 1991.

[7] Kasabov, N. and Q. Song. DENFIS: Dynamic, evolving neural-fuzzy inference systems and its application for time-series prediction. *IEEE Trans. on Fuzzy Systems*, 10(2):144-154, 2002.

[8] Lin, Cheng-Jian and Chi-Feng Wu. An efficient symbiotic particle swarm optimization for recurrent functional neural fuzzy network design. *International Journal of Fuzzy Systems*, 11(4), 2009.

[9] Nauk, D. and R. Kruse. A Neuro-Fuzzy Controller Learning by Fuzzy Error Propagation. *In Proceedings of the North American Fuzzy Information Processing Society NAFIPSвЂ™92, Mexico*, :388-397, 1992.

[10] M. Nogaard and O. Ravn and N. Poulsen and L. K. Hansen. Springer-Verlag, 2003.

[11] Phansalkar, V. V. and P. S. Sastry. Analysis of the back-propagation algorithm with momentum. *IEEE Trans. on Neural Networks*, 5(3):505-506, 1994.

[12] Rankovic, V. and J. Radulovic and N. Grujovic and D. Divac. Neural Network Model Predictive Control of Nonlinear Systems Using Genetic Algorithms. *Int. Journal of Computers and Communications*, 7(3):540-549, 2012.

[13] M. Riedmiller and H. Braun. A direct adaptive method for faster backpropagation learning: The RPROP algorithm. *In Proc. of International Conference on Neural Networks*, :586-591, 1993.

[14] Socha, K. and C. Blum. An ant colony optimization algorithm for continuous optimization: application to feed-forward neural network training. *Neural Computing and Applications*, 16(3):235-247, 2007.

[15] SГёrensen, P. H. and M. NГёrdaard and O. Ravn and N. K. Poulsen. Implementation of neural network based non-linear predictive control. *Neurocomputing*, 28:37-51, 1999.

[16] T. Takagi and M. Sugeno. Fuzzy identification of systems and its application to modeling and control, вЂќ IEEE Transactions on Systems, Man, Cybernetics. 15(1):116-132, 1985.

[17] Terzyiska, M. and Todorov, Y. and Petrov, M. Nonlinear Model Predictive Controller with Adaptive Learning rate Scheduling of an internal model. *In. Proc of the Int. Conf. вЂ?Modern Trends in ControlвЂ™*, :289-298, 2006.

[18] Fok Hing Chi Tivive and A. Bouzerdoum. Efficient training algorithms for a class of shunting inhibitory convolutional neural networks. *IEEE Trans. on Neural Networks*, 16(3):541-556, 2005.

[19] Y. Todorov and M. Terziyska and S. Ahmed and M. Petrov. Fuzzy-Neural Predictive Control using Levenberg-Marquardt optimization approach. *Proceedings of International IEEE conference on Innovations in Intelligent Systems and Applications*, :1-5, 2013.

[20] Y. Todorov and M. Terzyiska and M. Petrov. Recurrent Fuzzy-Neural Network with Fast Learning Algorithm for Predictive Control. *Lecture Notes on Computer Science вЂ" ICANN 2013*, pages 459-466, 2013. Springer-Verlag.

[21] Tiantian Xie and Hao Yu and J. Hewlett and P. RГizycki and B. Wilamowski. Fast and Efficient Second-Order Method for Training Radial Basis Function Networks. *IEEE Transactions on Neural Networks and Learning Systems*, 23(4), 2012.

[22] Yao, J. and Mao, J. and Zhang, W. Application of Fuzzy Tree on Chaotic Time Series Prediction. *In. IEEE Proc. of Int. Conf. on Aut. and Logistics*, :326-330, 2008.

CREATING GUI AND USING NEURAL NETWORK TOOL TO STUDY THE STATIC EQUATION OF LINEAR CNN

VICTORIA RASHKOVA

Introduction

Cellular Neural Networks (CNN) is a new class of neural networks, first introduced by Leon Chua and Lin Yang in the year 1988. They use a grid of non-linear dynamic circuits which are connected to each other. As a result, it is possible to transmit a large amount of information in real time.

The basic circuit unit of CNNs is called a cell. It contains linear and nonlinear circuit elements, which typically are linear capacitors, linear resistors, linear and nonlinear controlled sources, and independent sources. All the cells of a CNN have the same circuit structure and element values.

One of the key features of a CNN is that the individual cells are nonlinear dynamical systems, but that the coupling between them is linear. Roughly speaking, one could say that these arrays are nonlinear but have a linear spatial structure, which makes the use of techniques for their investigation common or attractive in engineering and physics [1].

Here the dynamical systems describing CNNs are presented. For a general CNN whose cells are made up of time-invariant circuit elements, each cell $C(i, j)$ is characterized by its CNN cell dynamics:

$$\dot{x}_{ij}(t) = -x_{ij}(t) + \sum_{C(k,l)\in N_r(i,j)} \tilde{A}_{ij,kl}(y_{kl}(t), y_{ij}(t)) + \sum_{C(k,l)\in N_r(i,j)} \widetilde{B}_{ij,kl}(u_{kl}, u_{ij}) + I_{ij} \quad (1)$$

$$1 \le i \le M, 1 \le j \le M$$

where x_{ij}, y_{ij} and u_{ij} refer to the state, output and input voltage of a cell $C(i,j)$: $C(i,j)$ refers to a grid point associated with a cell on the 2D grid. $C(k,l) \in N_r(i,j)$ is a grid point (cell) in the neighborhood within a radius r of the cell $C(i,j)$. I_{ij} is an independent current source. $\tilde{A}$ and $\tilde{B}$ are nonlinear cloning templates, which specify the interactions between each cell and all its neighbor cells in terms of their input, state and output variables [1]. In the linear case instead of $\tilde{A}$ and $\tilde{B}$ we have the following templates:

$$\sum_{C(k,l)\in N_r(i,j)} A_{ij,kl}y_{kl}(t) + \sum_{C(k,l)\in N_r(i,j)} B_{ij,kl}u_{kl} \tag{2}$$

When the templates are spatially independent, each cell is described by two real matrices A and B. In other words, the linear CNN has the following static equation:

$$\dot{x}_{ij}(t) = -x_{ij}(t) + A * y_{ij}(t) + B * u_{ij} + I_{ij} \tag{3}$$

where A and B are called templates for feedback and control templates and $*$ is the convolution operator.

The symmetry of the templates for feedback A is necessary to demonstrate the complete stability of a CNN.

The purpose of this paper is to present some features of the graphical user interface in MATLAB. GUI facilitates the work of users as they explore a variety of parameters of different systems and equations, for example using the static equation CNN (3).

Creating a Graphical User Interface and Options

An interface to study the static equation of linear CNNs can be developed using the built-in MATLAB graphics editor GUIDE. MATLAB functions with additional software packages called "Toolbox". These are designed for a wide range of tasks in different areas such as: Signal Processing Tools – for processing data (signals); Control System Tools - for analysis and synthesis of systems management; System Identification Tools – for identification of dynamical systems; Optimization Tools - for solving optimization problems; Neural Network Tools - for work with neural networks; Pattern Recognition Tools – for sinister transformation; etc. The graphical user interface provides the user with a familiar environment in which to work. This environment contains: pushbuttons, toggle buttons, lists, menus, text boxes, and so forth (described below). However, GUIs are harder to program because a GUI-based program must be prepared for executing each different event. Such inputs are known as events, and a program that responds to events is said to be event driven [5].

To start GUIDE it is necessary to select the GUIDE icon from the MATLAB tool menu. In the GUIDE Quick Start dialog box, select the Blank GUI (Default) template, and then click OK (shown in Figure 1).

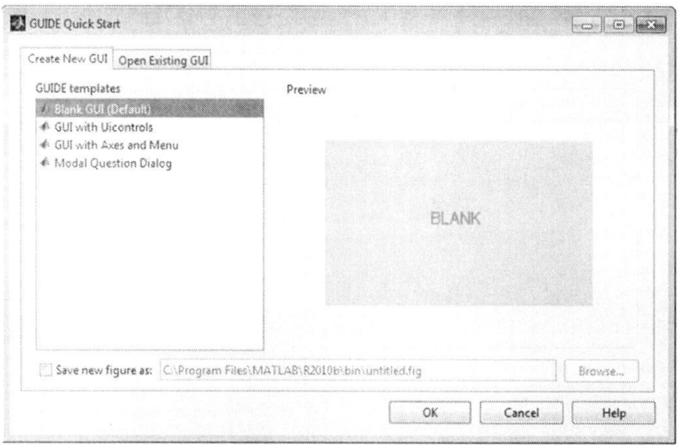

Figure 1. GUIDE quick start

This displays a dialog box to create a user interface that contains a set of tools called controls. The Layout Editor then appears as shown in the following figure (Figure 2).

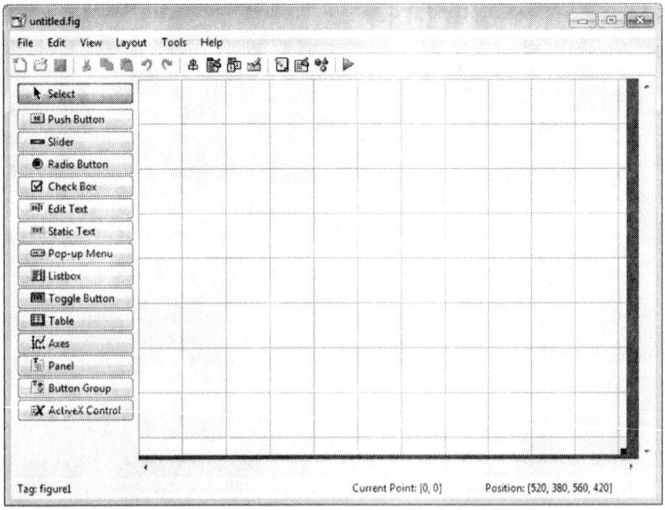

Figure 2. Layout Editor

The GUI controls are described here.

- **Push Button** - Push buttons generate an action when clicked.
- **Slider** - Sliders accept numeric input within a specific range by enabling the user to move a sliding bar, which is called a slider or thumb.
- **Radio Button** - Radio buttons are similar to check boxes, but are typically mutually exclusive within a group of related radio buttons. That is, you can select only one button at any given time.
- **Check Box** - Check boxes generate an action when checked and indicate their state as checked or not checked. Check boxes are useful for multiple choices.
- **Edit Text** - Edit text controls are fields that enable users to enter or modify text strings.
- **Static Text** - Static text is typically used to label other controls, provide directions to the user, or indicate values associated with a slider. Users cannot change static text interactively and there is no way to invoke the callback routine associated with it.
- **Pop-Up Menu** - Pop-up menus open to display a list of choices when users click the arrow.

- **List Box** - List boxes display a list of items and enable users to select one or more items.
- **Toggle Button** - Toggle buttons generate an action and indicate whether they are turned on or off. When you click a toggle button, it appears depressed, showing that it is on. When you release the mouse button, the toggle button's callback executes.
- **Table** - For work with table data.
- **Axes** - Axes enable your GUI to display graphics.
- **Panel** - Panels group GUI components. Panels can make a user interface easier to understand by visually grouping related controls. A panel can have a title and various borders.
- **Button Group** - Button groups are like panels but can be used to manage exclusive selection behavior for radio buttons and toggle buttons. A button group overwrites the callback properties of the radio buttons and toggle buttons that it manages.
- **ActiveX Component** - ActiveX components enable you to display ActiveX controls in your GUI.

When you save your GUI application, GUIDE automatically generates an M-file that you can use to control how the GUI works. This M-file provides code to initialize the GUI and contains a framework for the GUI callbacks - the routines that execute in response to user-generated events such as a mouse click. Using the M-file editor, you can add code to the callbacks to perform the functions you want [5]. Each control in GUI form has one or more user-written routines (executable MATLAB code) known as callbacks, named for the fact that they "call back" to MATLAB to ask it to do things. The execution of each callback is triggered by a particular user action such as pressing a screen button, clicking a mouse button, selecting a menu item, typing a string or a numeric value, or passing the cursor over a component. The GUI then responds to these events. You, as the creator of the GUI, provide callbacks which define what the components do to handle events. You can select, size, and position these components as you like. Using callbacks you can make the components do what you want when the user clicks or manipulates them with keystrokes. The graphical user interface for the study of static equations of linear CNNs has the form shown in Figure 3.

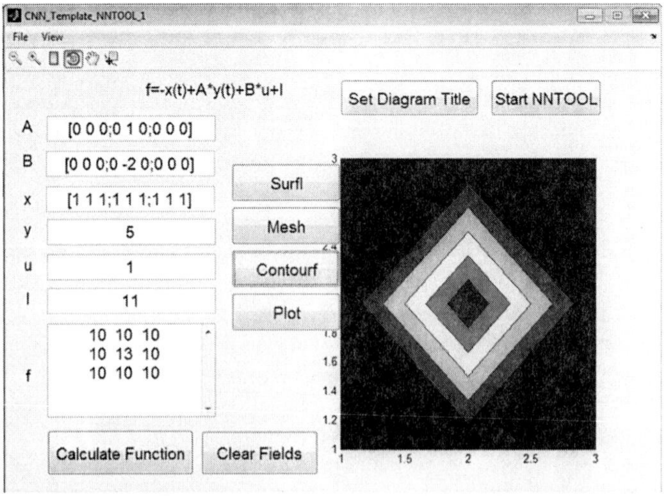

Figure 3. Main Menu

The GUI contains:

- Eight Static Text controls;
- Eight Push Button controls;
- Seven Edit Text controls;
- One Axes control.

The Graphic User Interface contains a toolbox menu - with icons for Zoom Out; Zoom In; Insert Color Bar; 3D Rotate; Pan and Data Cursor - shown in Figure 4. The Graphic User Interface contains a main menu too (shown in Figure 3):

Figure 4. Toolbox

After inserting values for the different parameters and selecting the "Calculate Function" button tool from the form, the function is calculated and displayed in the text box as shown in Figures 3 and 5.

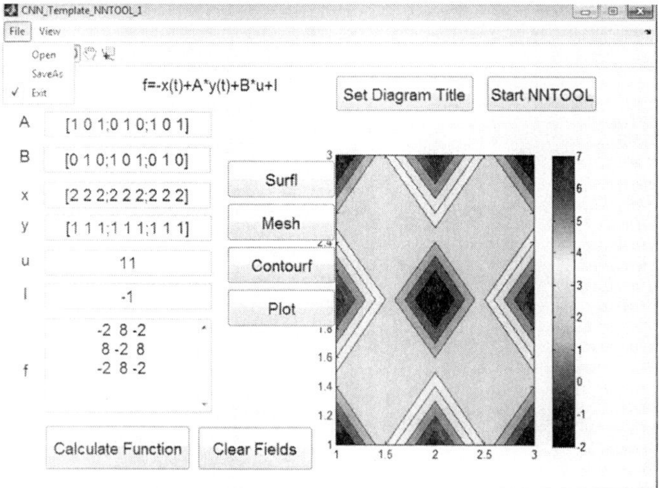

Figure 5. Calculated function

The Command Button "Clear Fields" clears all fields and allows the user to enter new values for the parameters of the function. The choice Command Buttons "Surfl", "Mesh", "Contour" and "Plot" show different kinds of plots: surface, mesh, contour and Plot respectively. The Command Button "Set Diagram Title" shows the title and marks a place along the axes x and y. The Command Button "Start NNTOOL" starts the Neural Network Tool that is embedded in MATLAB, The Network/Data Manager window is a basic part of the GUI. This window has its own work space distinct from the command line of MATLAB workspace. During its use we can import data from the command prompt of MATLAB, we can create new data and we can also export results from the Network Data Manager to the command prompt and store them in Tables [3], [4] (Figure 6).

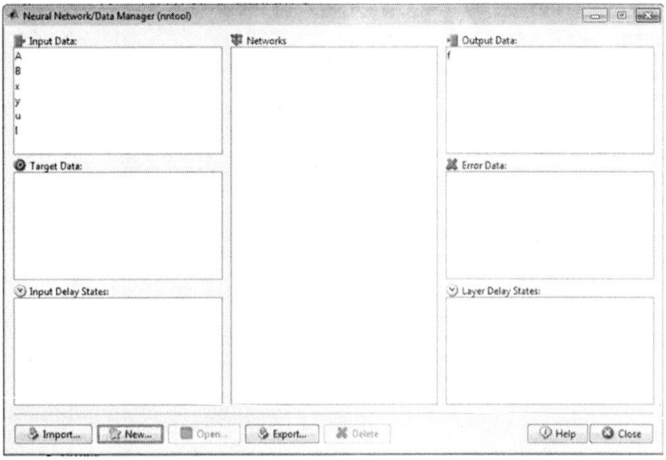

Figure 6. Network Data Manager

The choice of option "Data Type" is compulsory:

- Inputs;
- Targets;
- Input Delay States;
- Layer Delay States;
- Outputs; or
- Errors.

To create the network we chose the card "Network" in the "Create Network or Data" window. One must chose the variable from the catalog "Input data", and/or the variable from the catalogue "Target data" respectively [2].

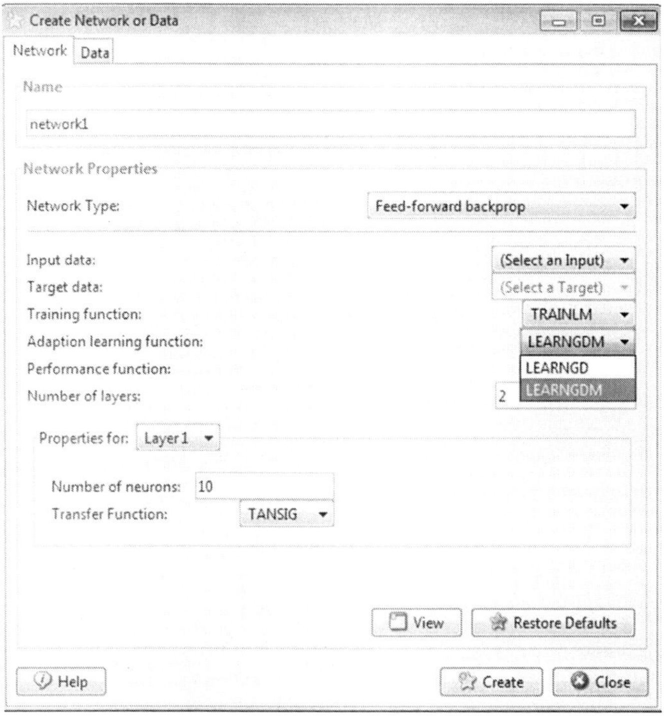

Figure 7. Create Network or Data

The possible Training functions are:

- **Trainbr** - Bayesian regularization;
- **Trainlm** - Levenberg-Marquardt back propagation;
- **Trainoss** - One step secant back propagation;
- **Trainr** - Random order incremental training with learning functions.

The possible Performance Functions are:

- **Mae** - Mean absolute error performance function;
- **Mse** - Mean squared error performance function;
- **Msne** - Mean squared normalized error performance function;

- **Msnereg** -Mean squared normalized error with regularization performance functions;
- **Msereg** - Mean squared error with regularization performance function;
- **Mseregec** - Mean squared error with regularization and economization performance function;
- **Sse** - Sum squared error performance function.

Some of the possible Transfer functions are:

- **Hardlim** - Hard limit transfer function;
- **Learnp** - Learning function;
- **Compet** - Competitive transfer function;
- **Netinv** - Inverse transfer function;
- **Poslin** - Positive linear transfer function;
- **Purelin** - Linear transfer function;
- **Radbas** - Radial basis transfer function;
- **Satlins** - Symmetric saturating linear transfer function;
- **Tansig** - Hyperbolic tangent sigmoid transfer function;
- **Tribas** - Triangular basis transfer function.

The simulation results for different values of A and B templates are shown in Figure 8.

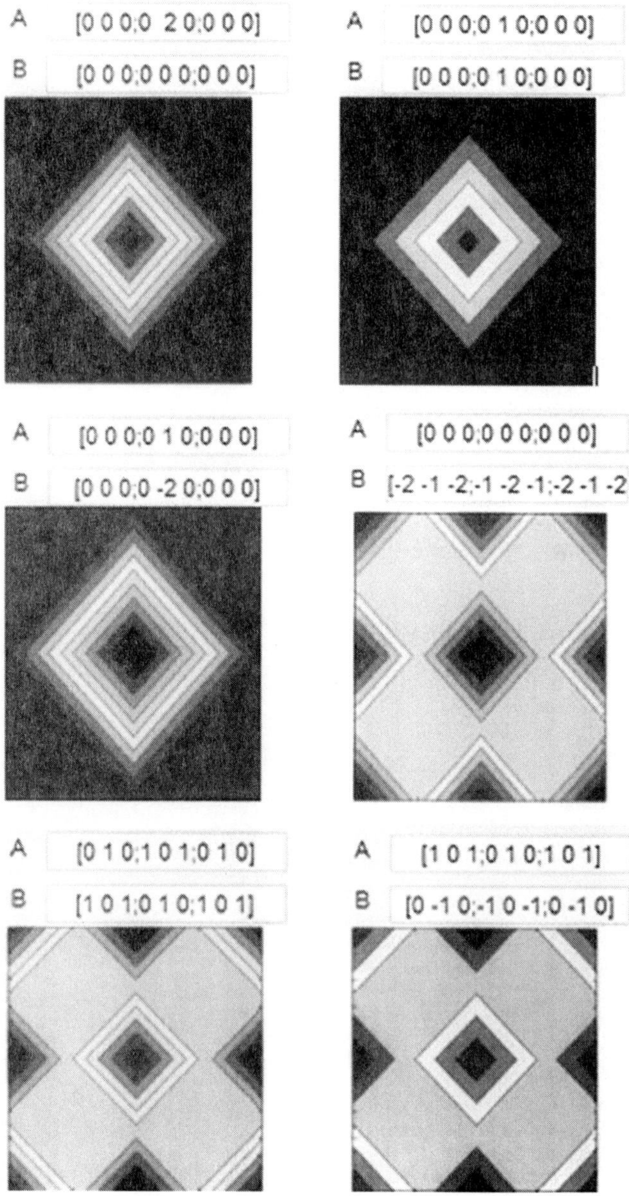

Figure 8. Simulation results

Part of the source code of the GUI Application is shown in Figure 9.

```
function pushbutton3_Callback(hObject, eventdata, handles)
% hObject    handle to pushbutton3 (see GCBO)
% eventdata  reserved - to be defined in a future version of MATLAB
% handles    structure with handles and user data (see GUIDATA)
set(handles.A,'string','');
set(handles.B,'string','');
set(handles.I,'string','');
set(handles.x,'string','');
set(handles.u,'string','');
set(handles.y,'string','');
set(handles.f,'string','');
delete xlabel
delete ylabel
delete title

function pushbutton5_Callback(hObject, eventdata, handles)
% hObject    handle to pushbutton5 (see GCBO)
% eventdata  reserved - to be defined in a future version of MATLAB
% handles    structure with handles and user data (see GUIDATA)
%hundles.axes1=handles.f
A=str2num(get(handles.A,'string'));
B=str2num(get(handles.B,'string'));
I=str2num(get(handles.I,'string'));
x=str2num(get(handles.x,'string'));
u=str2num(get(handles.u,'string'));
y=str2num(get(handles.y,'string'));
int=num2str(-x+A.*y+B.*u+I)
set(handles.f,'string',int);
f=-x+A.*y+B.*u+I;
axes(handles.axes1)
contourf(f)
```

Figure 9. Application source code

Conclusion

As GUIDE is built into MATLAB, it is designed to use all toolboxes available for MATLAB, and it is an effective tool for building graphical tools that are used in different areas.

The advantages of GUI are as follows:

- Code is produced efficiently. A relative large size of CNN can be built;

- Easy and straightforward design;
- Flexible for the user to set the inputs and initial states of variables;
- Ability to have an interactive dialogue with the user;
- Graphical display of data;
- Opportunity for rapid calculation of any type of mathematical problem;
- Ability to import data;
- Does not require the user to have special programming skills;
- Removes the step of continuously introducing the same programming code for the study of the same equation for various values of the parameters.

Disadvantages:

- The main disadvantage of using a GUI is that a much more complicated programming code is used to create the application. The user must have programming skills if he wants to add or change the parameters.

Acknowledgement

The present document has been produced with the financial assistance of the European Social Fund under the Operational Programme "Human Resources Development". The contents of this document are the sole responsibility of the "Angel Kanchev" University of Rousse and can under no circumstances be regarded as reflecting the position of the European Union or the Ministry of Education and Science of the Republic of Bulgaria.

Project B„– BG051PO001-3.3.06-0008 "Supporting Academic Development of Scientific Personnel in Engineering and Information Science and Technologies".

Bibliography

[1] A. Slavova, Cellular Neural Networks: Dynamics and Modeling, Kluwer Academic Publishers, 2003

[2] Hamada Ahmed Khadragy, Cellular Neural Networks, Praveena Annadurai, 2013

[3] Howard Demuth, Mark Beale, Neural Network Toolbox, Users Guide

[4] Lazaros Iliadis, The Graphical Interface of Neural Network Toolkit in MATLAB and Applications

[5] MATLAB, Creating Graphical User Interface, Release 2007b, 2007, book

CHAPTER SEVEN

NONLINEAR WAVES AND SIMULATIONS

GLOBAL SOLVABILITY AND FINITE TIME BLOW UP OF THE SOLUTION TO SIXTH ORDER BOUSSINESQ EQUATION

N. KUTEV, N. KOLKOVSKA AND M. DIMOVA

Introduction

The aim of this paper is to study the Cauchy problem for the generalized sixth order Boussinesq equation

$$\begin{aligned} &U_{tt} - U_{xx} - \beta_1 U_{ttxx} + \beta_2 U_{xxxx} + \beta U_{ttxxxx} = f(U)_{xx} \\ &X \in \mathbb{R}, \quad t \in [0,T), \quad T \leq \infty \end{aligned} \quad , \tag{1}$$

$$U(X,0) = U_0(X), \quad U_t(X,0) = U_1(X), \quad x \in \mathbb{R}. \tag{2}$$

Here $\beta_1 \geq 0$, $\beta_2 > 0$ and $\beta \geq 0$ are dispersive coefficients. The nonlinear term $f(U)$ has the form

$$f(U) = a|U|^p U + b|U|^{2p} U, \quad p \geq 1, \quad a, b = const \neq 0. \tag{3}$$

For $\beta_2 = 1$, $\beta_1 = \beta = 0$, Eq.(1), referred to as the "good" Boussinesq equation, is a universal model for nonlinear wave dynamics in weakly dispersive media. For example it models surface waves in shallow waters. For $\beta_1, \beta_2 \neq 0$, $\beta = 0$, Eq. (1) is the Boussinesq paradigm equation and is derived from the full Boussinesq model in [1]. When $\beta_2 = 1$, $\beta_1 = 0$, $\beta = 1$ Eq.(1) is transformed into the Rosenau equation which describes the dynamics of the nonlinear lattice [9].

Here we focus on Eq. (1) with non-zero dispersive coefficients. In this case Eq.(1) occurs in the water wave problem with non-zero surface tension, see [10].

In [16, 17, 20] the authors consider the long-time behavior of solutions to (1), (2) for small initial data using the contraction mapping theorem. Finite time blow-up and nonlinear scattering are established under certain hypotheses on the nonlinearities $f(U)$. This is even though these results cannot give the global existence for typical nonlinearities, e.g. $f(U) = U^2$.

In [13, 18] Eq.(1) has been studied with nonlinear terms

$$f(U) = a|U|^p \quad \text{or} \quad f(U) = a|U|^{p-1}U, \qquad a \neq 0, \;\; p > 1. \tag{4}$$

In these papers the global existence or finite time blow-up of the weak solutions with subcritical or critical initial energy $E(0) \leq d$ was proved by means of the potential well method. The supercritical case $E(0) > d$ is considered in [11].

In [12, 15, 19] the potential well method is applied to the Rosenau equation (Eq.(1) with $\beta_1 = 0$) and nonlinear terms (4).

Combined power-type nonlinearities are investigated in [5, 14] for the fourth order generalized Boussinesq equation (Eq.(1) with $\beta = 0$).

The special nonlinearity (3) is well-known as generalized Bernoulli (or Lienard) type nonlinearity. The cubic-quintic nonlinearities, i.e., $p = 2$ in (3), arise in a number of mathematical models of physical processes, e.g. in some models with significance in the theory of atomic chains [7] and shape-memory alloys [2].

In the present paper we study problem (1), (2) for all constants a and b in (3) by means of different methods. The main results are formulated in Theorem 3.1 from Section 3.

In the case $b < 0$ we give a complete answer to the question about the global existence or finite time blow-up of the solution by the well-known potential well method. In the case $b > 0$, $a < 0$, $a^2 - \frac{(p+2)^2}{(p+1)} b > 0$ we apply the new nonstandard potential well method, proposed in [5]. This method is based on new invariant sets and a new critical energy constant d_+, which is an analog of the critical energy constant d. For $b > 0$, $a > 0$ or $b > 0$, $a < 0$, $a^2 - \frac{(p+2)^2}{(p+1)} b \leq 0$ we prove the global existence of the solutions by the method of the law of conservation, without any restriction on the initial energy $E(0)$. The critical energy constants d and d_+, crucial for the global solvability or finite time blow-up in the framework of the potential well methods, are calculated explicitly using the ground state solutions of Eq. (1) at the end of Section 3. A conservative finite difference scheme for the numerical solution of (1), (2) is proposed and studied in Section 4. The numerical experiments which we performed illustrate and support our theoretical results.

Preliminaries

Throughout the paper we denote $L^2(\mathbb{R})$ and $H^s(\mathbb{R})$ by L^2 with norm $\| u \|$ and H^s with norm $\| u \|_s$ respectively, and define the inner product (u, v) as $(u, v) = \int_{\mathbb{R}} uv\, dx$.

After the change of the variable $x = X/\sqrt{\beta_2}$, problem (1), (2) can be rewritten in the following form

$$\begin{array}{l} \beta_2 u_{tt} - u_{xx} - \beta_1 u_{ttxx} + u_{xxxx} + \beta_3 u_{ttxxxx} = f(u)_{xx} \\ x \in \mathbb{R}, \quad t \in [0, T), \quad T \leq \infty \end{array}, \tag{5}$$

$$u(x, 0) = u_0(x), \quad u_t(x, 0) = u_1(x), \qquad x \in \mathbb{R}, \tag{6}$$

where $u(x, t) = U(\sqrt{\beta_2}x, t)$, $u_0(x) = U_0(\sqrt{\beta_2}x)$, $u_1(x) = U_1(\sqrt{\beta_2}x)$ and $\beta_3 = \beta/\beta_2$.

We study problem (5), (6) with initial data

$$u_0 \in H^1, \quad u_1 \in H^1, \quad (-\Delta)^{-1/2} u_1 \in L^2, \tag{7}$$

where $(-\Delta)^{-s} u = \mathcal{F}^{-1}(|\xi|^{-2s} \mathcal{F}(u))$ for $s > 0$ and $\mathcal{F}(u)$, $\mathcal{F}^{-1}(u)$ are the Fourier and the inverse Fourier transform. Under regularity assumptions (7), problem (5), (6) has a unique solution $u \in C^1([0, T_m); H^1)$ with maximal existence time interval $[0, T_m)$, $T_m \leq \infty$ (see [18], Theorem 2.3). Moreover $(-\Delta)^{-1/2} u_t \in L^2$ for every $t \in (0, T_m)$ and the solution satisfies the conservation law

$$E(t) = E(0) \quad \text{for every} \quad t \in [0, T_m), \tag{8}$$

where

$$E(t) = \frac{1}{2}\left(\beta_2\left\|(-\Delta)^{-1/2}u_t\right\|^2 + \beta_1 \parallel u_t \parallel^2 + \beta_3 \parallel u_{tx} \parallel^2 + \parallel u \parallel_1^2\right)$$

$$+ \int_{\mathbb{R}} F(u)\, dx, \qquad F(u) = \int_0^u f(s)\, ds \qquad (9)$$

(for more details see [18], Theorem 2.4).

In the framework of the potential well method we introduce the potential energy functional $J(u)$, the Nehari functional $I(u)$, the Nehari manifold $\mathcal{N}$, and the sets W, V as

$$J(u) = \frac{1}{2} \parallel u \parallel_1^2 + \int_{\mathbb{R}} F(u)\, dx, \quad I(u) = J'(u)u = \parallel u \parallel_1^2 + \int_{\mathbb{R}} uf(u)\, dx,$$

$$W = \{u \in \mathrm{H}^1 : I(u) > 0\} \cup \{0\}, V = \{u \in \mathrm{H}^1 : I(u) < 0\},$$

$$\mathcal{N} = \{u \in \mathrm{H}^1 :\ I(u) = 0,\ \parallel u \parallel_1 \neq 0\}.$$

Let us mention some differences in the definition of the Nehari manifold $\mathcal{N}$ depending on the constants a and b in (3).

In the case $\mathbf{b} < \mathbf{0}$ the set W is simply connected and contains the origin. The Nehari manifold $\mathcal{N}$ is on a positive distance to zero and divides H^1 into two sets W and V (see [5], Section 5). In this case the critical energy constant d is defined as

$$d = \inf_{u \in \mathcal{N}} J(u), \qquad 0 < d < \infty.$$

If $\mathbf{b} > \mathbf{0}$, $\mathbf{a} < \mathbf{0}$ and $\mathbf{a}^2 - \frac{(\mathbf{p}+2)^2}{\mathbf{p}+1}\mathbf{b} > \mathbf{0}$, then Nehari manifold $\mathcal{N}$ has a more complicated structure (see [5], Section 6), i.e.

$$\mathcal{N} = \mathcal{N}_+ \cup \mathcal{N}_- \cup \mathcal{N}_0, \quad \text{where}$$

$$\mathcal{N}_\pm = \{\lambda_\pm(u)u:\ u \in \mathrm{H}^1,\ \parallel u \parallel_{\mathrm{H}^1} \neq 0,\ G(u) > 0,\ I(\lambda_\pm(u)u) = 0\},$$

$$\mathcal{N}_0 = \{\lambda_0(u)u:\ u \in \mathrm{H}^1,\ \parallel u \parallel_{\mathrm{H}^1} \neq 0,\ G(u) = 0,\ I(\lambda_0(u)u) = 0\}.$$

Here $G(u)$ is the discriminant of the equation

$$I(\lambda u) = \lambda^2\left(\| u \|_{\mathrm{H}^1}^2 + a|\lambda|^p \int_{\mathbb{R}} |u|^{p+2}\, dx + b\lambda^{2p} \int_{\mathbb{R}} |u|^{2p+2}\, dx\right) = 0 \tag{10}$$

with respect to $|\lambda|^p$, namely

$$G(u) = \| u \|_{\mathrm{L}^{p+2}}^{2(p+2)} - \frac{4b}{a^2} \| u \|_1^2 \| u \|_{\mathrm{L}^{2p+2}}^{2p+2}. \tag{11}$$

With $\lambda_{\pm}(u)$, $\lambda_{+}(u) < \lambda_{-}(u)$ we denote the positive roots of (10) for $G(u) > 0$ and with $\lambda_0(u)$ we denote the unique positive root of (10) for $G(u) = 0$. In this case $d = -\infty$ (see [5]) and we introduce a new constant d_+ and a set $\widetilde{W}$, which are analogs of d and W:

$$d_+ = \inf_{u \in \mathcal{N}_+ \cup \mathcal{N}_0} J(u), \qquad 0 < d_+ < \infty,$$

$$\widetilde{W} = \mathrm{H}^1 \backslash \bar{K},$$

where $K = \{\lambda u\colon\ u \in \mathrm{H}^1,\ \ G(u) > 0 \ \ and\ \ \lambda > \lambda_+(u)\}$.

As in [5] we give an equivalent definition of the set $\widetilde{W}$ with easily checkable conditions, namely

$$\widetilde{W} = W_+ \cup W_0 \cup W_- \cup \{0\}, \quad \text{where}$$

$$W_- = \{u \in \mathrm{H}^1\colon\ G(u) < 0\},$$

$$W_+ \cup W_0 = \{u \in \mathrm{H}^1 \backslash \{0\}\colon\ G(u) \geq 0 \quad \text{and}$$

$$S(u) = \| u \|_{\mathrm{L}^{p+2}}^{p+2} + G^{1/2}(u) + (2/a) \| u \|_{\mathrm{H}^1}^2 < 0\}. \tag{12}$$

If $\mathbf{b > 0}$, $\mathbf{a > 0}$ then $I(\lambda u) \geq \lambda^2 \|u\|_{\mathrm{H}^1}^2 > 0$ for every $\lambda \neq 0$ and $\|u\|_{\mathrm{H}^1} \neq 0$, i.e. $\mathcal{N} = \emptyset$ and $W \equiv \mathrm{H}^1$.

Theoretical results

In this section we prove the global existence or finite time blow-up of the weak solutions of (5), (6) depending on the energy of the initial data and the values of a, b in (3).

Theorem 3.1 *Suppose* $u_0 \in H^1$, $u_1 \in H^1$ *and* $(-\Delta)^{-1/2}u_1 \in L^2$. *Then*

(i) for $b < 0$ *and* $E(0) < d$:

(a) if $u_0 \in W$ *then problem (5), (6) has a unique global solution defined for every* $t \in [0, \infty)$;

(b) if $u_0 \in V$ *then the solution of problem (5), (6) blows up in a finite time.*

(ii) for $b > 0$, $a < 0$, $a^2 - \frac{(p+2)^2}{p+1} b > 0$ *and* $E(0) < d_+$:

if $u_0 \in \widetilde{W}$ *then problem (5), (6) has a unique global solution defined for every* $t \in [0, \infty)$.

(iii) for $b > 0$:

(a) if $a > 0$ *or* $a < 0$, $a^2 - \frac{(p+2)^2}{p+1} b < 0$ *then problem (5), (6) has a unique global solution with uniformly bounded* H^1 *norm for every* $t \in [0, \infty)$.

(b) if $a < 0$ *and* $a^2 - \frac{(p+2)^2}{p+1} b = 0$ *then problem (5), (6) has a unique global solution which possibly blows up for* $t \to \infty$.

Sketch of the proof. The proofs of $(i)_a$ and (ii) follow the ideas of the standard and the non-standard potential well methods respectively. These methods are based on the invariance of the sets W, V and $\widetilde{W}$ under the flow of the equation (see the proof of Theorems 5.2, 5.3 and 6.3 in [5]). In case $(i)_a$, from the conservation law (8) we have $I(u(t)) > 0$ for every $t \in [0, T_m)$ and

$$E(0) = E(t) = \frac{1}{2}\left(\beta_2\left\|(-\Delta)^{-1/2}u_t\right\|^2 + \beta_1 \parallel u_t \parallel^2 + \beta_3 \parallel u_{xt} \parallel^2\right)$$

$$+\frac{p}{2(p+2)} \parallel u \parallel_1^2 - \frac{bp}{2(p+1)(p+2)} \parallel u \parallel_{L^{2p+2}}^{2p+2} + \frac{1}{p+2} I(u)$$

$$\geq \frac{p}{2(p+2)} \parallel u \parallel_1^2,$$

$$\text{i.e.} \quad \parallel u \parallel_1^2 \leq \frac{2(p+2)}{p} E(0) \text{ for every } t \in [0, T_m). \tag{13}$$

In case (ii) we get $I(\lambda^* u) > 0$ for $\lambda^* = (2/(p+2))^{1/p}$ (see the proof of Theorem 6.4 in [5]) and from (8) we have

$$E(0) \geq J(u) = \frac{1}{2(\lambda^*)^2} I(\lambda^* u) + \frac{bp^2}{2(p+1)(p+2)^2} \parallel u \parallel_{L^{2p+2}}^{2p+2}$$

$$\geq \frac{bp^2}{2(p+1)(p+2)^2} \parallel u \parallel_{L^{2p+2}}^{2p+2},$$

$$\text{i.e.} \quad \parallel u \parallel_{L^{2p+2}}^{2p+2} \leq \frac{2(p+2)^2(p+1)}{bp^2} E(0).$$

From the Hölder inequality we get

$$\frac{|a|}{p+2}\int_{\mathbb{R}} |u|^{p+2}\, dx \leq \frac{|a|}{p+2}\left(\int_{\mathbb{R}} u^2\, dx \int_{\mathbb{R}} |u|^{2p+2}\, dx\right)^{1/2}$$

$$\leq \frac{1}{4}\int_{\mathbb{R}} u^2\, dx + \frac{4a^2}{(p+2)^2}\int_{\mathbb{R}} |u|^{2p+2}\, dx.$$

Applying again the conservation law and the above estimate we obtain

$$E(0) \geq J(u) \geq \frac{1}{2} \parallel u \parallel_1^2 - \frac{1}{4} \parallel u \parallel_1^2 - \frac{4a^2}{(p+2)^2} \parallel u \parallel_{L^{2p+2}}^{2p+2}$$

$$\geq \frac{1}{4} \parallel u \parallel_1^2 - \frac{8(p+1)a^2}{p^2 b} E(0),$$

$$\text{i.e.} \quad \parallel u \parallel_1^2 \leq 4\left(1 + \frac{8(p+1)a^2}{p^2 b}\right) E(0) \text{ for every } t \in [0, T_m). \tag{14}$$

From (13) and (14) and the local existence result ([20], Theorem 2.3) it follows that the solutions in cases $(i)_a$ and (ii) are globally defined, i.e. $T_m = \infty$.

The proof of the finite time blow up in case $(i)_b$ is based on the concavity method of Levine by means of the function

$$\phi(t) = \beta_2\left\|(-\Delta)^{-1/2}u\right\|^2 + \beta_1 \parallel u \parallel^2 + \beta_3 \parallel u_x \parallel^2$$

(see for more details the proof of Theorem 5.3 in [5]).

In case $(iii)_a$, from the conservation law (8) we have

$$E(0) = E(t) \geq \frac{c^2-1}{2c^2} \parallel u \parallel_1^2, \quad \text{if} \quad a^2 - \frac{(p+2)^2}{p+1} b < 0,$$

where $c^2 = \frac{(p+2)^2 b}{(p+1)a^2} > 1$, and

$$E(0) = E(t) \geq \frac{1}{2} \parallel u \parallel_1^2, \quad \text{if} \quad a > 0, \; b > 0.$$

The rest of the proof follows from the local existence result ([20], Theorem 2.3).

In case $(iii)_b$, since $a^2 - \frac{(p+2)^2}{p+1} b = 0$ from (8) we have

$$E(0) = E(t) \geq \frac{1}{2}(\parallel u_t \parallel^2 + \parallel u_x \parallel^2).$$

From the inequality

$$\parallel u^2(t) \parallel^2 \leq \left(\parallel u_0 \parallel^2 + t^2 \sup_{0 \leq s \leq t} \parallel u_t(s) \parallel^2\right) \leq 2(\parallel u_0 \parallel^2 + 2t^2 E(0))$$

for every $T > 0$ and for every $t \in (0, T)$ it follows that $\parallel u(t) \parallel_1^2 \leq c_1 + c_2 T^2$ with constants c_1, c_2 independent of t. Thus, in case (iii), a finite time blow-up of the solution is impossible from the local existence result. Theorem 3.1 is proved.

From Theorem 3.1 it is clear that the explicit value of the constants d and d_+ is crucial for the theoretical and numerical analysis of problem (5),(6).

In our previous paper [5] we evaluate the critical energy constants d and d_+ by means of the ground state solutions $\psi(x)$ of (5), i.e. the solution to the stationary problem

$$\psi''(x) = \psi(x) + a|\psi(x)|^p\psi(x) + b|\psi(x)|^{2p}\psi(x), \tag{15}$$

$$\psi(x) \to 0 \quad \text{as} \quad |x| \to \infty.$$

If $b < 0$ or $b > 0$, $a < 0$, $a^2 - \frac{(p+2)^2}{p+1}b > 0$ then the unique (up to the sign and translation of the coordinate system) solution $\psi(x)$ of (15) is defined by

$$\psi(x) = (p+2)^{1/p}\left(\sqrt{a^2 - \frac{(p+2)^2}{p+1}b}\ \cosh(px) - a\right)^{-1/p} \tag{16}$$

(see Section 4 in [5]). Since $d = J(\psi) = J(\psi) - \frac{1}{2}I(\psi)$ for $p = 2$ we get the following explicit expression for d:

$$\begin{aligned} d|_{p=2} = \; & -\frac{384ab}{(3a^2-16b)^2} - \frac{3a}{8b(3a^2-16b)^2}(9a^4 - 256b^2 - 192a^2b) \\ & -\frac{12a}{(3a^2-16b)} + \frac{\sqrt{3}(3a^2-16b)}{32(-b)^{3/2}}\left(\frac{\pi}{2} + \arctan\frac{a}{4}\sqrt{\frac{3}{-b}}\right). \end{aligned} \tag{17}$$

If

$$b > 0, \quad a < 0 \quad and \quad a^2 - \frac{4(p+2)(p+1)}{3p+2}b \geq 0, \tag{18}$$

then $\psi \in \mathcal{N}_+$ and the explicit value of d_+ is given by:

$$\begin{aligned} d_+|_{p=2} = \; & -\frac{384ab}{(3a^2-16b)^2} - \frac{3a}{8b(3a^2-16b)^2}(9a^4 - 256b^2 - 192a^2b) \\ & -\frac{12a}{(3a^2-16b)} + \frac{\sqrt{3}(3a^2-16b)}{64b^{3/2}}\ln\frac{\sqrt{3}a+4\sqrt{b}}{\sqrt{3}a-4\sqrt{b}}. \end{aligned} \tag{19}$$

Numerical results

We discretize (5), (6) using a regular mesh with step h in a sufficiently large space interval $[-L_1, L_2]$: $x_i = -L_1 + ih,\ h = (L_1 + L_2)/N,\ i = 0, \dots, N$. For approximation of Eq. (5) we propose the following family of finite difference schemes depending on the real parameter θ:

$$B\left(\frac{v_i^{n+1}-2v_i^n+v_i^{n-1}}{\tau^2}\right) - \Lambda v_i^n + \Lambda^2 v_i^n = \Lambda\left(\frac{F(v_i^{n+1})-F(v_i^{n-1})}{v_i^{n+1}-v_i^{n-1}}\right), \qquad (20)$$

$$B = \beta_2 I - (\beta_1 + \theta\tau^2)\Lambda + (\beta_3 + \theta\tau^2)\Lambda^2.$$

Here: τ is a time-step; $t^n = n\tau$; v_i^n is a discrete approximation to u at (x_i, t^n); I is the identity operator; Λ and Λ^2 are the standard three and five-point discretizations of the second and fourth derivatives respectively; function $F(u)$ is defined in (9). The approximations of initial conditions (6) and asymptotic boundary conditions are given by

$$v_i^0 = u_0(x_i),$$

$$v_i^1 = u_0(x_i) + \tau u_1(x_i) \qquad (21)$$

$$+\frac{\tau^2}{2}(\beta_2 I - \beta_1\Lambda + \beta_3\Lambda^2)^{-1}(\Lambda u_0 - \Lambda^2 u_0 + \Lambda f(u_0))(x_i),$$

$$v_i^{n+1} = 0,\ \Lambda v_i^{n+1} = 0 \quad \text{for} \quad i = 0, N.$$

The nonlinear scheme with respect to v_i^{n+1} (20) is linearized using the method of successive iterations. In our calculations we consider the already computed function v^n as an initial approximation to the sought function v^{n+1}. The iterations stop when the relative error between two successive iterations is less than a given tolerance ε. Usually 3-8 nonlinear iterations are sufficient for convergence with tolerance 10^{-13}. The resulting systems of linear algebraic equations are five-diagonal with constant matrix coefficients. To solve them we apply a special kind of non-monotonic Gaussian elimination with pivoting.

Following the ideas and technique from [3] we can prove that the scheme (20), (21) is unconditionally stable for $\theta \geq 1/4$ and has second order of convergence in space and time. Another important feature of the proposed numerical scheme is its conservativeness. We introduce the discrete energy functional $E_h(v^n)$ which approximates to the energy functional $E(t)$ in (8) (the subscript i is omitted):

$$E_h(v^n) = -\beta_2\langle\Lambda^{-1}v_t^n, v_t^n\rangle + (\beta_1 + \tau^2(\theta - 1/4))\langle v_t^n, v_t^n\rangle$$

$$-(\beta_3 + \tau^2(\theta - 1/4))\langle\Lambda v_t^n, v_t^n\rangle$$

$$+1/4\langle v^n + v^{n+1} - \Lambda(v^n + v^{n+1}), v^n + v^{n+1}\rangle$$

$$+\langle F(v^{n+1}) + F(v^n), 1\rangle\rangle,$$

where $\langle\cdot,\cdot\rangle$ is a standard discrete scalar product at fixed time and $v_t^n = (v^{n+1} - v^n)/\tau$. In a similar way to [3] we prove that the discrete energy $E_h(v^n)$ is conserved in time:

$$E_h(v^n) = E_h(v^0), \quad n = 1,2,....$$

Later on we present numerical experiments for nonlinearity $f(u) = au^3 + bu^5$ and initial conditions

$$u_0(x) = -\delta\psi(x), \qquad \psi(x) = 2\left(\sqrt{a^2 - \frac{16}{3}b}\,\cosh(2x) - a\right)^{-1/2},$$
$$u_1(x) = 0. \tag{22}$$

Here $\psi(x)$ is the ground state solution defined by (16) with $p = 2$, and δ is a positive constant. The numerical experiments are performed for $\beta_i = 1$, $i = 1,2,3$. We use a regular mesh defined in [-300,300] with space step $h = 0.01$ and time step $\tau = 0.01$. In addition we perform mesh refinement analysis. In (20) we set the parameter $\theta = 1/2$.

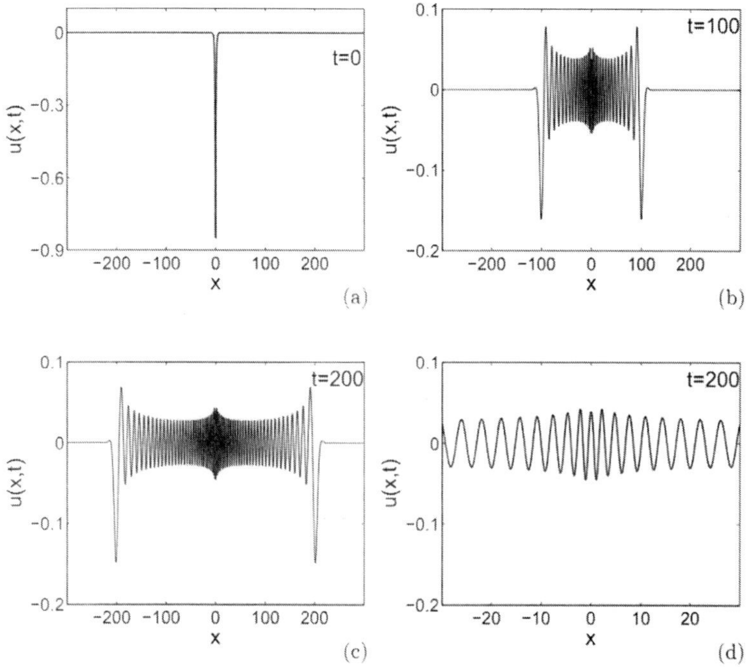

Figure 1: Profiles of the numerical solution $u(x,t)$ of (5), (22) computed for $a=2$, $b=-5$, $\delta=0.8$ at different evolution times: (a) t=0; (b) t=100; (c) t=200; (d) zoom of the graph in (c) around the origin.

Example 1. The constants a, b, δ are fixed to $a=2$, $b=-5$, $\delta=0.8$. Since $b<0$ from formula (17) we get the exact value of the critical energy constant d, namely $d\approx 1.01419855$. For this initial data $E_h(0)\approx 0.81846496<d$, while the computed value of $I(u_0)$, $\bar{I}(u_0)\approx 1.25865583>0$. The graphics of the numerical solution at different evolution times are presented in Fig. 1(a) - Fig. 1(c). Figure 1(d) shows a zoom of the graphic in Fig. 1(c) around the origin. We see that the numerical solution stays bounded within a large fixed time interval, i.e. the behavior of the numerical solution is fully consistent with the statements of Theorem $3.1(i)_a$.

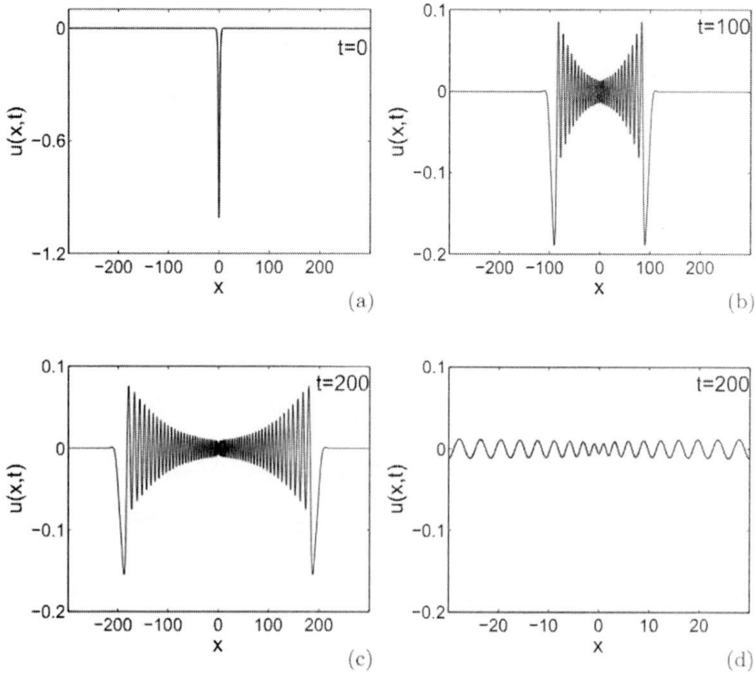

Figure 2: Profiles of the numerical solution $u(x,t)$ of (5), (22) computed for $a=-2, b=0.65$, $\delta=0.8$ at different evolution times: (a) t=0; (b) t=100; (c) t=200; (d) zoom of the graph in (c) around the origin.

Example 2. We choose $a=-2$, $b=0.65$, $\delta=0.8$. The constants a and b satisfy the conditions in case (ii) of Theorem 3.1, i.e. $b>0$, $a<0$, $a^2-\frac{(p+2)^2}{p+1}b>0$. In order to apply Theorem 3.1(ii) we need to check both conditions: $E(0)<d_+$ and $u_0\in\widetilde{W}$. Since the constants a and b satisfy the additional conditions (18) we obtain the exact value of the new critical energy constant d_+ by formula (19), i.e. $d_+\approx 0.87861623$. For this choice of constants we have: $E_h(0)\approx 0.80376024<d_+$, and the computed values of $G(u)$ and $S(u)$ are: $\overline{G}(u_0)\approx 0.66437578>0$, and $\overline{S}(u_0)\approx -1.02680152<0$. From (12) it follows that $u_0\in\widetilde{W}$ and therefore the initial data fulfill the conditions of Theorem 3.1(ii). Fig. 2 shows plotting of the graphics of the numerical solution at different evolution times. It is clear that through the evolution the numerical solutions stay bounded within a large fixed time interval.

We have performed numerical tests for a variety of values of the parameters β_1, β_2, β_3, a, and b. The behaviour of all numerical solutions is in full agreement with Theorem 3.1.

Acknowledgments

This work is partially supported by the Bulgarian Science Fund under grant DDVU 02/71.

Bibliography

[1] C. I. Christov, *An energy-consistent dispersive shallow-water model*, Wave motion **34** (2001), 161-174.

[2] F. Falk, E.W. Laedke, K. H. Spatschek, Stability of solitary-wave pulses in shape-memory alloys, Phys. Rev. B **36**(1987), no. 6, 3031-3041.

[3] N. Kolkovska, M. Dimova, *A new conservative finite difference scheme for Boussinesq paradigm equation*, Cent. Eur. J. Math. **10** (2012), no. 3, 1159-1171.

[4] N. Kutev, N. Kolkovska, M. Dimova, *Global existence of Cauchy problem to Boussinesq paradigm equation*, Comput. Math. Appl. **65** (2013), 500-511.

[5] N. Kutev, N. Kolkovska, M. Dimova, Global existence to generalized Boussinesq equation with combined power-type nonlinearities, J. Math. Anal. Appl. **410** (2014), 427-444.

[6] Y. Liu, R. Xu, Potential well method for Cauchy problem of generalized double dispersion equations, J. Math. Anal. Appl. **338** (2008), 1169–1187.

[7] G.A. Maugin, Nonlinear Waves in Elastic Crystals, Oxford University Press, 1999.

[8] A. Porubov, Amplification of nonlinear strain waves in solids, World Scientific, 2003

[9] P. Rosenau, *Dynamics of dense discrete systems*, Progress of Theoretical Physics **79** (1988), no. 5, 1028-1042

[10] G. Schneider, C.E. Wayne, *Kawahara dynamics in dispersive media*, Phys D, **152--153** (2001), 384-394.

[11] H. Taskesen, N. Polat, *Global existence for a double dispersive sixth order Boussinesq equation*, Contemporary Analysis and Applied Mathematics **1** (2013), no. 1, 60-69.

[12] H. Taskesen, N. Polat, A. Ertas, *On global solutions for the Cauchy problem of a Boussinesq-type equation*, Abstract and Applied Analysis, **2012** (2012), Article ID 535031, 10 p.

[13] Jiang Xiaoli, Ding Yunhua, Liu Yacheng and Xu Runzhang, Global well-posedness of Cauchy problem for 1D generalized Boussinesq equations at both low initial energy level and critical initial energy level, AIP Conf. Proc. **1389** (2011), 1640-1643.

[14] R. Xu, Chauchy problem of generalized Boussinesq equation with conbined power-type nonlinearities, Math. Methods in Appl. Sci. **34** (2011), 2318-2328.

[15] R. Xu, Y. Liu, B. Liu, The Cauchy problem for a class of the multidimensional Boussinesq-type equation, Nonlinear Analysis **74** (2011), 2425-2437.

[16] Y. Wang, C. Mu, *Blow-up and scattering of solution for a generalized Boussinesq equation*, Applied Mathematics and Computation **188** (2007), 1131-1141.

[17] Y. Wang, C. Mu, Y. Wu, *Decay and scattering of solutions for a generalized Boussinesq equation*, J. Differential Equations **247** (2009), 2380-2394.

[18] S. Wang, H. Xue, *Global solution for a generalized Boussinesq equation*, Applied Mathematics and Computation **204** (2008), 130-136.

[19] S. Wang, G. Xu, *The Cauchy problem for the Rosenau equation*, Nonlinear Analysis **71** (2009), 456-466.

[20] S. Xia, J. Yuan, Existence and scattering of small solutions to a Boussinesq type equation of sixth order, Nonlinear Analysis **73** (2010), 1015-1027.

INVESTIGATION OF TWO NUMERICAL SCHEMES FOR THE 2D BOUSSINESQ PARADIGM EQUATION IN A MOVING FRAME COORDINATE SYSTEM

DANIELA VASILEVA AND NATALIA KOLKOVSKA

Introduction

Different versions of the Boussinesq Equation (BE) are used to model surface waves in a shallow fluid layer. One important feature of BE is the balance between nonlinearity and dispersion, which leads to solutions of the type "permanent waves" (solitons) [1]. Accurate derivation of the Boussinesq system combined with an approximation, which reduces the full model to a single equation, leads to the Boussinesq Paradigm Equation (BPE) [2]:

$$u_{tt} = \Delta[u - F(u) + \beta_1 u_{tt} - \beta_2 \Delta u], \quad F(u) := \alpha u^2, \tag{1}$$

where u is the surface elevation of the wave, $\beta_1, \beta_2 > 0$ are two dispersion coefficients, and $\alpha > 0$ is an amplitude parameter. The main difference between (1) and the original Boussinesq Equation is the presence of a term proportional to $\beta_1 \neq 0$ called ``rotational inertia".

It has been recently shown that the 2D BPE admits stationary translating localized solutions [3, 4, 5], which can be obtained approximately using finite differences, a perturbation technique, or the Galerkin spectral method. Results for their time behavior and structural stability are presented in [6, 7, 8, 9, 10] using different numerical methods. All results are in good agreement and show that the 2D localized soliton solutions with initial data from [5] are not stable -- they either disperse in the form of ring-waves or blow up in finite time (depending on the parameters). In [11] we continued our investigations using a moving frame coordinate system. This allows us to keep the localized structure in the center of the coordinate system, to use a small computational box and to compute the solution for larger times. The same unstable behavior of the 2D localized soliton solutions was demonstrated there. This motivated us to investigate in this paper the time behavior of known stable 1D solitons, for when they are taken as initial data for the 2D problem.

In the present work we study the properties of the finite difference schemes (FDS) proposed in [11]. First the equation (1) is transformed in order to keep the soliton in the center of the new coordinate system. We set $z := x - ct$, where c is the velocity of the stationary propagating soliton. Then we obtain the following equation for $U(z, y, t) := u(z + ct, y, t)$ in the moving frame coordinate system

$$(I - \beta_1 \widetilde{\Delta}) \frac{\partial^2 U}{\partial t^2} - 2c \frac{\partial^2 U}{\partial t\, \partial z} + 2c\beta_1 \frac{\partial^2}{\partial t\, \partial z} \widetilde{\Delta} U = -\beta_2 \frac{\partial^4 U}{\partial y^4} - (\beta_2 - \beta_1 c^2) \frac{\partial^4 U}{\partial z^4}$$

$$-(2\beta_2 - \beta_1 c^2) \frac{\partial^4 U}{\partial y^2\, \partial z^2} + \frac{\partial^2 U}{\partial y^2} + (1 - c^2) \frac{\partial^2 U}{\partial z^2} - \alpha \widetilde{\Delta} F(U). \qquad (2)$$

Here I is the identity operator and $\widetilde{\Delta}$ stands for the Laplace operator with respect to variables z and y.

The fourth order spatial derivatives on the right hand side of (2) constitute a fourth order elliptic operator if $c^2 < \beta_2 / \beta_1$. In a similar way the second order derivatives generate a second order elliptic operator if $c^2 < 1$. Therefore we suppose in the following that the velocity c satisfies the restriction $c^2 < min(1, \beta_2/\beta_1)$.

In the next section we describe two numerical schemes for solving BPE in the moving frame coordinate system. The first uses central finite differences for the mixed $(\frac{\partial^2}{\partial t\, \partial z})$ derivative, while the second uses upwind finite differences. The grid is quasi-uniform and the truncation error of both FDS is second order in space and time. The properties of the numerical methods corresponding to the linearized BPE are studied in Section 3. It is proved that the proposed FDS are stable with respect to initial data, if $c^2 < min(1, \beta_2/\beta_1)$.

In Section 4 some known unstable and stable 1D solutions are investigated numerically in 1D and 2D settings. The results demonstrate the second order of convergence of the schemes. The stable 1D solutions preserve their shape for very large times. The corresponding numerical solutions of the 2D problem are stable in relatively narrow domains in the y −direction, but seem to be not stable in relatively wide domains in the y −direction.

Numerical method for solving BPE in the moving frame coordinate system

We introduce the new dependent function W,

$$W(z,y,t) := U - \beta_1 \widetilde{\Delta} U \tag{3a}$$

and substituting it in Eq. (2) we get the following equation

$$W_{tt} - 2cW_{tz} + c^2 W_{zz} = \frac{\beta_2}{\beta_1}\widetilde{\Delta} W + \frac{\beta_1 - \beta_2}{\beta_1^2}(U - W) - \Delta F(U). \tag{3b}$$

Thus we obtain a system consisting of an equation for U, Eq. (3a), and an equation for W: Eq. (3b).

The following implicit time stepping can be designed for the system (3):

$$\frac{W_{ij}^{n+1} - 2W_{ij}^{n} + W_{ij}^{n-1}}{\tau^2} - c\frac{V^z[W_{ij}^{n+1} - W_{ij}^{n-1}]}{\tau} + \frac{c^2}{2}\Lambda^{zz}[W_{ij}^{n+1} + W_{ij}^{n-1}]$$

$$= \frac{\beta_2}{2\beta_1}\Lambda[W_{ij}^{n+1} + W_{ij}^{n-1}] + \frac{\beta_1 - \beta_2}{2\beta_1^2}[U_{ij}^{n+1} - W_{ij}^{n+1} + U_{ij}^{n-1} - W_{ij}^{n-1}]$$

$$-\alpha\Lambda G(U_{ij}^{n+1}, U_{ij}^{n}, U_{ij}^{n-1}), \tag{4a}$$

$$U_{ij}^{n+1} - \beta_1 \Lambda U_{ij}^{n+1} = W_{ij}^{n+1}, \quad i = 0, \dots, N_x + 1, \ j = 0, \dots, N_y + 1. \tag{4b}$$

Here τ is the time increment, the nonlinear term U^2 is approximated by

$$G(U_{ij}^{n+1}, U_{ij}^{n}, U_{ij}^{n-1}) = (U_{ij}^{n})^2, \tag{5}$$

or

$$G(U_{ij}^{n+1}, U_{ij}^{n}, U_{ij}^{n-1}) = \left[(U_{ij}^{n+1})^2 + U_{ij}^{n+1}U_{ij}^{n-1} + (U_{ij}^{n-1})^2\right]/3, \quad (6)$$

$\Lambda = \Lambda^{zz} + \Lambda^{yy}$ stands for the difference approximation of the Laplace operator $\widetilde{\Delta}$ on a non-uniform grid, for example

$$\Lambda^{zz}W_{ij} = \frac{2W_{i-1j}}{h_{i-1}^{z}(h_i^z + h_{i-1}^z)} - \frac{2W_{ij}}{h_i^z h_{i-1}^z} + \frac{2W_{i+1j}}{h_i^z(h_i^z + h_{i-1}^z)},$$

and V^z is the central difference approximation of $\frac{\partial}{\partial z}$ defined by

$$V_z W_{ij} = \frac{h_{i-1}^z W_{i+1j}}{h_i^z(h_i^z + h_{i-1}^z)} - \frac{h_i^z W_{i-1j}}{h_{i-1}^z(h_i^z + h_{i-1}^z)} + \frac{(h_i^z - h_{i-1}^z)W_{ij}}{h_i^z h_{i-1}^z}.$$

Another way to approximate W_{zt} for $c > 0$ is by the following "upwind" approximation

$$W_{zt} \approx \frac{W_{i+1j}^{n+1} - W_{ij}^{n+1} - W_{i+1j}^{n} + W_{ij}^{n}}{2\tau h_i^z} + \frac{W_{ij}^{n} - W_{i-1j}^{n} - W_{ij}^{n-1} + W_{i-1j}^{n-1}}{2\tau h_{i-1}^z}.$$

The values of the functions sought at the $(n-1)$-st and n-th time stages are considered as known when computing the $(n+1)$-st stage. When the approximation G of the nonlinear term U^2 is obtained using (6), the system (4) is linearized using internal (Picard) iterations [12], i.e. we perform successive iterations for W and U on the $(n+1)$-st stage, starting with initial data from the already computed n-th stage. Usually 5-10 nonlinear iterations are sufficient for convergence with tolerance 10^{-14}. This kind of linearization for the Boussinesq equation was proposed and investigated in [13] in order to conserve the energy of the numerical solution.

The following quasi-uniform grid is used in the z −direction

$$z_i = \sinh[\hat{h}_z(i - n_z)], \; z_{N_z+1-i} = -z_i, i = n_z + 1, \dots, N_z + 1, \; z_{n_z} = 0,$$

where N_z is an odd number, $n_z = (N_z + 1)/2$, $\hat{h}_z = D_z/N_z$, and D_z is selected so as to have a large enough computational region. The grid in the y −direction is uniform.

In order to test the properties of the numerical method, we take the known one-dimensional solutions of the problem (see [2]) as initial data:

$$U(z,y,0) = U^{sech}(z) = (1-c^2)\frac{1.5}{\alpha}\, sech^{\,2}\!\left(\frac{z}{2}\sqrt{(1-c^2)/(\beta_2-\beta_1 c^2)}\right). \tag{7}$$

The second initial condition is chosen as $\frac{\partial}{\partial t}U(z,y,0) = 0$.

Because of the localization of the wave profile in the z −direction, the boundary conditions in this direction can be set equal to zero, when the size of the computational domain in the z −direction is large enough. Neumann $(\frac{\partial U}{\partial y} = \frac{\partial W}{\partial y} = 0)$ or Dirichlet $(U = U^{sech},\ W = U - \beta_1 \widetilde{\Delta} U^{sech})$ boundary conditions are imposed in the y −direction.

The coupled system of equations (4) is solved by the Bi-Conjugate Gradient Stabilized Method with ILU preconditioner [14]. In most examples we set the tolerance for the iterative solution of the linear systems to be 10^{-14}.

Analysis of the finite difference schemes

In this section we study the stability of the linear schemes corresponding to both FDS. We analyze the first scheme, with central finite difference approximation to the first space derivative. The analysis of the scheme with "upwind" approximation leads to stability results similar to the results formulated in Theorem 3.1.

First, we eliminate function W from (3a) and (3b) and obtain one FDS for the discrete function U:

$$(I-\beta_1\Lambda)\left(\frac{U_{ij}^{n+1}-2U_{ij}^{n}+U_{ij}^{n-1}}{\tau^2}\right) - c(I-\beta_1\Lambda)V^z\frac{(U_{ij}^{n+1}-U_{ij}^{n-1})}{\tau} =$$

$$-\Lambda\left(\beta_2\Lambda^{yy}\overline{U}_{ij}^{n} + (\beta_2-\beta_1 c^2)\Lambda^{zz}\overline{U}_{ij}^{n}\right) + (1-c^2)\Lambda^{zz}\overline{U}_{ij}^{n} + \Lambda^{yy}\overline{U}_{ij}^{n} \tag{8}$$

$$-\alpha\Lambda G(U_{ij}^{n+1}, U_{ij}^{n}, U_{ij}^{n-1}).$$

Here $\overline{U}_{ij}^{n} = 0.5(U_{ij}^{n+1} + U_{ij}^{n-1})$. Further, we omit the notion (ij) whenever possible.

We consider the space of discrete functions, which vanish on the computational boundary with Dirichlet boundary conditions, and define the discrete scalar product in this space in the standard way. Note that the operator $-\Lambda$ is a self-adjoint and positive definite operator in this space. We perform the analysis of the numerical method supposing that the mesh is uniform in both directions.

We rewrite FDS (8) in the operator form

$$B\left(\frac{U_{ij}^{n+1}-2U_{ij}^{n}+U_{ij}^{n-1}}{\tau^2}\right) + R\left(\frac{U_{ij}^{n+1}-U_{ij}^{n-1}}{2\tau}\right) + AU_{ij}^{n}$$

$$= -\alpha\Lambda G(U_{ij}^{n+1}, U_{ij}^{n}, U_{ij}^{n-1}),$$

where operators B, R and A are defined as

$$B = I - \beta_1\Lambda + 0.5\tau^2(-(1-c^2)\Lambda^{zz} - \Lambda^{yy} + \beta_2\Lambda\Lambda^{yy} + (\beta_2 - \beta_1 c^2)\Lambda\Lambda^{zz}),$$

$$R = -2c(I - \beta_1\Lambda)V^{z},$$

$$A = -(1-c^2)\Lambda^{zz} - \Lambda^{yy} + \beta_2\Lambda\Lambda^{yy} + (\beta_2 - \beta_1 c^2)\Lambda\Lambda^{zz}.$$

We study the properties of operators included in this FDS. It is straightforward to prove that operators A and B are self-adjoint and positive definite operators -- $(AU, U) > 0$ and $(BU, U) > 0$. The discrete energy of the operator R is equal to zero -- $(RU, U) = 0$. The operator $B - \frac{\tau^2}{4}A$ is also positive definite.

We analyze the linear problem corresponding to (8) :

$$B\frac{U_{ij}^{n+1}-2U_{ij}^{n}+U_{ij}^{n-1}}{\tau^2} + R\left(\frac{U_{ij}^{n+1}-U_{ij}^{n-1}}{2\tau}\right) + AU_{ij}^{n} = g_{ij}^{n}, \quad (9)$$

with a given function g independent of U. Using the stability theory from Chapter 6 in [15], we get the following theorem:

Theorem 3.1 *Let $c^2 < min(1, \beta_2/\beta_1)$. Then the finite difference scheme (9) is stable with respect to the initial data and the function g. Moreover, the following estimate holds*

$$((-\Lambda)U^{(n)}, U^{(n)}) \leq C[((-\Lambda)U^{(0)}, U^{(0)}) + ((-\Lambda)^{-1}BU_t^{(0)}, U_t^{(0)})$$

$$+ \textstyle\sum_{m=1}^{n} \tau(g^m, g^m)]$$

with constant C independent of U, h and τ.

Using the stability estimates from Theorem 3.1 we can prove convergence results for the numerical solution of a nonlinear scheme. These investigations are very similar to those given in [16] and we omit them here.

For the upwind finite difference scheme the operator B is

$$B = (I - c\frac{\tau h}{2}\Lambda^{zz})(I - \beta_1\Lambda) + \frac{\tau^2}{2}(-(1 - c^2)\Lambda^{zz} - \Lambda^{yy} + \beta_2\Lambda\Lambda^{yy} +$$

$$(\beta_2 - \beta_1 c^2)\Lambda\Lambda^{zz}).$$

The operators A and R do not change. The operators B and $B - \frac{\tau^2}{4}A$ are self-adjoint and positive definite, and Theorem 3.1 holds in this case as well.

Numerical experiments

In [17] it is proved that the solution (7) of the 1D generalized Boussinesq equation ($\beta_2 = \alpha = 1,\ \beta_1 = 0$) is stable when $0.25 < c^2 < 1$. In [18] nonlinear instability is obtained when $c^2 \leq 0.25$. In our numerical experiments we observe a similar behavior for the 1D BPE, i.e. when $\beta_1 = 1$. That is why in the next examples solutions for $\beta_1 = \beta_2 = \alpha = 1$ are investigated.

Example 1. The first example is for the phase speed $c = 0.4$, i.e. the 1D soliton solution (7) should be unstable. First we compute the numerical solution using (5) for the approximation of the nonlinear term and Neumann boundary conditions in the y −direction. The basic grid for the 2D solution has 256×16 cells in the region $[-50,50] \times [-1,1]$, the time step is $\tau = 0.2$. We compare the solution in the 2D setting with the 1D solution, computed on a grid with 256 cells in the interval $[-50,50]$ and with the same time step $\tau = 0.2$. The maximum of the difference between the numerical and the exact solution $\delta(U) := \max|U - U^{\text{sech}}|$ is shown in Table 1. Both numerical solutions (2D and 1D) blow-up at time $t \approx 76$. The central difference and upwind approximations of W_{tz} lead to practically the same values in the numerical solution for $t \leq 72$. After that the solutions start to grow very fast and then the upwind approximation of W_{tz} leads to larger errors in the numerical solution. The comparison between the 2D and 1D settings shows that both produce the same errors even near the blow-up time.

Table 1: The difference $\delta(U) := \max|U - U^{\text{sech}}|$ between the exact and the numerical solution for $c = 0.4$

	central differences		upwind differences	
t	2D solution	1D solution	2D solution	1D solution
8	1.57e-3	1.57e-3	1.57e-3	1.57e-3
16	5.59e-3	5.59e-3	5.60e-3	5.60e-3
24	1.43e-2	1.43e-2	1.44e-2	1.44e-2
32	3.15e-2	3.15e-2	3.16e-2	3.16e-2
40	6.45e-2	6.45e-2	6.49e-2	6.49e-2
48	1.29e-1	1.29e-1	1.30e-1	1.30e-1
56	2.63e-1	2.63e-1	2.66e-1	2.66e-1
64	5.93e-1	5.93e-1	6.01e-1	6.01e-1
72	2.41	2.41	2.51	2.51
75	56.61	56.61	83.54	83.54
76	1.22e+11	1.22e+11	1.48e+14	1.48e+14

The evolution of the 1D solution is shown in Fig.1. The evolution of the cross-sections of the 2D solution is the same, because the 2D solution keeps its constant behavior in the y −direction.

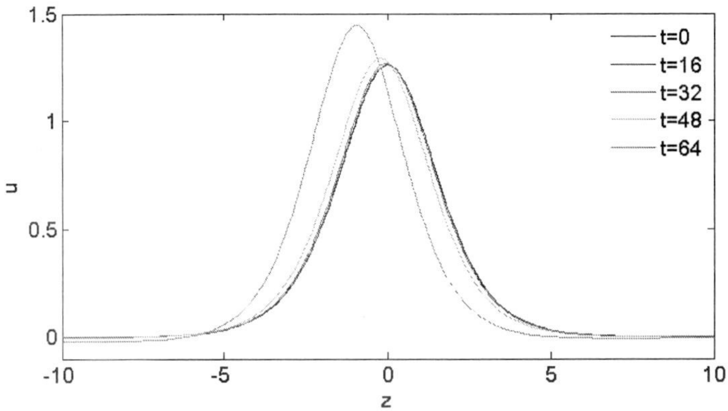

Figure 1: Evolution of the 1D solution for $c = 0.4$.

Example 2. For the phase speed $c = 0.6$ we investigate the 1D and 2D solutions, on the same grids and with the same boundary conditions as in the previous example. Both solutions are stable (we computed them up to time $t = 10^6$, see Table 2). The difference between the exact and the numerical solution is one and the same for the 2D and the 1D solution of the problem. The central difference and upwind approximations of W_{tz} lead to practically the same values in the numerical solution. Both approximations of the nonlinear term (see (5) and (6)), as well as a stronger tolerance for the iterative solution of the linear systems (10^{-28}) also lead to practically the same results.

Table 2: The difference $\delta(U) := \max|U - U^{\text{sech}}|$ between the exact and the numerical solution for $c = 0.6$

	central differences		upwind differences	
t	2D solution	1D solution	2D solution	1D solution
10^2	5.15e-3	5.15e-3	5.16e-2	5.16e-3
10^3	9.73e-3	9.73e-3	9.66e-3	9.66e-3
10^4	1.80e-2	1.80e-2	1.83e-2	1.83e-2
10^5	1.05e-2	1.05e-2	6.94e-3	6.94e-3
10^6	1.01e-2	1.01e-2	8.49e-3	8.49e-3

We also investigate the convergence of the 2D solution on three grids -- the basic grid has 256×8 cells, $\tau = 0.2$, the finer has 512×16 cells, $\tau = 0.1$, the finest has 1024×32 cells, $\tau = 0.05$. The order of convergence l is computed as $l = \log_2 \frac{\delta(U_{k-1})}{\delta(U_k)}$, where k is the number of the corresponding grid. The results in Table 3 show second order of convergence of the numerical solution.

Table 3: The difference $\delta(U)$ between the exact and the numerical solution, and the order of convergence l for $c = 0.6$

		$t = 100$		$t = 200$		$t = 400$	
τ	$N_x \times N_y$	$\delta(U)$	l	$\delta(U)$	l	$\delta(U)$	l
central differences, approximation (5) of the nonlinear term							
0.2	256×8	5.15e-3		9.50e-3		1.67e-2	
0.1	512×16	1.30e-3	1.99	2.46e-3	1.95	4.83e-3	1.79
0.05	1024×32	3.26e-4	2.00	6.20e-4	1.99	1.25e-3	1.95
central differences, approximation (6) of the nonlinear term							
0.2	256×8	5.15e-3		9.50e-3		1.67e-2	
0.1	512×16	1.30e-3	1.99	2.46e-3	1.95	4.83e-3	1.79
0.05	1024×32	3.26e-4	2.00	6.20e-4	1.99	1.25e-3	1.95
upwind differences, approximation (5) of the nonlinear term							
0.2	256×8	5.16e-3		9.52e-3		1.67e-2	
0.1	512×16	1.30e-3	1.99	2.46e-3	1.95	4.83e-3	1.79
0.05	1024×32	3.26e-4	2.00	6.20e-4	1.99	1.25e-3	1.95
upwind differences, approximation (6) of the nonlinear term							
0.2	256×8	5.15e-3		9.52e-3		1.67e-2	
0.1	512×16	1.30e-3	1.99	2.46e-3	1.95	4.83e-3	1.79
0.05	1024×32	3.26e-4	2.00	6.20e-4	1.99	1.25e-3	1.95

The evolution of the 1D solution is shown in Fig.2. The evolution of the cross-sections of the 2D solution is the same, because the 2D solution keeps its constant behavior in the y −direction.

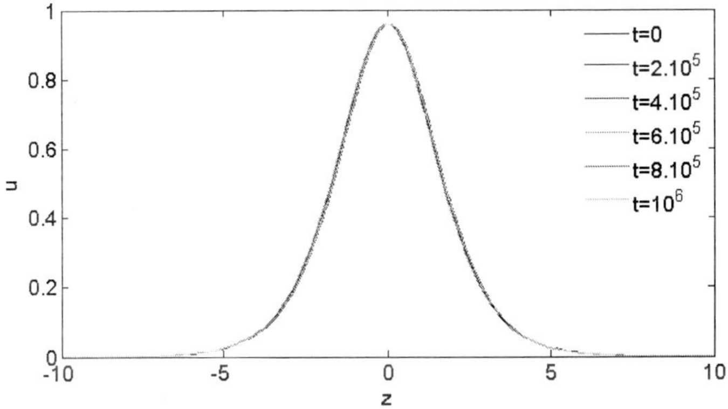

Figure 2: Evolution of the 1D solution for $c = 0.6$.

Finally, let us note that size of the domain in the y –direction is very important for the stability of the 2D solution. For example, if $y \in [-10,10]$, the solution of the 2D problem preserves its shape up to $t \approx 140$, but after that the wave develops a nonconstant behaviour in the y-direction, starts to grow and the solution blows up at $t \approx 175$. The number of maxima, which appear in the y –direction, strongly depends on the size of the domain in this direction. This behavior of the 2D solutions will be examined in detail in a forthcoming article.

Example 3. Results for $c = 0.9$ and $(x, y) \in [-50,50] \times [-1,1]$ are shown in Table 4, Table 5 and Fig.3. The behaviour of the solution is quite similar to that for $c = 0.6$: second order convergence of the solution is demonstrated; there is no practical difference between both discretizations of the mixed derivative W_{tz}, both approximations of the nonlinear term; and the solution does not depend on the tolerance prescribed for the solution of the linear systems, arising after the discretisation. The difference between the exact and the approximate solution $\delta(U)$ is one and the same for 1D and 2D settings of the problem. The solutions also preserve their shape for very large times ($t = 10^6$).

Table 4: The difference $\delta(U) := \max|U - U^{\text{sech}}|$ between the exact and the numerical solution for $c = 0.9$

	central differences		upwind differences	
t	2D solution	1D solution	2D solution	1D solution
10^2	2.09e-4	2.09e-4	2.09e-4	2.09e-4
10^3	1.64e-3	1.64e-3	1.63e-3	1.63e-3
10^4	3.27e-3	3.27e-3	3.29e-3	3.29e-3
10^5	3.42e-3	3.42e-3	3.33e-3	3.33e-3
10^6	1.81e-3	1.81e-3	2.86e-3	2.86e-3

Table 5: The difference $\delta(U)$ between the exact and the numerical solution, and the order of convergence l for $c = 0.9$

		$t = 400$		$t = 800$		$t = 1200$	
τ	$N_x \times N_y$	$\delta(U)$	l	$\delta(U)$	l	$\delta(U)$	l
0.4	128×8	2.86e-3		5.37e-3		7.05e-3	
0.2	256×16	7.40e-4	1.95	1.33e-3	2.01	2.05e-3	1.78
0.1	512×32	1.88e-4	1.98	3.37e-4	1.99	5.33e-4	1.94

In order to show second order convergence for larger times, we need either to use very fine grids in the x −direction or to impose Dirichlet boundary conditions in the y −direction. That is why in the next table we present results with Dirichlet boundary conditions in the y −direction. As can be seen, the errors in this case are much lower and second order convergence is demonstrated up to time $t = 10^6$.

Table 6: The difference $\delta(U)$ between the exact and the numerical solution, and the order of convergence l for $c = 0.9$, in the case of Dirichlet boundary conditions

		$t = 10^2$		$t = 10^4$		$t = 10^6$	
τ	$N_x \times N_y$	$\delta(U)$	l	$\delta(U)$	l	$\delta(U)$	l
central differences							
0.8	64×4	2.56e-4		2.46e-4		2.50e-4	
0.4	128×8	6.31e-5	2.02	6.33e-5	1.96	6.48e-5	1.92
0.2	256×16	1.59e-5	1.99	1.60e-5	1.98	1.55e-5	2.06
upwind differences							
0.8	64×4	2.52e-4		2.42e-4		2.49e-4	
0.4	128×8	6.35e-5	1.99	6.23e-5	1.97	6.31e-5	1.98
0.2	256×16	1.59e-5	2.00	1.58e-5	1.98	1.57e-5	2.01

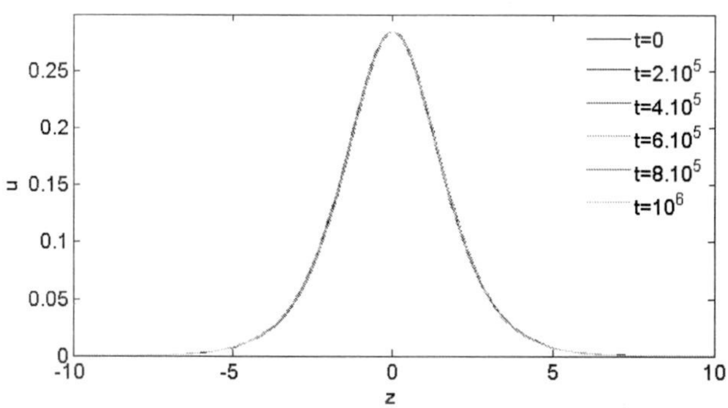

Figure 3: Evolution of the 1D solution for $c = 0.9$.

As in the previous example, the stability of the 2D solution strongly depends on the size of the domain in the y −direction. For example, when $y \in [-10,10]$ and Neumann boundary conditions are imposed in the y −direction, the solution loses its constant behaviour in the y −direction at $t \approx 1600$, begins to grow and blows up for $t \approx 2200$.

Conclusions

The moving frame coordinate system helps us to keep the soliton in the center of the coordinate system, where the grid is much finer. It also reduces the effects of reflection from the boundaries and allows us to use a small computational box and to compute the solution for very large times.

We have investigated two finite difference schemes for the time evolution of the solutions of BPE in a moving frame coordinate system. We have proved that the proposed finite difference schemes for the linearized BPE are stable with respect to initial data, if the velocity c satisfies $c^2 < min(1, \beta_2/\beta_1)$.

The numerical experiments we have presented demonstrate the second order of convergence of the schemes. Both discretizations of the mixed derivative W_{tz}, as well as both approximations of the nonlinear term, lead to practically one and the same results. The stable 1D solutions preserve themselves for very large times. The solutions of the 2D problem for the same parameters and in small intervals for y also preserve their shape for very large times.

But the solutions of the 2D problem in large intervals for y seem to be unstable: the waves preserve their shape for relatively long intervals of time (depending on the parameters), but after that the waves lose their constant behavior in the y −direction, the solutions start to grow and blow up. Most probably this effect is due to the instability of the exact solution of the 2D differential problem in wide domains, even when the corresponding 1D solution is stable. As was mentioned, this behavior of the 2D solutions will be a subject of future research.

Acknowledgment. This work has been partially supported by Grant DDVU02/71 from the National Science Fund provided by the Ministry of Education, Youth, and Science of the Republic of Bulgaria.

Bibliography

[1] Boussinesq, J.V.: Théorie des ondes et des remous qui se propagent le long d'un canal rectangulaire horizontal, en communiquant au liquide contenu dans ce canal des vitesses ensiblement pareilles de la surface au fond. *Journal de Mathématiques Pures et Appliquées*, **17** (1872), 55-108

[2] Christov, C.I.: An energy-consistent Galilean-invariant dispersive shallow-water model. *Wave Motion*, **34** (2001), 161-174

[3] Christou, M.A., Christov, C.I.: Fourier-Galerkin method for 2D solitons of Boussinesq equation. *Math. Comput. Simul.*, **74** (2007), 82-92

[4] Christov, C.I.: Numerical implementation of the asymptotic boundary conditions for steadily propagating 2D solitons of Boussinesq type equations. *Math. Comp. Simulat.*, **82** (2012), 1079-1092

[5] Christov, C.I., Choudhury, J.: Perturbation solution for the 2D Boussinesq Equation. *Mech. Res. Commun.*, **38** (2010), 274-281

[6] Chertock, A., Christov, C.I., Kurganov, A.: Central-upwind schemes for the Boussinesq paradigm equation. *Proc. 4th Russian-German Advanced Research Workshop on Comp. Science and High Performance Computing, NNFM*, **113** (2011), 267-281

[7] Christov, C.I., Kolkovska, N., Vasileva, D.: On the numerical simulation of unsteady solutions for the 2D Boussinesq paradigm equation, *Lecture Notes Computer Science*, **6046** (2011), 386-394

[8] Christov, C.I., Kolkovska, N., Vasileva, D.: Numerical investigation of unsteady solutions for the 2D Boussinesq paradigm equation, *5th Annual Meeting of the Bulgarian Section of SIAM, Sofia, Bulgaria, BGSIAM'10 Proceedings*, 2011, 11-16

[9] Dimova, M., Vasileva, D.: Comparison of two numerical approaches to Boussinesq paradigm equation, *Lecture Notes Computer Science*, **8236** (2013), 255-262

[10] Kolkovska, N., Angelow, K.: A multicomponent alternating direction method for numerical solving of Boussinesq paradigm equation, *Lecture Notes Computer Science*, **8236** (2013), 371-378

[11] Vasileva, D., Christov, C.I.: On the numerical investigation of unsteady solutions for the 2D Boussinesq paradigm equation in a moving frame coordinate system, *6th Annual Meeting of the Bulgarian Section of SIAM, Sofia, Bulgaria, BGSIAM'11 Proceedings*, 2012, 103-108

[12] Ames, W.F.: Nonlinear Partial Differential Equations in Engineering. Academic Press (1965)

[13] Christov, C.I., Velarde, M.G.: Inelastic interaction of Boussinesq solitons. *J. Bifurcation & Chaos*, **4** (1994), 1095-1112

[14] van der Vorst, H.: Iterative Krylov methods for large linear systems. *Cambridge Monographs on Appl. and Comp. Math.*, **13** (2009)

[15] Samarsky, A.: *The Theory of Difference Schemes*. Marcel Dekker Inc., New York, (2001)

[16] Kolkovska, N, Dimova, M.: A new conservative finite difference scheme for Boussinesq paradigm equation. *Central Eur. J. Math.*, **10** (2012), 1159-1171

[17] Bona, J., Sachs, R.: Global existence of smooth solutions and stability of solitary waves for a generalized Boussinesq equation, *Comm. Math. Phys.*, **118** (1988), 15-29

[18] Liu, Y: Instability of solitary waves for generalized Boussinesq equations, *J. Dynamics Differential Equations* **5** (1993), 537-558

CHAPTER EIGHT

MATHEMATICAL PHYSICS EQUATIONS AND APPLICATIONS IN INDUSTRY

EXAMPLES OF G-STRAND EQUATIONS

DARRYL D. HOLM AND ROSSEN I. IVANOV

Introduction

The G-strand equations for a map $\mathbb{R} \times \mathbb{R}$ into a Lie group G are associated with a G-invariant Lagrangian. The Lie group manifold is also the configuration space for the Lagrangian. The G-strand itself is the map $g(t,s): \mathbb{R} \times \mathbb{R} \to G$, where t and s are the independent variables of the G-strand equations. The Euler-Poincaré reduction of the variation principle leads to a formulation where the dependent variables of the G-strand equations take values in the corresponding Lie algebra $\mathfrak{g}$ and its co-algebra, $\mathfrak{g}^*$, with respect to the pairing provided by the variation derivatives of the Lagrangian.

We review examples of two G-strand constructions, including matrix Lie groups and the Diffeomorphism group. In some cases the G-strand equations are completely integrable $1+1$ Hamiltonian systems that admit soliton solutions.

Our presentation is based on our previous works [14, 8, 9, 12, 10] and aims to illustrate the G-strand construction with two simple but instructive examples:

(i) $SO(3)$-strand integrable equations for Lax operators, quadratic in the spectral parameter;

(ii) $\mathrm{D}iff(\mathbb{R})$-strand equations. These equations are in general non-integrable; however they admit solutions in $2+1$ space-time with singular support (e.g., peakons). The one- and two-peakon equations obtained from the $\mathrm{D}iff(\mathbb{R})$-strand equations can be solved analytically, and potentially they can be applied in the theory of image registration. Our example is with a system which is a $2+1$ generalization of the Hunter-Saxton equation.

Ingredients of Euler-Poincaré theory for left G-Invariant Lagrangians

Let G be a Lie group. A map $g(t,s)\colon \mathbb{R}\times\mathbb{R}\to G$ has two types of tangent vectors, $\dot{g} := g_t \in TG$ and $g' := g_s \in TG$. Assume that the Lagrangian density function $L(g,\dot{g},g')$ is left G-invariant. The left G-invariance of L permits us to define $l\colon \mathfrak{g}\times\mathfrak{g}\to\mathbb{R}$ by

$$L(g,\dot{g},g') = L(g^{-1}g, g^{-1}\dot{g}, g^{-1}g') \equiv l(g^{-1}\dot{g}, g^{-1}g').$$

Conversely, this relation defines for any reduced Lagrangian $l = l(u,v)\colon \mathfrak{g}\times\mathfrak{g}\to\mathbb{R}$ a left G-invariant function $L\colon TG\times TG\to\mathbb{R}$ and a map $g(t,s)\colon \mathbb{R}\times\mathbb{R}\to G$ such that

$$u(t,s) := g^{-1}g_t(t,s) = g^{-1}\dot{g}(t,s) \quad and \quad v(t,s) := g^{-1}g_s(t,s) = g^{-1}g'(t,s).$$

Lemma 2.1 *The left-invariant tangent vectors* $u(t,s)$ *and* $v(t,s)$ *at the identity of G satisfy*

$$v_t - u_s = -\,ad_u v\,. \tag{1}$$

Proof. The proof is standard and follows from equality of cross derivatives $g_{ts} = g_{st}$.

Equation (1) is usually called a *zero-curvature relation.*

Theorem 2.2 (Euler-Poincare theorem for left-invariant Lagrangians)

With the preceding notation, the following two statements are equivalent:

1. Variation principle on $TG\times TG$ $\delta\int_{t_1}^{t_2} L(g(t,s),\dot{g}(t,s),g'(t,s))\,ds\,dt = 0$ holds, for variations $\delta g(t,s)$ of $g(t,s)$ vanishing at the end points in t and s. The function $g(t,s)$ satisfies Euler-Lagrange equations for L on G, given by

$$\frac{\partial L}{\partial g} - \frac{\partial}{\partial t}\frac{\partial L}{\partial g_t} - \frac{\partial}{\partial s}\frac{\partial L}{\partial g_s} = 0.$$

2. The constrained variation principle[8]

$$\delta \int_{t_1}^{t_2} l(u(t,s), v(t,s))\, ds\, dt = 0$$

holds on $\mathfrak{g} \times \mathfrak{g}$, using variations of $u := g^{-1}g_t(t,s)$ and $v := g^{-1}g_s(t,s)$ of the forms

$$\delta u = \dot{w} + \mathrm{ad}_u w \quad and \quad \delta v = w' + \mathrm{ad}_v w \,,$$

where $w(t,s) := g^{-1}\delta g \in \mathfrak{g}$ vanishes at the endpoints. The *Euler-Poincaré* equations hold on $\mathfrak{g}^* \times \mathfrak{g}^*$ (*G-strand equations*)

$$\frac{d}{dt}\frac{\delta l}{\delta u} - ad_u^* \frac{\delta l}{\delta u} + \frac{d}{ds}\frac{\delta l}{\delta v} - ad_v^* \frac{\delta l}{\delta v} = 0, \qquad \partial_s u - \partial_t v = [\, u, v\,] = \mathrm{ad}_u v$$

where $(\mathrm{ad}^* : \mathfrak{g} \times \mathfrak{g}^* \to \mathfrak{g}^*)$ is defined via $(\mathrm{ad} : \mathfrak{g} \times \mathfrak{g} \to \mathfrak{g})$ in the dual pairing $\langle \cdot , \cdot \rangle : \mathfrak{g}^* \times \mathfrak{g} \to \mathbb{R}$ by,

$$\left\langle \mathrm{ad}_u^* \frac{\delta \ell}{\delta u}, v \right\rangle_{\mathfrak{g}} = \left\langle \frac{\delta \ell}{\delta u}, \mathrm{ad}_u v \right\rangle_{\mathfrak{g}}.$$

In 1901 Poincaré in his famous work proves that, when a Lie algebra acts locally transitively on the configuration space of a Lagrangian mechanical system, the well known Euler-Lagrange equations are equivalent to a new system of differential equations defined on the product of the configuration space with the Lie algebra. These equations are now called in his honor Euler-Poincaré equations. In modern language the content of Poincaré's article [13] is presented for example in [7, 5]. An English translation of the article [13] can be found as Appendix D in [7].

G-strand equations on matrix Lie algebras

Denoting $m := \delta\ell/\delta u$ and $n := \delta\ell/\delta v$ in $\mathfrak{g}^*$, the *G*-strand equations become

---

[8]As with the basic Euler-Poincaré equations, this is not strictly a variational principle in the same sense as the standard Hamilton's principle. It is more like the Lagrange d'Alembert principle, because we impose the stated constraints on the variations allowed.

$$m_t + n_s - \mathrm{ad}^*_u m - \mathrm{ad}^*_v n = 0 \quad and \quad \partial_t v - \partial_s u + \mathrm{ad}_u v = 0.$$

For G a semisimple *matrix Lie group* and $\mathfrak{g}$ its *matrix Lie algebra* these equations become

$$\begin{aligned} m_t^T + n_s^T + \mathrm{ad}_u m^T + \mathrm{ad}_v n^T = &\quad 0, \\ \partial_t v - \partial_s u + \mathrm{ad}_u v = &\quad 0 \end{aligned} \tag{2}$$

where the ad-invariant pairing for semi-simple matrix Lie algebras is given by

$$\langle m\,, n\rangle = \frac{1}{2}\mathrm{tr}(m^T n),$$

the transpose gives the map between the algebra and its dual $(\,\cdot\,)^T : \mathfrak{g} \to \mathfrak{g}^*$. For semisimple matrix Lie groups, the adjoint operator is the matrix commutator. Examples are studied in [14, 9, 12].

Lie-Poisson Hamiltonian formulation

Legendre transformation of the Lagrangian $\ell(u, v)$: $\mathfrak{g} \times \mathfrak{g} \to \mathbb{R}$ yields the Hamiltonian $h(m, v)$: $\mathfrak{g}^* \times \mathfrak{g} \to \mathbb{R}$

$$h(m, v) = \langle m\,, u\rangle - \ell(u, v)\,. \tag{3}$$

Its partial derivatives imply

$$\frac{\delta l}{\delta u} = m\,, \quad \frac{\delta h}{\delta m} = u \quad and \quad \frac{\delta h}{\delta v} = -\,\frac{\delta \ell}{\delta v} = v.$$

These derivatives allow one to rewrite the Euler-Poincaré equation solely in terms of momentum m as

$$\begin{aligned} \partial_t m &= \mathrm{ad}^*_{\delta h/\delta m}\, m + \partial_s \frac{\delta h}{\delta v} - \mathrm{ad}^*_v\, \frac{\delta h}{\delta v}\,, \\ \partial_t v &= \partial_s \frac{\delta h}{\delta m} - \mathrm{ad}_{\delta h/\delta m}\, v\,. \end{aligned} \tag{4}$$

Assembling these equations into Lie-Poisson Hamiltonian form gives,

$$\frac{\partial}{\partial t}\begin{bmatrix} m \\ v \end{bmatrix} = \begin{bmatrix} ad^*_{(.)}m & \partial_s - \mathrm{ad}^*_v \\ \partial_s + \mathrm{ad}_v & 0 \end{bmatrix} \begin{bmatrix} \delta h/\delta m \\ \delta h/\delta v \end{bmatrix}. \tag{5}$$

The Hamiltonian matrix in equation (5) also appears in the Lie-Poisson brackets for Yang-Mills plasmas, for spin glasses and for perfect complex fluids, such as liquid crystals.

Example: Integrable $SO(3)$ G-strands with Lax operator, quadratic in the spectral parameter

The integrable $SO(3)$ G-strand system was studied in [14] by linking it to the integrable P-chiral model of [15, 1, 4]. The Lax operator of the system in [14] is linear in the spectral parameter. In this example we will provide the zero curvature representation of a $SO(3)$ G-Strand equation and thereby prove its integrability, where the Lax operator is quadratic in the spectral parameter.

The hat map $\hat{}: \ (\boldsymbol{so}(\mathbf{3}), [\cdot,\cdot]) \to (\mathbb{R}^{\mathbf{3}},\times)$

The Lie algebra $(\mathfrak{so}(3), [\cdot,\cdot])$ with matrix commutator bracket $[\,\cdot\,,\cdot\,]$ maps to the Lie algebra $(\mathbb{R}^3,\times)$ with vector product $\times$, by the linear isomorphism

$$\mathbf{u}:=(u^1,u^2,u^3)\in\mathbb{R}^3 \mapsto \hat{u}:=\begin{bmatrix} 0 & -u^3 & u^2 \\ u^3 & 0 & -u^1 \\ -u^2 & u^1 & 0 \end{bmatrix} \in so(3)\,.$$

In matrix and vector components, the linear isomorphism is $\hat{u}_{ij}:= -\varepsilon_{ijk}u^k$. Equivalently, this isomorphism is given by $\hat{u}\mathbf{v}=\mathbf{u}\times\mathbf{v}$ for all $\mathbf{u},\mathbf{v}\in\mathbb{R}^3$. This is the hat map $\hat{}: (so(3),[\cdot,\cdot]) \to (\mathbb{R}^3,\times)$, which holds for the skew-symmetric 3×3 matrices in the matrix Lie algebra $\mathfrak{so}(3)$.

One may verify the following useful formulas for $\mathbf{u},\mathbf{v},\mathbf{w}\in\mathbb{R}^3$:

$$(\mathbf{u}\times\mathbf{v})\hat{} = \hat{u}\,\hat{v} - \hat{v}\,\hat{u} =: [\hat{u},\hat{v}]\,,$$

$$[\hat{u},\hat{v}]\,\mathbf{w} = (\mathbf{u}\times\mathbf{v})\times\mathbf{w}\,,$$

$$((\mathbf{u}\times\mathbf{v})\times\mathbf{w})\,\widehat{}\; = [[\hat{u},\hat{v}]\,,\hat{w}]\,,$$

$$\mathbf{u}\cdot\mathbf{v} = -\tfrac{1}{2}trace(\hat{u}\,\hat{v}) =: \langle\,\hat{u}\,,\hat{v}\,\rangle\,,$$

in which the dot product of vectors is also the natural pairing of 3×3 skew-symmetric matrices.

The $SO(3)$ G-Strand system in $\mathbb{R}^3$ vector form

By using the hat map, $\mathfrak{so}(3)\to\mathbb{R}^3$, the matrix G-Strand system for $SO(3)$ [14] may be written in $\mathbb{R}^3$ vector form by following the analogy with the Euler rigid body in standard notation,

$$\begin{aligned} \partial_t\Pi + \Omega\times\Pi - \partial_s\Xi - \Gamma\times\Xi &= 0\,,\\ \partial_t\Gamma - \partial_s\Omega - \Gamma\times\Omega &= 0\,, \end{aligned} \tag{6}$$

where $\Omega := O^{-1}\,\partial_t O\in\mathfrak{so}(3)$ and $\Pi := \partial\ell/\,\partial\Omega\in\mathfrak{so}(3)^*$ are the body angular velocity and momentum, while $\Gamma := O^{-1}\,\partial_s O\in\mathfrak{so}(3)$ and $\Xi = -\,\partial\ell/\,\partial\Gamma\in\mathfrak{so}(3)^*$ are the body angular strain and stress. These G-Strand equations for $\mathfrak{g}=\mathfrak{so}(3)$ may be expressed in Lie-Poisson Hamiltonian form in terms of vector operations in $\mathfrak{so}(3)\times\mathbb{R}^3$ as,

$$\frac{\partial}{\partial t}\begin{bmatrix}\Pi\\ \Gamma\end{bmatrix} = \begin{bmatrix}\Pi\times & \partial_s+\Gamma\times\\ \partial_s+\Gamma\times & 0\end{bmatrix}\begin{bmatrix}\delta h/\delta\Pi = \Omega\\ \delta h/\delta\Gamma = \Xi\end{bmatrix}. \tag{7}$$

This Hamiltonian matrix yields a *Lie--Poisson bracket* defined on the dual of the semidirect-product Lie algebra $\mathfrak{so}(3)\rtimes\mathbb{R}^3$ with a two-cocycle given by ∂_s. Namely,

$$\begin{aligned} \{f,h\} &= \int\,[\delta f/\delta\Pi\,,\quad \delta f/\delta\Gamma]\cdot\begin{bmatrix}\Pi\times & \partial_s+\Gamma\times\\ \partial_s+\Gamma\times & 0\end{bmatrix}\begin{bmatrix}\delta h/\delta\Pi\\ \delta h/\delta\Gamma\end{bmatrix}ds\\ &= \int\,\Big(-\Pi\cdot\frac{\delta f}{\delta\Pi}\times\frac{\delta h}{\delta\Pi} - \Gamma\cdot\Big(\frac{\delta f}{\delta\Pi}\times\frac{\delta h}{\delta\Gamma} - \frac{\delta h}{\delta\Pi}\times\frac{\delta f}{\delta\Gamma}\Big)\\ &+ \qquad \frac{\delta f}{\delta\Pi}\partial_s\frac{\delta h}{\delta\Gamma} + \frac{\delta f}{\delta\Gamma}\partial_s\frac{\delta h}{\delta\Pi}\Big)\;ds. \end{aligned} \tag{8}$$

Dual variables are Π dual to $\mathfrak{so}(3)$ and Γ dual to $\mathbb{R}^3$. For more information about Lie-Poisson brackets, see [11].

The $\mathbb{R}^3$ G-Strand equations (6) combine two classic ODEs due separately to Euler and Kirchhoff into a single PDE system. The $\mathbb{R}^3$ vector representation of $\mathfrak{so}(3)$ implies that $ad^*_\Omega\Pi = -\,\Omega\times\Pi = -\mathrm{ad}_\Omega\Pi$, so the corresponding Euler-Poincaré equation has a ZCR. To find its integrability conditions, we set

$$L := \lambda^2 A + \lambda\Pi + \Gamma \quad and \quad M := \lambda^2 B + \lambda\Xi + \Omega\,, \tag{9}$$

and compute the conditions in terms of Π Ω, Ξ, Γ and the constant vectors A and B that are required to write the vector system (6) in zero-curvature form,

$$\partial_t L - \partial_s M - L\times M = 0\,. \tag{10}$$

By direct substitution of (9) into (10) and equating the coefficient of each power of λ to zero, one finds

$$\begin{array}{ll}\lambda^4 & : A\times B = 0\\ \lambda^3 & : A\times\Xi - B\times\Pi = 0\\ \lambda^2 & : A\times\Omega - B\times\Gamma + \Pi\times\Xi = 0\\ \lambda^1 & : \Pi\times\Omega + \Gamma\times\Xi = \partial_t\Pi - \partial_s\Xi \quad (EP\ equation)\\ \lambda^0 & : \Gamma\times\Omega = \partial_t\Gamma - \partial_s\Omega \quad (compatibility)\end{array} \tag{11}$$

where A and B are taken as constant nonzero vectors. These imply the following relationships

$$\begin{array}{ll}\lambda^4 & : A = \alpha B\\ \lambda^3 & : A\times(\Xi - \Pi/\alpha) = 0 \Rightarrow \Xi - \Pi/\alpha = \beta A\\ \lambda^2 & : A\times(\Omega - \Gamma/\alpha) = \Xi\times\Pi = \beta A\times\Pi\end{array}\,. \tag{12}$$

We solve equations (12) for the diagnostic variables Ξ and Ω, as

$$\Xi - \frac{1}{\alpha}\Pi = \beta A \quad and \quad \Omega - \frac{\Gamma}{\alpha} - \beta\Pi = \gamma A\,, \tag{13}$$

where α, β, γ are real scalars.

The conserved quantities can be evaluated from the Lax representation:

$$\begin{aligned} H_{-1} &= \int (A\cdot\Pi)ds \\ H_0 &= \int \left(\frac{(A\times\Pi)^2}{2|A|^2} + A\cdot\Gamma\right)ds \\ H_1 &= \int \left(\Pi\cdot\Gamma - (A\cdot\Pi)\left(\frac{(A\times\Pi)^2}{2|A|^4} + \frac{(A\cdot\Gamma)}{|A|^2}\right)\right)ds. \end{aligned} \tag{14}$$

Let us now try to find a Hamiltonian h as a linear combination of H_{-1}, H_0 and H_1, i.e. $h = c_{-1}H_{-1} + c_0H_0 + c_1H_1$ for some numerical constants c_k. We need to satisfy the two relations

$$\begin{aligned} \frac{\delta h}{\delta\Pi} &= c_{-1}\frac{\delta H_{-1}}{\delta\Pi} + c_0\frac{\delta H_0}{\delta\Pi} + c_1\frac{\delta H_1}{\delta\Pi} = \frac{1}{\alpha}\Gamma + \beta\Pi + \gamma A \equiv \Omega, \\ \frac{\delta h}{\delta\Gamma} &= c_{-1}\frac{\delta H_{-1}}{\delta\Gamma} + c_0\frac{\delta H_0}{\delta\Gamma} + c_1\frac{\delta H_1}{\delta\Gamma} = \frac{1}{\alpha}\Pi + \beta A \equiv \Xi \end{aligned} . \tag{15}$$

Comparing the scalar coefficients arising in front of the vectors A, Π and Γ from both sides of (15) we obtain

$$\begin{aligned} \alpha &= c_1 = 1, \\ \beta &= c_0 - \frac{A\cdot\Pi}{|A|^2}, \\ \gamma &= c_{-1} - c_0\frac{A\cdot\Pi}{|A|^2} - \frac{\Pi^2}{2|A|^2} + \frac{3(A\cdot\Pi)^2}{2|A|^4} - \frac{A\cdot\Gamma}{|A|^2}, \end{aligned} \tag{16}$$

wherec_{-1} and c_0 are arbitrary real constants. The most general Hamiltonian of course could contain combinations of all conserved quantities, with coefficients c_k where possibly $k > 1$. In such cases the expressions for α, β and γ will of course contain terms, related to the higher conserved quantities, entering the Hamiltonian.

The $\mathbf{Diff}(\mathbb{R})$-strand system

The constructions described briefly in the previous sections can be easily generalized in cases where the Lie group is the group of the Diffeomorphisms. Consider a Hamiltonian which is a right-invariant bilinear form $H(u, v)$. The manifold $\mathcal{M}$ where u and v are defined is $\mathbb{S}^1$ or in the case when the class of smooth functions vanishing rapidly at $\pm\infty$ is considered, we will allow $\mathcal{M} \equiv \mathbb{R}$. Let us introduce the notation

$u(g(x)) \equiv u \circ g$. Let us further consider a one-parametric family of diffeomorphisms, $g(x,t) \in Diff(\mathcal{M})$, by defining the t - evolution as

$$\dot{g} = u(g(x,t),t), \qquad g(x,0) = x, \qquad \text{i.e.} \qquad \dot{g} = u \circ g \in T_g G; \tag{17}$$

$u = \dot{g} \circ g^{-1} \in \mathfrak{g}$, where $\mathfrak{g}$, the corresponding Lie-algebra, is the algebra of vector fields, Vect$(\mathcal{M})$. Now we recall the following result:

Theorem 6.1 *(A. Kirillov, 1980, [2, 3]) The dual space of* $\mathfrak{g}$ *is a space of distributions but the subspace of local functionals, called the regular dual* $\mathfrak{g}^*$ *is naturally identified with the space of quadratic differentials* $m(x)dx^2$ *on* $\mathcal{M}$. *The pairing is given for any vector field* $u\partial_x \in Vect(\mathcal{M})$ *by*

$$\langle m dx^2, u\partial_x \rangle = \int_{\mathcal{M}} m(x)u(x)dx.$$

The co-adjoint action coincides with the action of a diffeomorphism on the quadratic differential:

$$Ad_g^*: \quad m dx^2 \mapsto m(g) g_x^2 dx^2$$

and

$$ad_u^* = 2u_x + u\partial_x.$$

Indeed, a simple computation shows that

$$\langle \mathrm{ad}^*_{u\,\partial_x} m dx^2, v\,\partial_x \rangle = \langle m dx^2, [u\,\partial_x, v\,\partial_x] \rangle = \int_{\mathcal{M}} m(u_x v - v_x u)dx =$$

$$\int_{\mathcal{M}} v(2mu_x + um_x)dx = \langle (2mu_x + um_x)dx^2, v\,\partial_x \rangle,$$

i.e. $\mathrm{ad}^*_u m = 2u_x m + u m_x$.

The $Diff(\mathbb{R})$-strand system arises when we choose $G = Diff(\mathbb{R})$. For a two-parametric group we have two tangent vectors

$$\partial_t g = u \circ g \quad and \quad \partial_s g = v \circ g \,,$$

where the symbol $\circ$ denotes composition of functions.

In this right-invariant case, the G-strand PDE system with reduced Lagrangian $\ell(u, v)$ takes the form,

$$\begin{aligned} \frac{\partial}{\partial t}\frac{\delta \ell}{\delta u} + \frac{\partial}{\partial s}\frac{\delta \ell}{\delta v} &= -\operatorname{ad}^*_u \frac{\delta \ell}{\delta u} - \operatorname{ad}^*_v \frac{\delta \ell}{\delta v} \,, \\ \frac{\partial v}{\partial t} - \frac{\partial u}{\partial s} &= \operatorname{ad}_u v \,. \end{aligned} \tag{18}$$

Of course, the distinction between the maps $(u, v): \mathbb{R} \times \mathbb{R} \to \mathfrak{g} \times \mathfrak{g}$ and their pointwise values $(u(t,s), v(t,s)) \in \mathfrak{g} \times \mathfrak{g}$ is clear. Likewise for the variational derivatives $\delta\ell/\delta u$ and $\delta\ell/\delta v$.

The $\mathrm{D}\boldsymbol{iff}(\mathbb{R})$-strand Hamiltonian structure

Upon setting $m = \delta\ell/\delta u$ and $n = \delta\ell/\delta v$, the right-invariant $\mathrm{D}iff(\mathbb{R})$-strand equations in (18) for maps $\mathbb{R} \times \mathbb{R} \to G = \mathrm{D}iff(\mathbb{R})$ in one spatial dimension may be expressed as a system of two 1+2 PDEs in (t, s, x),

$$\begin{aligned} m_t + n_s &= -\operatorname{ad}^*_u m - \operatorname{ad}^*_v n = -(um)_x - m u_x - (vn)_x - n v_x \,, \\ v_t - u_s &= -\operatorname{ad}_v u = -u v_x + v u_x \,. \end{aligned} \tag{19}$$

The Hamiltonian structure for these $\mathrm{D}iff(\mathbb{R})$-strand equations is obtained by Legendre transforming to

$$h(m, v) = \langle m, u \rangle - \ell(u, v) \,.$$

One may then write the equations (19) in Lie-Poisson Hamiltonian form as

$$\frac{d}{dt}\begin{bmatrix} m \\ v \end{bmatrix} = \begin{bmatrix} -ad^*_{(.)} m & \partial_s + \operatorname{ad}^*_v \\ \partial_s - \operatorname{ad}_v & 0 \end{bmatrix} \begin{bmatrix} \delta h/\delta m = u \\ \delta h/\delta v = -n \end{bmatrix}. \tag{20}$$

Singular solutions of the $\mathrm{Di}ff(\mathbb{R})$-strand equations

For simplicity we continue with the following choice of Lagrangian,

$$\ell(u,v) = \frac{1}{2}\int (u_x^2 + v_x^2)dx. \tag{21}$$

The $\mathrm{Di}ff(\mathbb{R})$-strand equations (19) admit peakon solutions in *both* momentum

$$m = -u_{xx} \quad \text{and} \quad n = -v_{xx},$$

with continuous velocities u and v. This is a two-component generalization of the Hunter-Saxton equation [18, 17].

Theorem 8.1 *The $\mathrm{Di}ff(\mathbb{R})$-strand equations (19) admit singular solutions expressible as linear superpositions summed over $a \in \mathbb{Z}$*

$$\begin{aligned} m(s,t,x) &= \textstyle\sum_a M_a(s,t)\delta(x - Q^a(s,t)), \\ n(s,t,x) &= \textstyle\sum_a N_a(s,t)\delta(x - Q^a(s,t)), \\ u(s,t,x) &= K * m = \textstyle\sum_a M_a(s,t)K(x,Q^a), \\ v(s,t,x) &= K * n = \textstyle\sum_a N_a(s,t)K(x,Q^a), \end{aligned} \tag{22}$$

where $K(x,y) = -\frac{1}{2}|x - y|$ is the Green function of the operator $-\partial_x^2$:

$$-\partial_x^2 K(x,0) = \delta(x).$$

The solution parameters $\{Q^a(s,t), M_a(s,t), N_a(s,t)\}$ with $a \in \mathbb{Z}$ that specify the singular solutions (22) (which we call 'peakons' for simplicity, although the Green function in this case is unbounded) are determined by the following set of evolutionary PDEs in s and t, in which we denote $K^{ab} := K(Q^a, Q^b)$ with integer summation indices $a, b, c, e \in \mathbb{Z}$:

$$\begin{aligned}
\partial_t Q^a(s,t) &= u(Q^a,s,t) = \textstyle\sum_b M_b(s,t)K^{ab}\,,\\
\partial_s Q^a(s,t) &= v(Q^a,s,t) = \textstyle\sum_b N_b(s,t)K^{ab}\,,\\
\partial_t M_a(s,t) &= -\,\partial_s N_a - \textstyle\sum_c (M_aM_c + N_aN_c)\frac{\partial K^{ac}}{\partial Q^a} \quad (no\ sum\ on\ a),\\
\partial_t N_a(s,t) &= \partial_s M_a + \textstyle\sum_{b,c,e} (N_bM_c - M_bN_c)\frac{\partial K^{ec}}{\partial Q^e}(K^{eb} - K^{cb})(K^{-1})_{ae}\,.
\end{aligned} \tag{23}$$

The last pair of equations in (23) may be solved as a system for the momentum, i.e., Lagrange multipliers (M_a, N_a), then used in the previous pair to update the support set of positions $Q^a(t,s)$.

Example: Two-peakon solution of a $Diff(\mathbb{R})$-strand

enote the relative spacing $X(s,t) = Q^1 - Q^2$ for the peakons at positions $Q^1(t,s)$ and $Q^2(t,s)$ on the real line and the Green's function $K = K(X)$. Then the first two equations in (23) imply

$$\begin{aligned}
\partial_t X &= -(M_1 - M_2)K(X)\,,\\
\partial_s X &= -(N_1 - N_2)K(X).
\end{aligned} \tag{24}$$

The second pair of equations in (23) may then be written as

$$\begin{aligned}
\partial_t M_1 &= -\,\partial_s N_1 - (M_1M_2 + N_1N_2)K'(X)\,,\\
\partial_t M_2 &= -\,\partial_s N_2 + (M_1M_2 + N_1N_2)K'(X)\,,\\
\partial_t N_1 &= \partial_s M_1 - (N_1M_2 - M_1N_2)K'(X)\,,\\
\partial_t N_2 &= \partial_s M_2 - (N_1M_2 - M_1N_2)K'(X)\,.
\end{aligned} \tag{25}$$

Assuming $X > 0$, $K'(X) = -\frac{1}{2}\operatorname{sgn}(X) = -\frac{1}{2}$. Introducing for convenience $L_{1,2} = M_{1,2} + iN_{1,2}$ we can rewrite (25) as

$$\begin{aligned}
(\partial_t - i\,\partial_s)L_1 &= \tfrac{1}{2}L_1\bar{L}_2\,,\\
(\partial_t - i\,\partial_s)L_2 &= -\tfrac{1}{2}\bar{L}_1L_2\,.
\end{aligned} \tag{26}$$

The solution for X can be expressed formally via $L_{1,2}$ from (24) as

$$X = \exp\left(\tfrac{1}{2}\Delta^{-1}\Re(L_1\bar{L}_2)\right),$$

where $\Delta = \partial_t^2 + \partial_s^2$ and $\Re(z)$ is the real part of z.

From the system (26) we obtain

$$\Delta \ln L_1 = -\frac{1}{4} L_1 \bar{L}_2, \qquad \Delta \ln L_2 = -\frac{1}{4} \bar{L}_1 L_2, \tag{27}$$

thus $\Delta \ln L_1 = \Delta \ln \bar{L}_2$ and $L_1 = \bar{L}_2 e^h$ where $h(s,t)$ is an arbitrary harmonic function: $\Delta h = 0$. Then for the variable $\tilde{Y} = \ln L_1$ we have the equation

$$\Delta \tilde{Y} = -\frac{1}{4} e^{2\tilde{Y}-h}, \tag{28}$$

and for $Y = \ln L_1 - \frac{1}{2}h - 2\ln 2 + \pi i$ we arrive at the Liouville's 2D equation

$$\Delta Y = e^{2Y}. \tag{29}$$

Solutions of (29) are known in the form

$$Y = \frac{1}{2} \ln \frac{w_s^2 + w_t^2}{f(w)}$$

where the function $f(w)$ could be w^2, $\cos^2 w$, $\sin^2 w$ or $\sinh^2 w$ with w being an arbitrary harmonic function, $\Delta w = 0$ see e.g. [6, 16, 20]. Thus the solutions $L_{1,2}$ depend on two arbitrary complex harmonic functions h, w. Hence the four peakon parameters $M_{1,2}$ and $N_{1,2}$ can be given in terms of four real arbitrary harmonic functions.

Other examples including complexification of the Camassa-Holm equation [19] are given in [8].

Conclusions

The G-strand equations comprise a system of PDEs obtained from the Euler-Poincaré (EP) variational equations for a G-invariant Lagrangian, coupled to an auxiliary *zero-curvature* equation. Once the G-invariant Lagrangian has been specified, the system of G-strand equations in (2) follows automatically in the EP framework. For matrix Lie groups, some of the G-strand systems are integrable. The singular solution of the $\mathrm{D}iff(\mathbb{R})$-strand equations (19) can also be obtained explicitly in some simple situations, and the freedom in the solution is given by several arbitrary harmonic functions of the variables s, t. The *complex* $\mathrm{D}iff(\mathbb{R})$-strand equations and their peakon collision solutions have also been solved by elementary means [8]. The stability of the single-peakon solution under perturbations into the full solution space of equations (19) would be an interesting problem for future work.

Acknowledgments

We are grateful for enlightening discussions of this material with F. Gay-Balmaz, T. S. Ratiu and C. Tronci. Work by RII was supported by the Science Foundation Ireland (SFI), under Grant No. 09/RFP/MTH2144. Work by DDH was partially supported by Advanced Grant 267382 FCCA from the European Research Council.

Bibliography

[1] Bordag L. A. and Yanovski A. B. Polynomial Lax pairs for the chiral $O(3)$ field equations and the Landau-Lifshitz equation. *J. Phys. A: Math. Gen.*, 28:4007-4013, 1995.

[2] Kirillov A. Orbits of the group of diffeomorphisms of a circle and local Lie superalgebras. *Funct. Anal. Appl. (English transl.)*, 15:135-137, 1981.

[3] Kirillov A. The orbit method. II. Infinite-dimensional Lie groups and Lie algebra. *Contemp. Math. (Providence, RI: Amer. Math. Soc.)*, 145:33-63, 1993.

[4] Yanovski A. B. Bi-Hamiltonian formulation of the $O(3)$ chiral fields equations hierarchy via a polynomial bundle. *J. Phys. A: Math. Gen.*, 31:8709-8726, 1998.

[5] Marle C.-M. On Henri Poincaré's Note "Sur une forme nouvelle des équations de la Mécanique". *J. Geom. Symm. Phys.*, 29:1-38, 2013.

[6] Crowdy D. General solutions to the 2D Liouville equation. *Int. J. Engng Sci.*, 35:141-149, 1997.

[7] Holm D. D. *Geometric Mechanics II: Rotating, Translating and Rolling*. World Scientific - Imperial College Press, Singapore, 2011.

[8] Holm D. D. and Ivanov R. I. G-Strands and Peakon Collisions on $\mathrm{D}iff(\mathbb{R})$. *SIGMA*, 9:027 (14 pages), 2013.

[9] Holm D. D. and Ivanov R. I. Matrix G-Strands. Nonlineariry 27:1445,2014. Preprint math-ph 1305.4010, available at

http://arxiv.org/pdf/1305.4010v1.pdf.

[10] Holm D. D. and Lucas A. M. Toda lattice G-Strands. Preprint nlin.SI 1306.2984, available at http://arxiv.org/pdf/1306.2984v1.pdf.

[11] Marsden J. E. and Ratiu T. S. *Introduction to Mechanics and Symmetry*. Springer, Berlin, 1999.

[12] Gay-Balmaz F., Holm D. D. and Ratiu T. S. Integrable G-Strands on semisimple Lie groups. J.Phys.A:Math.Theor. 47:075201, 2014. Preprint math-ph 1308.3800, available at http://arxiv.org/pdf/1308.3800v1.pdf.

[13] Poincaré H. Sur une forme nouvelle des équations de la Mécanique. *C. R. Acad. Sci. Paris*, CXXXII:369, 1901.

[14] Holm D. D., Ivanov R. I. and Percival J. R. G-Strands. *Journal of Nonlinear Science*, 22:517-551, 2012.

[15] Cherednik I. On the integrability of the 2-dimensional asymmetric chiral $O(3)$ field equations and their quantum analogue. *J. Nuc. Phys. (in Russian)*, 33:278-282, 1981.

[16] Ibragimov N. H. (ed.). CRC Handbook of Lie Group to Differential Equations. Vol. II. Applications in engineering and physical sciences. CRC Press, Boca Raton, Fl., 1995.

[17] Hunter J. and Zheng Y. On a Completely Integrable Nonlinear Hyperbolic Variational Equation. *Physica D*, 79:361-386, 1994.

[18] Hunter J. K. and Saxton R. A. Dynamics of Director Fields. *SIAM J. Appl. Math.*, 51:1498-1521, 1991.

[19] Camassa R. and Holm D. An integrable shallow water equation with peaked solitons. *Phys. Rev. Lett.*, 71:1661-1664, 1993.

[20] Kiselev A. V. On the geometry of Liouville equation: symmetries, conservation laws, and Bäcklund transformations. *Acta Appl. Math.*, 72:33-49, 2002.

SOME PROPERTIES OF THE SPECTRAL DECOMPOSITIONS FOR THE RECURSION OPERATORS RELATED TO THE CAUDREY-BEALS-COIFMAN SYSTEM IN THE PRESENCE OF $\mathbb{Z}p$ REDUCTIONS OF MIKHAILOV TYPE

A. B. YANOVSKI

Introduction

It is known that the characteristic property of the so-called soliton equations is that they admit the so-called Lax representation $[L, A] = 0$. In the last expression L, A are linear operators on ∂_x, ∂_t of degree 1, depending also on some functions $q_\alpha(x,t)$, $1 \leq \alpha \leq s$ (called `potentials') and a spectral parameter λ. Fixing L, the Lax representations for the nonlinear evolution equations (NLEEs), or soliton equations, associated with (related to) L or $L\psi = 0$ are equivalent to a system (in the case where$A$ depends linearly on ∂_t) of the type $(q_\alpha)_t = F_\alpha(q, q_x, \ldots)$, where $q = (q_\alpha)_{1 \leq \alpha \leq s}$. The linear problem $L\psi = 0$ is called an auxiliary linear problem. The schemes to find solutions to the NLEEs related to L could be quite different but what is essential is that the Lax representation permits one to pass from the original evolution defined by the equations $(q_\alpha)_t = F_\alpha$ to the evolution of some spectral data related to the problem $L\psi = 0$, see [7].

The Caudrey-Beals-Coifman system (CBC system), also called the Generalized Zakharov-Shabat system (GZS system) when the element J is real, is one of the best known auxiliary linear problems. It can be written as follows

$$L\psi = (\mathrm{i}\,\partial_x + q(x) - \lambda J)\psi = 0. \tag{1}$$

In its most general form $q(x)$ and J belong to a fixed simple Lie algebra $\mathfrak{g}$ in some finite dimensional irreducible representation. The element J

should be regular; that is, the kernel of ad_J ($\mathrm{ad}_J(X) \equiv [J,X]$, $X \in \mathfrak{g}$) is a Cartan subalgebra. We shall denote it by $\mathfrak{h}$ and shall assume that it is fixed, so we shall call it `the Cartan subalgebra'. The potential $q(x)$ takes values in the orthogonal complement $\mathfrak{h}^{\perp} = \bar{\mathfrak{g}}$ of $\mathfrak{h}$ with respect to the Killing form $\langle X,Y\rangle = \mathrm{tr}\ \mathrm{ad}_X \mathrm{ad}_Y$ and therefore $q(x) = \sum_{\alpha\in\Delta} q_\alpha E_\alpha$ where E_α are the root vectors; Δ is the root system of $\mathfrak{g}$ with respect to $\mathfrak{h}$. The scalar functions $q_\alpha(x)$ (the `potentials') are defined on $\mathbb{R}$, are complex valued, smooth, and tend to zero as $x \to \pm\infty$. We assume the properties of the simple Lie algebras are known, and our notation and normalizations are as in [7].

The system (1) is a generalization of the classical Zakharov Shabat system, passed through generalizations on $sl\ (n)$ for J real. It is then complex and has finally acquired the form in which we present it here. For bibliography see [4].

Because of the form of the Lax representation it is easy to understand that the spectral theory of the operator L in adjoint representation is important for the study of the NLEEs associated with L. Further, if L is considered an arbitrary faithful representation of the algebra $\mathfrak{g}$, and ψ is a fundamental solution $L\psi = 0$, then any $w = \psi X \psi^{-1}$ where X is a constant element from $\mathfrak{g}$ (adjoint solutions) satisfies $[L,w] = 0$. Thus finding the fundamental analytic solutions of $L\chi(x,\lambda) = 0$ is important for the spectral theory of the problem L (both in some typical representation and in adjoint representation). Through them is constructed the resolvent of the operator $\Psi \mapsto [L,\Psi]$, which gives rise to the so-called expansions over adjoint solutions (or Generalized Exponents), an important approach that started with the seminal work [1]; see [4] for extended bibliography. Recently there has been substantial interest in the theory of L in the presence of reductions, and the expansions have some specific properties here, but the part related to the discrete spectrum has not been given proper attention. We intend to fill this gap in the present work.

Completeness relations for the CBC system

First we are going to describe briefly the analytic solutions of the CBC system $L\psi = (\mathrm{i}\,\partial_x - \lambda J + q)\psi = 0$, on some simple Lie algebra $\mathfrak{g}$ in a typical representation, [3]. Let Δ be the root system of $\mathfrak{g}$ (defined by $\mathfrak{h} = \ ker\ \mathrm{ad}_J$) and let $\Sigma = \cup_{\alpha\in\Delta} \zeta_\alpha$ be a bunch of straight lines ζ_α where $\zeta_\alpha = \{\lambda : \mathrm{Im}(\lambda\alpha(J)) = 0\}$. The connected components of the set

$\mathbb{C}\backslash\Sigma$ are open sectors in the λ-plain. We shall denote these sectors by Ω_ν ordering them anti-clockwise. Clearly ν takes values from 1 to some even number $2M$. Then the boundary of the sector Ω_ν consists of two rays: l_ν and $l_{\nu+1}$ (l_ν comes before $l_{\nu+1}$ when we turn anti-clockwise) so that $\overline{\Omega}_\nu \cap \overline{\Omega}_{\nu-1} = l_\nu$. (The `bar' here denotes the closure). We shall understand the number ν modulo $2M$. Naturally,

$$\mathbb{C}\backslash\Sigma = \cup_{\nu=1}^{2M} \Omega_\nu, \qquad \Omega_\nu \cap \Omega_\mu = \emptyset, \quad \nu \neq \mu. \tag{2}$$

If $\alpha, \beta \in \Delta$, $\alpha \neq \beta$ in every sector Ω_ν either $\mathrm{Im}[\lambda(\alpha-\beta)(J)]$ does not change sign which permits one to define ν-ordering of the roots $\alpha >_\nu \beta$ if $\mathrm{Im}\lambda(\alpha-\beta)(J) > 0$, and consequently in each Ω_ν we have splitting into the sets of positive and negative roots $\Delta = \Delta_\nu^+ \cup \Delta_\nu^-$.

Limiting ourselves to the typical representation of $\mathfrak{g}$, for a large class of potentials $q(x)$ (in fact they form a dense open set in the space of the absolutely integrable potentials) it can be shown that in each of the sectors Ω_ν there exists a fundamental solution (FAS) $\chi_\nu(x,\lambda)$ of $L\chi = 0$ with the properties:

(a) $\chi_\nu(x,\lambda)$ is meromorphic in Ω_ν and has only a finite number of poles. The poles define the discrete spectrum of the problem and for simplicity below we shall assume that $\chi_\nu(x,\lambda)$ is analytic in Ω_ν.

(b) $\chi_\nu(x,\lambda)$ allows extension by continuity to the boundary of Ω_ν (to the rays l_ν and $l_{\nu+1}$).

(c) For $\lambda \in \Omega_\nu$ the function $\chi_\nu(x,\lambda)e^{i\lambda Jx}$ is bounded and we have

$$\lim_{x\to-\infty}\chi_\nu(x,\lambda)e^{i\lambda Jx} = \mathbb{1} \text{ and } \lim_{\lambda\to+\infty}\chi_\nu(x,\lambda)e^{i\lambda Jx} = \mathbb{1}.$$

Below we shall have frequently a situation when some set of functions $f_\nu(\lambda)$ is such that each $f_\nu(\lambda)$ is analytic in Ω_ν and allows extension to the boundary of this sector $l_\nu \cup l_{\nu+1}$. Then say on l_ν we have the extensions from the left and from the right. We shall denote the extension from the left by $f_\nu^+(\lambda)$ and from the right by $f_\nu^-(\lambda)$ (of course $\lambda \in l_\nu$). Thus for example on each l_ν we have the solutions $\chi_\nu^\pm(\lambda, x)$ of the CBC problem. We are not going to present the resolvent kernel of $\Psi \mapsto [L, \Psi]$ (it can be found in [3]), we simply remind readers of the completeness relations that are related to it.

Let π_0 be the orthogonal projector on $\bar{\mathfrak{g}}$, let E_α, $\alpha \in \Delta$ be the root vectors. Define the following functions, called Generalized Exponents or adjoint solutions.

$$e_\alpha^\nu(x,\lambda) = \pi_0(\chi_\nu(x,\lambda)E_\alpha\chi_\nu^{-1}(x,\lambda)), \quad \lambda \in \bar{\Omega}_\nu. \tag{3}$$

As we agreed for $\lambda \in l_\nu$ we shall write $e_\alpha^{(+;\nu)}(x,\lambda)$ if the solution is extended from the sector Ω_ν and $e_\alpha^{(-;\nu)}(x,\lambda)$ if the solution is extended from the sector $\Omega_{\nu-1}$. Then we have:

Theorem 2.1 *The completeness relation for the Generalized Exponents (without discrete spectrum terms) can be cast into the following form:*

$$\Pi_0\delta(x-y) =$$
$$\frac{1}{2\pi}\sum_{\nu=1}^{M}\int_{l_\nu} d\lambda\{\sum_{\alpha\in\delta_\nu^+}(e_\alpha^{(+;\nu)}(x)\otimes e_{-\alpha}^{(+;\nu)}(y) - e_{-\alpha}^{(-;\nu)}(x)\otimes e_\alpha^{(-;\nu)}(y))\} \tag{4}$$

where

$$\Pi_0 = \sum_{\gamma\in\Delta}\frac{E_\gamma\otimes E_{-\gamma}}{\gamma(J)} \tag{5}$$

$$\delta_\nu^\pm = \Delta_\nu^\pm \cap \delta_\nu, \quad \delta_\nu = \{\alpha\in\Delta : Im(\lambda\alpha(J)) = 0 \; for \; \lambda\in l_\nu\}. \tag{6}$$

In the above it is assumed that the rays are oriented from 0 to ∞ and for shortness we have omitted the dependence on λ. The formula itself must be understood in the following way. First, it is assumed that $\mathfrak{g}^*$ is identified with $\mathfrak{g}$, the pairing between them being given by the Killing form. So, for example, for $X, Y, Z \in \mathfrak{g}$ making a contraction of $X \otimes Y$ with Z on the right we obtain $X\langle Y, Z\rangle$, and making a contraction from the left we get $\langle Z, X\rangle Y$. Next, the formula for Π_0 implies that making a contraction with Π_0 on the right we get $\Pi_0 X = \mathrm{ad}_J^{-1}\pi_0 X$ and similarly from the left $X\Pi_0 = -\,\mathrm{ad}_J^{-1}\pi_0 X$. (On the space $\bar{\mathfrak{g}}$ the operator ad_J is invertible). When the discrete spectrum is taken into account, to the right hand side of this formula must be written a term which we denote by DSC.

Suppose that we have a L^1-integrable function $g: \mathbb{R} \mapsto \bar{\mathfrak{g}}$. Then we have the expansions (for $\varepsilon = +1$ and $\varepsilon = -1$ respectively):

$$g(x) = \frac{1}{2\pi}\sum_{\nu=1}^{2M}\int_{l_\nu}\{(\sum_{\alpha\in\delta_\nu^+} e_{\varepsilon\alpha}^{(+;\nu)}(x)\langle\langle e_{-\varepsilon\alpha}^{(+;\nu)},[J,g]\rangle\rangle - e_{-\varepsilon\alpha}^{(-;\nu)}(x)\langle\langle e_{\varepsilon\alpha}^{(-;\nu)},[J,g]\rangle\rangle\}) \mathrm{d}\lambda. \quad (7)$$

(When the discrete spectrum terms are taken into account, to the right side of this formula must be added a term $\mathrm{DSC}_{\pm}(g)$). In the above we used the following notation: for two functions $f_1(x), f_2(x)$ with values in $\mathfrak{g}$ we put

$$\langle\langle f_1, f_2\rangle\rangle = \int_{-\infty}^{+\infty}\langle f_1(x), f_2(x)\rangle \mathrm{d}x. \quad (8)$$

It can be shown that for either of the choices for ε the expansion (7) converges in the same sense as the Fourier expansions for $g(x)$. Finally, if one introduces the operators

$$\Lambda_{\pm}(X(x)) = \mathrm{ad}_J^{-1}\left(\mathrm{i}\,\partial_x X + \pi_0[q,X] + \mathrm{i}\,\mathrm{ad}_q\int_{\pm\infty}^{x}(id - \pi_0)[q(y),X(y)]\mathrm{d}y\right) \quad (9)$$

one can see that

$$(\Lambda_- - \lambda)e_{\alpha}^{(+;\nu)} = 0, \qquad (\Lambda_- - \lambda)e_{-\alpha}^{(-;\nu)} = 0, \quad \alpha \in \delta_\nu^+ \quad (10)$$

$$(\Lambda_+ - \lambda)e_{-\alpha}^{(+;\nu)} = 0, \qquad (\Lambda_+ - \lambda)e_{\alpha}^{(-;\nu)} = 0, \quad \alpha \in \delta_\nu^+. \quad (11)$$

The operators $\Lambda_{\pm}$ are the Generating Operators for the expansions (7); they play for these expansions the role that $\mathrm{i}\,\partial_x$ plays for the Fourier expansion. The importance of the above expansions for the theory of NLEEs (they appeared first in the famous work of [1] for the ZS system and later were generalized) is based on the fact that when one expands over the Generalized Exponents the potential $q(x)$ and $\delta q(x)$ one gets as coefficients a set of minimal scattering data and its variation. Moreover, the coefficients in these expansions have linear evolution for the NLEEs related to L. Through them can be obtained the NLEEs, their conservation laws, the hierarchy of symplectic structures, etc, see [4]. In fact the theory of the Recursion Operators is a theoretical tool which apart from explicit solutions can give most of the information about the NLEEs associated with L, [4]. They also have an interesting geometric interpretation as dual objects to Nijenhuis tensors on the manifold of potentials, on which it is defined as a special geometric structure, the Poisson-Nijenhuis structure. Then the NLEEs related to L could be interpreted as fundamental fields of that structure. This interpretation has been given by Magri, [8]; see [4] for the full theory.

Completeness relations in the presence of reductions defined by automorphisms

Assume that for the CBC system on the algebra $\mathfrak{g}$ we have a reduction group G_M generated by one element g acting as

$$g(\psi(x,\lambda)) = \mathcal{K}(\psi(x,\omega^{-1}\lambda)), \quad \omega = \exp\frac{2\pi\mathrm{i}}{p} \tag{12}$$

where $\mathcal{K}$ is an automorphism of order p of the Lie group G corresponding to the algebra $\mathfrak{g}$. It generates an automorphism of $\mathfrak{g}$ which we shall denote by the same letter $\mathcal{K}$. Since this immediately leads to $\mathcal{K}J = \omega J$ and $\mathcal{K}q = q$ the automorphism $\mathcal{K}$ preserves the Cartan subalgebra. Since $g^p = \ id\ $, G_M is isomorphic to $\mathbb{Z}_p$. The automorphism $\mathcal{K}$ acts also on the set of roots (that action we shall denote by the same letter again), and then for the root vectors we have $\mathcal{K}(E_\alpha) = \zeta(\alpha)E_{\mathcal{K}\alpha}$, where $\zeta(\alpha)$ are numbers such that $\zeta(\alpha)\zeta(-\alpha) = 1$ and $\zeta(\alpha)q(\beta) = \zeta(\alpha+\beta)$ if $\alpha+\beta \in \delta$ (see [7] for the details).

Since the automorphisms of $\mathfrak{g}$ we consider here leave $\mathfrak{h}$ invariant, we have the splittings:

$$\mathfrak{g} = \oplus_{s=0}^{p-1} \mathfrak{g}^{[s]}, \quad [\mathfrak{g}^{[k]}\mathfrak{g}^{[l]}] \subset \mathfrak{g}^{[k+l]}, \quad \bar{\mathfrak{g}} = \oplus_{s=0}^{p-1} \bar{\mathfrak{g}}^{[s]}, \quad \mathfrak{h} = \oplus_{s=0}^{p-1} \mathfrak{h}^{[s]}. \tag{13}$$

For each $X \in \mathfrak{g}^{[s]}, \mathfrak{h}^{[s]}, \bar{\mathfrak{g}}^{[s]}$ we have $\mathcal{K}X = \omega^s X$, the spaces with upper indexes s and k are orthogonal with respect to the Killing form unless $k+s \neq 0 (mod\ \ p)$, and the spaces $\bar{\mathfrak{g}}^{[s]}$, $\mathfrak{h}^{[k]}$ are always orthogonal.

The invariance of the set of fundamental solutions can be additionally specified if we take the FAS $\chi_\nu(x,\lambda)$ defined in Ω_ν, $\nu = 1,2,\ldots 2M$. Then one easily sees that $\mathcal{K}(\chi_\nu(x,\lambda))$ must be the analytic solution in the sector containing $\omega\lambda$, and one has an action of $\mathcal{K}$ (multiplication by ω) on the sectors Ω_ν and on the rays l_ν. One can see that if $\omega\Omega_\nu = \Omega_{\nu+a}$ then $\omega l_\nu = l_{\nu+a}$, $\delta^\pm_{\nu+a} = \delta^\pm_\nu$. Using these relations one has:

Theorem 3.1 *In case we have $\mathbb{Z}_p$ reductions the completeness can be put into the form:*

$$\Pi_0\delta(x-y) = \frac{1}{2\pi p}\sum_{\nu=1}^{2M}\sum_{k=1}^{p}\int_{l_\nu}\{[\sum_{\alpha\in\delta_\nu^+}\omega^k\mathcal{K}^k\otimes\mathcal{K}^k(e_\alpha^{(+;\nu)}(x)\otimes e_{-\alpha}^{(+;\nu)}(y)) - \omega^k\mathcal{K}^k\otimes\mathcal{K}^k(e_{-\alpha}^{(-;\nu)}(x)\otimes e_\alpha^{(-;\nu)}(y))]\}d\lambda. \qquad (14)$$

Note that the numbers $\zeta(\alpha)$ don't appear any more: this occurs because we apply $\mathcal{K}$ always on products of the type $E_\alpha\otimes E_{-\alpha}$. The expansions of a function $g(x)$ over the adjoint solutions can be simplified further, if for arbitrary x the value $g(x)\in \mathfrak{g}^{[s]}$

$$g(x) = \frac{\varepsilon}{2\pi p}\sum_{\nu=1}^{2M}\int_{l_\nu}\{\sum_{\alpha\in\delta_\nu^+}[\sum_{k=1}^{p}\omega^{-ks}\mathcal{K}^k(e_{\varepsilon\alpha}^{(+;\nu)}(x,\lambda))\langle\langle e_{-\varepsilon\alpha}^{(+;\nu)},[J,h]\rangle\rangle - -\sum_{k=1}^{p}\omega^{-ks}\mathcal{K}^k(e_{-\varepsilon\alpha}^{(-;\nu)}(x,\lambda))\langle\langle e_{\varepsilon\alpha}^{(-;\nu)},[J,g]\rangle\rangle]\}d\lambda. \qquad (15)$$

In the above are written two expansions, one for $\varepsilon = +1$ and the other for $\varepsilon = -1$. Making a contraction one can see that $g(x)$ is actually expanded over the functions:

$$e_\alpha^{(\pm;\nu;s)}(x,\lambda) = \sum_{k=1}^{p}\omega^{-ks}K^k(e_\alpha^{(\pm;\nu)}(x,\lambda)),\quad \nu = 1,2,\dots,a \qquad (16)$$

which are up to a multiplier p the orthogonal projections of $e_\alpha^{(\pm;\nu)}(x,\lambda)$ on $\bar{\mathfrak{g}}^{[s]}$.

The operators $\Lambda_\pm$ map functions with values in $\bar{\mathfrak{g}}^{[s]}$ into functions with values in $\bar{\mathfrak{g}}^{[s-1]}$. In particular, for the new Generalized Exponents we have

$$\Lambda_- e_\alpha^{(+;\nu;s)} = \lambda e_\alpha^{(+;\nu,s-1)}, \quad \Lambda_- e_{-\alpha}^{(-;\nu,s)} = \lambda e_{-\alpha}^{(-;\nu.s-1)}, \qquad \alpha \in \delta_\nu^+ \tag{17}$$

$$\Lambda_+ e_{-\alpha}^{(+;\nu,s)} = \lambda e_{-\alpha}^{(+;\nu,s-1)}, \quad \Lambda_- e_\alpha^{(-;\nu,s)} = \lambda e_\alpha^{(-;\nu,s-1)}, \qquad \alpha \in \delta_\nu^+. \tag{18}$$

Therefore $e_\alpha^{(\pm;\nu,s)}$ are not eigenfunctions of $\Lambda_\pm$. However, we obviously have:

$$\Lambda_-^p e_\alpha^{(+;\nu;s)} = \lambda^p e_\alpha^{(+;\nu,s)}, \quad \Lambda_-^p e_{-\alpha}^{(-;\nu,s)} = \lambda^p e_{-\alpha}^{(-;\nu.s)}, \qquad \alpha \in \delta_\nu^+ \tag{19}$$

$$\Lambda_+^p e_{-\alpha}^{(+;\nu,s)} = \lambda^p e_{-\alpha}^{(+;\nu,s)}, \quad \Lambda_+^p e_\alpha^{(-;\nu,s)} = \lambda^p e_\alpha^{(-;\nu,s)}, \qquad \alpha \in \delta_\nu^+. \tag{20}$$

This is important, so let us formulate it separately:

Theorem 3.2 *For the expansions (15) the role of the Recursion Operators are played by the p-th powers of the operators $\Lambda_\pm$.*

We want to see now what happens with the discrete spectrum in case we have $\mathbb{Z}_p$ reduction and if the above conclusion also holds in some sense.

The discrete spectrum

To take into account the contribution of the discrete spectrum is in general quite a complicated task, since its structure unlike the continuous spectrum depends on the representation of the algebra $\mathfrak{g}$ in which we consider the problem $L\psi = 0$. We shall skip discussion of this issue, as it can be found in [3]. Here we shall remind readers only of the main facts in the case that $q(x)$ is regular potential. First, the discrete and the continuous spectrum do not overlap and the discrete spectrum consists of a finite number of poles of the functions χ_ν and χ_ν^{-1} which do not depend on x. As will be seen below, in adjoint representation we need to know the singularities of $e_\alpha^\nu(x, \lambda)$ which of course depend on the singularities of χ_ν and χ_ν^{-1}. The information about the singularities of χ_ν is essential in some questions (for example if one wants to perform Inverse Scattering Transform reducing it to a suitable Riemann-Hilbert problem). However, since our task is different, we shall not attempt to track down these singularities to χ_ν and χ_ν^{-1}; all we need to know is that for regular potentials $e_\alpha^\nu(x,\lambda)$ has singularities at $\lambda_{\nu;k} \in \Omega_\nu$, $k = 1,2,\dots N_\nu$. (For some particular α some of the singularities could be removable).

Then the discrete spectrum contribution $\mathrm{DSC}_\pm(g)$ to the expansions (7), according to [3], is given making contractions to the right and to the left and integrating over $\mathbb{R}$ (we identify $\mathfrak{g}$ and $\mathfrak{g}^*$ via Killing form) of the expression DSC given below with $[J, g(x)]$ where $g(x)$ is an absolutely integrable function on the line taking values in $\bar{\mathfrak{g}}$. DSC has the form

$$DSC = -\mathrm{i}\sum_{\nu=1}^{2M}\sum_{\alpha\in\Delta_\nu^+}\sum_{k=1}^{N_\nu}\mathrm{Res}(Q_{\nu,\alpha}(x,y,\lambda);\lambda_{\nu,k}) \qquad (21)$$

where $Q_{\nu;\alpha}(x,y,\lambda) = e_\alpha^\nu(x,\lambda) \otimes e_{-\alpha}^\nu(y,\lambda)$.

In the case of $\mathbb{Z}_p$ reduction the expression (21) can be cast in another form. The starting point for our considerations will be the equations $\mathcal{K}(\chi_\nu(x,\lambda)) = \chi_{\nu+a}(x,\omega\lambda)$ and $\mathcal{K}(e_\alpha^\nu(x,\lambda)) = \varepsilon(\alpha)e_{\mathcal{K}\alpha}^{\nu+a}(x,\omega\lambda)$. They show that if $e_\alpha^\nu(x,\lambda)$ has a pole of some order at $\lambda = \lambda_0$ in Ω_ν then $e_{\mathcal{K}\alpha}^{\nu+a}(x,\lambda)$ will have the same type of singularity at $\omega\lambda_0$ in $\Omega_{\nu+a}$. Thus the reduction group acts also on the poles of the functions $e_\alpha^\nu(x,\lambda)$ (on the discrete spectrum of L) dividing it into orbits in which the poles are obtained by multiplication by some power of ω. The orders of poles in different orbits could of course be different but since they are a finite number we can assume that these orders are not higher than some number d. Since working with the formula (21) is cumbersome, let us consider the contribution from just one $Q_{\nu;\alpha}$ and one pole $\lambda = \lambda_0$ (in fact from the poles from the orbit of λ_0) located in Ω_ν. We note that necessarily $\lambda_0 \neq 0$.

Assume at $\lambda = \lambda_0$ we have a pole of order d then in some discs centered at λ_0 and $\omega\lambda_0$ respectively we shall have:

$$e_\alpha^\nu(x,\lambda) = \sum_{s=-d}^{\infty} A_{\alpha;s}^\nu(x)(\lambda-\lambda_0)^s,\; e_{\mathcal{K}\alpha}^{\nu+a}(x,\lambda) = \sum_{s=-d}^{\infty} A_{\mathcal{K}\alpha;s}^{\nu+a}(x)(\lambda-\omega\lambda_0)^s \qquad (22)$$

where $A^{\nu}_{\alpha;s}(x)$ are some functions with values in $\mathfrak{g}$ and $A^{\nu}_{\alpha;-d}(x) \neq 0$. From the relation between $e^{\nu}_{\alpha}(x,\lambda)$ and $e^{\nu+a}_{\mathcal{K}\alpha}(x,\lambda)$ we get $\mathcal{K}A^{\nu}_{\alpha;s}(x) = \varepsilon(\alpha)\omega^{s}A^{\nu+a}_{\mathcal{K}\alpha;s}(x)$ and therefore

$$\mathrm{Res}(Q_{\nu;\alpha}(x,y,\lambda);\lambda_0) = \textstyle\sum_{s+l=-1} A^{\nu}_{\alpha;s}(x) \otimes A^{\nu}_{-\alpha;l}(y). \qquad (23)$$

Consequently,

$$\mathrm{Res}(Q_{\nu+a;\mathcal{K}\alpha}(x,y,\lambda);\omega\lambda_0) = \sum_{l+s=-1} A^{\nu+a}_{\mathcal{K}\alpha;s}(x) \otimes A^{\nu+a}_{-\mathcal{K}\alpha;l}(y) =$$

$$\sum_{l+s=-1} \omega^{-(s+l)}\mathcal{K}A^{\nu}_{\alpha;s}(x) \otimes \mathcal{K}A^{\nu}_{-\alpha;l}(y) = \omega\mathcal{K} \otimes \mathcal{K}(\mathrm{Res}(Q_{\nu;\alpha};\lambda_0))(x,y).$$

Let us now make a contraction from the right with $[J,g](y)$ where g is a smooth function defined on the line $\mathbb{R}$ with values in $\mathfrak{g}^{[k]}$. Then $[J,g]$ takes values in $\mathfrak{g}^{[s+1]}$ and we obtain

$$\mathrm{Res}(Q_{\nu+a;\mathcal{K}\alpha};\omega\lambda_0).\,[J,g](x,y) = \omega^{-s}\mathcal{K}(\mathrm{Res}(Q_{\nu;\alpha};\lambda_0).\,[J,g])(x,y).$$

Now summing up the terms of the above type over the poles belonging to the orbit defined by λ_0 and taking into account that for $X \in \mathfrak{g}$ the map $X \mapsto \frac{1}{p}\sum_{j=1}^{p} \omega^{-sj}\mathcal{K}^{j}X$ is a projector onto the subspace $\mathfrak{g}^{[s]}$, after some algebraic transformations we get the expression

$$\tfrac{1}{p}\textstyle\sum_{s+l=-1} A^{\nu;k}_{\alpha;s}(x)\langle A^{\nu;-(k+1)}_{-\alpha;l}(y), [J,g](y)\rangle \qquad (24)$$

where for $\beta \in \delta^{+}_{\nu}$ we defined $A^{\nu;s}_{\beta;l}(y) = \sum_{k=1}^{p} \omega^{-sk}\mathcal{K}^{k}A^{\nu}_{\beta;l}(y)$.

If instead of contraction from the right we perform contraction from the left we obtain similar expressions; so finally, integrating from $-\infty$ to $+\infty$ over y (over x for the expressions obtained by left contraction) as is necessary in order to obtain expansions over the adjoint solutions, we obtain the expressions (for $\varepsilon = +1$ and $\varepsilon = -1$):

$$\tfrac{1}{p}\textstyle\sum_{s=-d}^{d-1} A^{\nu;k}_{\varepsilon\alpha;s}(x)\langle\langle A^{\nu;-(k+1)}_{-\varepsilon\alpha;-(s+1)}, [J,h]\rangle\rangle, \quad \varepsilon = \pm 1. \qquad (25)$$

Let us denote by $\lambda_{\nu,j}$, $\nu = 1,2, \dots a$ and $j = 1,2 \dots N_\nu$ the different poles in the fundamental sectors $\Omega_1, \Omega_2, \dots, \Omega_a$. Then we need to write an additional index to the functions $A^{\nu}_{\alpha;s}$; they become $A^{\nu}_{\alpha;s;j}$. Consequently, the functions $A^{\nu;k}_{\alpha;s}$ also aquire an additional index to become $A^{\nu;k}_{\alpha;s;j}$. Then, in the case where we expand a function g with values in $\overline{\mathfrak{g}}^{[k]}$, the discrete spectrum contributions $DSC_{\pm}(g)$ to the right hand side of formula (7) will be given by

$$DSC_{\pm}(g) = \frac{-\mathrm{i}}{p} \sum_{\alpha \in \Delta_\nu^+} \sum_{\nu=1}^{a} \sum_{j=1}^{N_\nu} \sum_{s=-d}^{d-1} A^{\nu;k}_{\varepsilon\alpha;s;j}(x) \langle\langle A^{\nu;-(k+1)}_{-\varepsilon\alpha,-(s+1);j}, [J, g] \rangle\rangle \tag{26}$$

for $\varepsilon = \pm 1$. The choice $\varepsilon = +1$ corresponds to the expressions obtained using contraction from the right while $\varepsilon = -1$ corresponds to the expressions obtained using contraction from the left. Since this formula is too cumbersome, in what follows we shall consider the contribution due to only one pole, located at $\lambda = \lambda_0$ in one of the fundamental sectors.

Now we consider the action of the Recursion Operators on the discrete spectrum (compare to (10), (11)). As mentioned, it will suffice to consider the contribution from only one pole $\lambda = \lambda_0$. From the fact that $(\Lambda_{\pm} - \lambda) e^{\nu}_{\alpha}(x, \lambda) = 0$ we easily get that

$$\Lambda_{\pm} A^{\nu}_{\beta;-d}(x) = \lambda_0 A^{\nu}_{\beta;-d}(x) \tag{27}$$

$$\Lambda_{\pm} A^{\nu}_{\beta;s}(x) = \lambda_0 A^{\nu}_{\beta;s}(x) + A^{\nu}_{\beta;s-1}(x), \quad -d < s < d$$

which shows that acting on the vectors $A^{\nu}_{\beta;s}(x)$, $s = d-1, d-2, \dots -d$ the operators $\Lambda_{\pm}$ have canonical single block Jordan form. (It is easy to show that if $\lambda_0 \neq 0$, which is our case, and $A^{\nu}_{\beta;-d}(x) \neq 0$ all these vectors must be linearly independent). We are interested now to see how this agrees with the splitting defined by $\mathcal{K}$. Since $\Lambda_{\pm} \circ \mathcal{K} = \omega \mathcal{K} \circ \Lambda_{\pm}$ we obtain

$$\Lambda_{\pm} A^{\nu;k}_{\beta;-d}(x) = \lambda_0 A^{\nu;k-1}_{\beta;-d}(x) \tag{28}$$

$$\Lambda_{\pm} A^{\nu;k}_{\beta;s}(x) = \lambda_0 A^{\nu;k-1}_{\beta;s}(x) + A^{\nu;k-1}_{\beta;s-1}(x), \quad -d < s < d.$$

The action does not have canonical Jordan form in any space of functions taking value in $g^{[s]}$, so it is not the representation we look for. From the other side, for an arbitrary k we have $A_{\beta;s}^{\nu;k+p}(x) = A_{\beta;s}^{\nu;k}(x)$, so using induction we get that

$$\Lambda_{\pm}^{p} A_{\beta;s}^{\nu;k}(x) = \sum_{l=0}^{p} C_{p}^{l} \lambda_0^{p-l} A_{\beta;s-l}^{\nu;k}(x) \tag{29}$$

where in order to be able to write nicely the formula we assume that $A_{\beta;s}^{\nu;k} \equiv 0$ for $s < -d$ and C_p^s are the binomial coefficients. (In fact from (27) we get the same formula for the functions $A_{\beta;s}^{\nu}(x)$.) If $A_{\beta;l}^{\nu;k}(x) = 0$ for $l = -d, -d+1 \dots c-1$ but $A_{\beta;c}^{\nu;k}(x) \neq 0$, $c < d-1$ we see that the formulae remain the same - one must simply start with the index c instead of $-d$. In the most degenerate case only $A_{\beta;d-1}^{\nu;k}(x) \neq 0$ and it will be an eigenfunction. If none of these options is true then $A_{\beta;l}^{\nu;k}(x) = 0$ for $-d \leq l < d$ and the terms corresponding to λ_0 will not contribute to the discrete spectrum part of the expansions. We shall call the poles for which this happens '$g^{[k]}$-removable'. Assume that λ_0 is not $g^{[k]}$-removable. As we explained, things in general are the same as in the case $A_{\beta;-d}^{\nu;k}(x) \neq 0$, so let us assume that this is true. Using induction it is not hard to prove that the the functions $\{A_{\beta;l}^{\nu;k}(x): d-1 \geq l \geq -d\}$ are linearly independent.

The subspace $V_{\beta}^{\nu;k}(\lambda_0)$ generated by the functions $\{A_{\beta;l}^{\nu;k}(x)\}_{l=-d}^{d-1}$ is invariant under the action of $\Lambda_{\pm}^{p}$. If one takes in it the basis

$$\{A_{\beta;-d}^{\nu;k}(x), A_{\beta;-d+1}^{\nu;k}(x), \dots, A_{\beta;d-1}^{\nu;k}\}$$

then to $\Lambda_{\pm}^{p}$ corresponds a $2d \times 2d$ matrix that is upper triangular, with constant entries and with λ_0^p on the diagonal, which is in fact the p-th power of the canonical Jordan matrix we had before. As we explained, if λ_0 is not $g^{[k]}$-removable the size of this matrix would be smaller and even reduced to a 1×1 'matrix'.

The discrete spectrum contribution from $\lambda_{\nu,j}$ in the case where we make a contraction with $[J, g]$ to the left (right) is a linear combination of terms belonging to the spaces $V_\beta^{\nu;k}(\lambda_{\nu,j})$. Thus after elimination of the '$\mathfrak{g}^{[k]}$-removable' poles we have:

Theorem 4.1 *The discrete spectrum contribution to the expansion of* L^1*-integrable function* $g\colon \mathbb{R} \mapsto \bar{\mathfrak{g}}$ *belongs to the direct sum of the subspaces* $V_\beta^{\nu;k}(\lambda_{\nu;j})$. *These subspaces are not invariant under the action of* $\Lambda_\pm$ *but under the action of* $\Lambda_\pm^p$. *On a suitable basis* $\Lambda_\pm^p$ *has block upper triangular form with* λ_j^p *on the diagonal and the blocks either have dimension* $2c_j \times 2c_j$, *where* c_j *are some positive integers less than or equal to the orders of the corresponding poles, or they are* 1×1 *blocks.*

Conclusion

The analysis of the completeness relations in the case we have considered, where we have $\mathbb{Z}_p$ reductions, shows that the role of the Recursion Operators $\Lambda_\pm$ (the operators for the system without reductions) is now played by the operators $\Lambda_\pm^p$. This completes the picture, since the fact was established earlier for the geometric aspects of the theory of the Recursion Operators (interpretation as Nijenhuis tensors for certain P-N structures) and its algebraic aspects - see [10]. Now we have it for the spectral aspect of the theory.

Acknowledgments

The author is grateful to NRF South Africa incentive grant 2013 for financial support.

Bibliography

[1] M. Ablowitz, D. Kaup, A. Newell and H. Segur, *The inverse scattering method -- Fourier analysis for nonlinear problems*, Studies in Appl. Math. **53** (1974), 249-315.

[2] R. Beals and R. Coifman, *Scattering and Inverse scattering for First Order Systems*, Comm. Pure & Appl. Math. **37** (1984), 39-90.

[3] V. Gerdjikov and A. Yanovski, *Completeness of the eigenfunctions for the Caudrey-Beals-Coifman system*, J. Math. Phys. **35** (1994), 3687-3725.

[4] V. Gerdjikov, G. Vilasi and A. Yanovski, *Integrable Hamiltonian Hierarchies - Spectral and Geometric Methods*, Springer, Heidelberg, 2008.

[5] V. Gerdjikov, N. Kostov and T. Valchev, *Generalized Zakharov-Shabat Systems and Nonlinear Evolution Equations with Deep Reductions*, In: BGSIAM'09, S. Margenov, S. Dimova and A. Slavova (Eds), Demetra, Sofia, 2010, pp 51-57.

[6] V. Gerdjikov, A. Yanovski, On Soliton Equations with $\mathbb{Z}_h$ and $\mathbb{D}_h$ Reductions: Conservation Laws and Generating Operators, J. Geom. Symm. Phys. (JSPG) **31** (2013), 57-92.

[7] M. Goto and F. Grosshans, *Semisimple Lie Algebras. Lecture Notes in Pure and Applied Mathematics* **38**, M. Dekker Inc., New-York & Basel, 1978.

[8] F. Magri, A simple model of the integrable Hamiltonian equations, J. Math. Phys. **19** (1978), 1156-1162.

[9] A. Mikhailov, The reduction problem and inverse scattering method, Physica **3D** (1981), 73-117.

[10] A. Yanovski, Geometry of the Recursion Operators for Caudrey-Beals-Coifman system in the presence of Mikhailov $\mathbb{Z}_p$ reductions, J. Geom. Symm. Phys. **25** (2012), 77-97.

[11] A. Yanovski, Recursion Operators and Expansions over adjoint solutions for the Caudrey-Beals-Coifman system with $\mathbb{Z}_p$ reductions of Mikhailov type, J. Geom. Symm. Phys. (JSPG) **30**, 105-119, (2013)

[12] V. Zakharov, S. Manakov, S. Novikov and L. Pitaevskii, *Theory of solitons: the inverse scattering method*, Consultants Bureau, branch of Plenum Publishing Company, New York, 1984.

MKDV-TYPE OF EQUATIONS RELATED TO $\mathfrak{sl}(N, \mathbb{C})$ ALGEBRA

V. S. GERDJIKOV, D. M. MLADENOV, A. A. STEFANOV AND S. K. VARBEV

1 Introduction

The general theory of nonlinear evolution equations (NLEE) allowing Lax representation is well developed [1, 2, 3, 4, 5, 6, 7]. This paper deals with NLEE that allow Lax representation with deep reductions. This means that they can be written as the commutativity condition of two ordinary differential operators of the type

$$\begin{aligned} L\psi &\equiv i\frac{\partial\psi}{\partial x} + U(x,t,\lambda)\psi = 0, \\ M\psi &\equiv i\frac{\partial\psi}{\partial t} + V(x,t,\lambda)\psi = \psi C(\lambda), \end{aligned} \tag{1}$$

where $U(x,t,\lambda)$, $V(x,t,\lambda)$ and $C(\lambda)$ are some polynomials of λ to be defined below. We request also that the Lax pair (1) possesses $\mathbb{Z}_N$-reduction groups [8]. For the case of $\mathbb{Z}_N$-reduction this means that we impose on (1) and (2) a $\mathbb{Z}_N$-reduction by [8]

$$C_1 U(x,t,\lambda) C_1^{-1} = U(x,t,\omega\lambda), \quad C_1 V(x,t,\lambda) C_1^{-1} = V(x,t,\omega\lambda), \tag{2}$$

where $C_1^N = \mathbb{1}$ is a Coxeter automorphism of the algebra $\mathfrak{sl}(N,\mathbb{C})$ and $\omega = \exp(2\pi i/N)$.

Below we consider only the simplest possible case, when the underlying algebra is $\mathfrak{sl}(N,\mathbb{C})$ and the group of reduction is $\mathbb{Z}_N$. The class of relevant NLEE may be considered as generalizations of the derivative NLS equations [9, 10] - see also [8, 11].

$$i\frac{\partial\psi_k}{\partial t} + \gamma\frac{\partial}{\partial x}\left(\cot(\pi k/N)\cdot\psi_{k,x} + i\sum_{p=1}^{N-1}\psi_p\psi_{k-p}\right) = 0, \tag{3}$$

$k = 1,2,\dots,N-1$, where γ is a constant and the index $k-p$ should be understood modulus N and $\psi_0 = \psi_N = 0$.

Section 2 contains preliminaries necessary to derive the NLEE. In particular we provide a convenient basis for $\mathfrak{sl}(N,\mathbb{C})$ which is compatible with the $\mathbb{Z}_N$-reduction. In Section 3 we derive MKdV equations for any N. In section 4 we derive the evolution equations for scattering matrix of the Lax operator. In Section 5 we show that additional $\mathbb{Z}_2$-reductions can be imposed on the MKdV equations. We also give several particular examples of these equations.

2 Preliminaries

Let us consider the Lax operator (1). To this end we will use a convenient basis in the Lie algebra $\mathfrak{sl}(N,\mathbb{C})$ which is compatible with the $\mathbb{Z}_N$-reduction. Here and below all indices are understood modulus N. The automorphism Ad_{C_1} ($Ad_{C_1}(Y) \equiv C_1 Y C_1^{-1}$ for every Y from $\mathfrak{g}$) defines a grading in the Lie algebra

$$\mathfrak{sl}(N,\mathbb{C}) = \bigoplus_{k=0}^{N-1} \mathfrak{g}^{(k)}, \tag{4}$$

where $\mathfrak{g}^{(k)}$ is the eigenspace of Ad_{C_1} corresponding to its eigenvalue ω^{-k}, $k = 0,1,\dots,N-1$. The calculations are much simpler if we introduce a convenient basis in $\mathfrak{g}^{(k)}$ compatible with the grading:

$$J_s^{(k)} = \sum_{j=1}^{N} \omega^{kj} E_{j,j+s}, \qquad C^{-1} J_s^{(k)} C = \omega^{-k} J_s^{(k)}, \tag{5}$$

where $E_{j,s}$ is an $N \times N$ matrix defined by $(E_{j,s})_{q,r} = \delta_{jq}\delta_{sr}$. Obviously, $J_s^{(k)}$ satisfies the commutation relations

$$[J_s^{(k)}, J_l^{(m)}] = (\omega^{ms} - \omega^{kl}) J_{s+l}^{(k+m)}. \tag{6}$$

3 Derivation of the equations

We choose $U(x,t,\lambda)$ and $V(x,t,\lambda)$ as follows:

$$U(x,t,\lambda) = Q(x,t) - \lambda J, \quad Q(x,t) = \sum_{j=1}^{N-1} \psi_j(x,t) J_j^{(0)}, \quad J = aJ_0^{(1)} \tag{7}$$

$$V(x,t,\lambda) = V_3(x,t) + \lambda V_2(x,t) + \lambda^2 V_1(x,t) - \lambda^3 K, \tag{8}$$

where

$$\begin{aligned} V_1(x,t) &= \sum_{k=1}^{N} v_k^1(x,t) J_k^{(2)}, \qquad & V_2(x,t) &= \sum_{l=1}^{N} v_l^2(x,t) J_l^{(1)}, \\ V_3(x,t) &= \sum_{j=1}^{N-1} v_j^3(x,t) J_j^{(0)}, & K &= bJ_0^{(3)}. \end{aligned} \tag{9}$$

The constants a and b determine the dispersion law of the MKdV equations.

The next step is to request that $[L, M] = 0$ identically with respect to λ. This leads to a set of recursion relations, generalizing the ones in [1], which allow us to express $V_k(x,t)$, $k = 1,2,3$ through $Q(x,t)$ and its x-derivatives. Skipping the details we get:

$$v_k^1(x,t) = \frac{b}{a}(\omega^{2k} + \omega^k + 1)\psi_k, \qquad k = 1, \dots, N-1, \tag{10}$$

and $v_N^1 = C(t)$ with $C(t)$ being an arbitrary function of time. For

$$\begin{aligned} v_l^2(x,t) &= \frac{b}{a^2} \sum_{j+k=l}^{N-1} \frac{\omega^{2l} + \omega^{2j+k} - \omega^k - 1}{1-\omega^l} \psi_j \psi_k \\ &\quad + i\frac{b}{a^2}\left(\frac{\omega^{2l} + \omega^l + 1}{1-\omega^l}\right)\frac{\partial \psi_l}{\partial x} - \frac{C}{a}(\omega^l + 1)\psi_l, \end{aligned} \tag{11}$$

for $l = 1, \dots, N-1$ and

$$v_N^2 = -\frac{b}{a^2} \sum_{j+l=0}^{N-1} \left(\cos\frac{2\pi j}{N} + \frac{1}{2}\right)\psi_j \psi_l + D(t), \tag{12}$$

with D(t) being another arbitrary function of time. And for

$$v_j^3 = \frac{b}{a^3} \cot\left(\frac{\pi j}{N}\right) \sum_{k+l=j}^{N-1} \frac{\partial}{\partial x}(\psi_k \psi_l) + \frac{C}{a^2} \sum_{m+l=j}^{N-1} (\psi_m \psi_l)$$

$$+\frac{b}{2a^3}\sum_{k+l=j}^{N-1}\frac{\cos\frac{\pi(k-l)}{N}}{\sin\frac{\pi j}{N}}\frac{\partial}{\partial x}(\psi_k\psi_l)-\frac{D}{a}\psi_j$$

$$+\frac{b}{a^3}\sum_{l+m=j}^{N-1}\sum_{i+k=l}^{N-1}(\psi_i\psi_k\psi_m)+\frac{3b}{2a^3}\sum_{l+m=j}^{N-1}\cot\left(\frac{\pi l}{N}\right)\frac{\partial\psi_l}{\partial x}\psi_m$$

$$+\frac{b}{a^3}\sum_{l+m=j}^{N-1}\sum_{i+k=l}^{N-1}\frac{\sin\frac{\pi(j-2k)}{N}-\sin\frac{\pi(j-2m)}{N}}{\sin\frac{\pi j}{N}}(\psi_i\psi_k\psi_m)$$

$$-\frac{b}{4a^3}\cot\left(\frac{\pi j}{N}\right)\sum_{l+m=j}^{N-1}\frac{\partial}{\partial x}(\psi_l\psi_m)+\frac{C}{a^2}\cot\left(\frac{\pi j}{N}\right)\frac{\partial\psi_j}{\partial x}$$

$$-\frac{b}{2a^3}\sum_{l+m=j}^{N-1}\frac{\cos\frac{\pi(l-m)}{N}}{\sin\frac{\pi j}{N}}\frac{\partial}{\partial x}(\psi_l\psi_m)+\frac{b}{a^3}\left(\cot^2\frac{\pi j}{N}-\frac{1}{4\sin^2\frac{\pi j}{N}}\right)\frac{\partial^2\psi_j}{\partial x^2}$$

$$+\frac{b}{a^3}\sum_{k=1}^{N-1}\left(\cos\frac{2\pi k}{N}+\frac{1}{2}\right)(\psi_k\psi_{N-k}\psi_j) \tag{13}$$

where j is running from 1 to N-1. We choose $C(t)=0$ and $D(t)=0$.

In the end for Q(x,t) we get

$$\alpha\frac{\partial\psi_j}{\partial t}=\left(\cot^2\frac{\pi j}{N}-\frac{1}{4\sin^2\frac{\pi j}{N}}\right)\frac{\partial^3\psi_j}{\partial x^3}+\sum_{l+m=j}^{N-1}\sum_{i+k=l}^{N-1}\frac{\partial}{\partial x}(\psi_i\psi_k\psi_m)$$

$$+\sum_{l+m=j}^{N-1}\sum_{i+k=l}^{N-1}\frac{\sin\frac{\pi(j-2k)}{N}-\sin\frac{\pi(j-2m)}{N}}{\sin\frac{\pi j}{N}}\frac{\partial}{\partial x}(\psi_i\psi_k\psi_m)$$

$$+\sum_{k=1}^{N-1}\left(\cos\frac{2\pi k}{N}+\frac{1}{2}\right)\frac{\partial}{\partial x}(\psi_k\psi_{N-k}\psi_j)+\frac{3}{4}\cot\left(\frac{\pi j}{N}\right)\sum_{k+l=j}^{N-1}\frac{\partial^2}{\partial x^2}(\psi_k\psi_l)$$

$$+\frac{3}{4}\sum_{k+l=j}^{N-1}\frac{\partial}{\partial x}\left(\cot\left(\frac{\pi l}{N}\right)\frac{\partial\psi_l}{\partial x}\psi_k+\cot\left(\frac{\pi k}{N}\right)\frac{\partial\psi_k}{\partial x}\psi_l\right) \tag{14}$$

where $\alpha=a^3/b$.

4 The evolution of the scattering matrix

Let us introduce the Jost solutions $\phi_{\pm}(x,t,\lambda)$ of the Lax pair by:

$$\lim_{x\to-\infty}\phi_-(x,t)e^{i\lambda J_0^{(1)}x} = \mathbb{1}, \quad \lim_{x\to\infty}\phi_+(x,t)e^{i\lambda J_0^{(1)}x} = \mathbb{1}, \tag{15}$$

They are related by:

$$\phi_-(x,t,\lambda) = \phi_+(x,t,\lambda)T(\lambda,t) \tag{16}$$

where $T(\lambda,t)$ is known as the scattering matrix. Both Jost solutions $\phi_{\pm}(x,t,\lambda)$ satisfy equations (1). Let us now calculate the limit

$$\lim_{x\to\infty} M\phi_+(x,t) \quad = (i\frac{\partial}{\partial t} - \lambda D - \lambda^2 C - \lambda^3 K)e^{-i\lambda J_0^{(1)}x} = e^{-i\lambda J_0^{(1)}x}C(\lambda). \tag{17}$$

Assuming that the definitions of the Jost solutions are t-independent we find that

$$C(\lambda,t) = -\lambda D - \lambda^2 C - \lambda^3 K. \tag{18}$$

Next we calculate

$$\begin{aligned}\lim_{x\to\infty} M\phi_-(x,t) \quad &= (i\frac{\partial}{\partial t} - C(\lambda))e^{-i\lambda J_0^{(1)}x}T(\lambda,t) \\ &= e^{-i\lambda J_0^{(1)}x}(i\frac{\partial T}{\partial t} - C(\lambda)T(\lambda,t)) = e^{-i\lambda J_0^{(1)}x}T(\lambda,t)C(\lambda)\end{aligned} \tag{19}$$

Therefore, if $Q(x,t)$ satisfies the MKdV equations (14) the scattering matrix $T(\lambda,t)$ must satisfy the following linear evolution equation:

$$i\frac{\partial T}{\partial t} - [C(\lambda), T(\lambda,t)] = 0. \tag{20}$$

In the particular case when $C = D = 0$ we get:

$$i\frac{\partial T}{\partial t} + \lambda^3[K, T(\lambda,t)] = 0, \qquad (21)$$

whose solution is

$$T_{ij}(\lambda,t) = e^{i\lambda^3(\omega^{3i}-\omega^{3j})t}T_{ij}(\lambda,0). \qquad (22)$$

Thus $T_{ij}(\lambda,0)$ is the Cauchy data for the initial conditions of the scattering matrix, therefore solving the MKdV equations. (14) is reduced to solving the direct and the inverse scattering problem for the Lax operator L: see [8, 10, 12].

5 Additional Involutions and Examples

Along with the $\mathbb{Z}_N$-reduction (2), we can introduce one of the following involutions ($\mathbb{Z}_2$-reductions):

$$\begin{array}{llll}
a) & K_0^{-1}U^\dagger(x,t,\kappa_1(\lambda))K_0 = U(x,t,\lambda), & \kappa_1(\lambda) = \omega^{-1}\lambda^* \\
b) & K_0^{-1}U^*(x,t,\kappa_1(\lambda))K_0 = -U(x,t,\lambda), & \kappa_1(\lambda) = -\omega^{-1}\lambda^* \\
c) & U^T(x,t,-\lambda) = -U(x,t,\lambda), &
\end{array}$$

(23)

where $K_0^2 = \mathbb{1}$. We choose

$$K_0 = \sum_{k=1}^{N} E_{k,N-k+1}.$$

The action of K_0 on the basis is as follows:

$$K_0\left(J_s^{(k)}\right)^\dagger K_0 = \omega^{k(s-1)}J_s^{(k)}, \qquad K_0\left(J_s^{(k)}\right)^* K_0 = \omega^{-k}J_{-s}^{(k)}, \qquad (24)$$

from which we derive the reductions below.

Immediate consequences of eq. (23) are the constraints on the potentials:

$$
\begin{array}{llll}
a) & K_0^{-1}Q^{\dagger}(x,t)K_0 = Q(x,t), & K_0^{-1}(J_0^{(1)})^{\dagger}K_0 & = \omega^{-1}J_0^{(1)}, \\
b) & K_0^{-1}Q^{*}(x,t)K_0 = -Q(x,t), & K_0^{-1}(J_0^{(1)})^{*}K_0 & = \omega^{-1}J_0^{(1)}, \\
c) & Q^{T}(x,t) = -Q(x,t), & (J_0^{(1)})^{T} & = J_0^{(1)}.
\end{array}
$$

(25)

More specifically from eq. (25) it follows that each of the algebraic relations below:

$$
\begin{array}{llll}
a) & \psi_j^*(x,t) = \psi_j(x,t), & \alpha = \alpha^*, \\
b) & \psi_j^*(x,t) = -\psi_{N-j}(x,t), & \alpha = \alpha^*, \\
c) & \psi_j(x,t) = -\psi_{N-j}(x,t), &
\end{array}
\tag{26}
$$

where $j = 1, \dots, N-1$, are compatible with the evolution of the MKdV eqs. (14).

6 Some particular cases

Special examples of DNLS systems of equations can be found in [10].

In the case of $\mathfrak{sl}(2,\mathbb{C})$ algebra we obtain the well-known MKdV equation

$$
\alpha \frac{\partial \psi_1}{\partial t} = -\frac{1}{4}\frac{\partial^3 \psi_1}{\partial x^3} - \frac{1}{2}\frac{\partial}{\partial x}(\psi_1^3). \tag{27}
$$

In the case of $\mathfrak{sl}(3,\mathbb{C})$ algebra we have the system of trivial equations $\partial_t \psi_1 = 0$ and $\partial_t \psi_2 = 0$.

And finally in the case of $\mathfrak{sl}(4,\mathbb{C})$ algebra we find a new system of exactly integrable nonlinear partial differential equations

$$
\alpha \frac{\partial \psi_1}{\partial t} = \frac{1}{2}\frac{\partial^3 \psi_1}{\partial x^3} + \frac{3}{2}\frac{\partial}{\partial x}\left(\frac{\partial \psi_2}{\partial x}\psi_3\right) + \frac{3}{2}\frac{\partial}{\partial x}(\psi_1\psi_2^2) + \frac{\partial}{\partial x}(\psi_3^3), \tag{28}
$$

$$
\alpha \frac{\partial \psi_2}{\partial t} = -\frac{1}{4}\frac{\partial^3 \psi_2}{\partial x^3} + \frac{3}{4}\frac{\partial^2}{\partial x^2}(\psi_1^2) - \frac{3}{4}\frac{\partial^2}{\partial x^2}(\psi_3^2)
$$

$$
+3\frac{\partial}{\partial x}(\psi_1\psi_2\psi_3) - \frac{1}{2}\frac{\partial}{\partial x}(\psi_2^3) \tag{29}
$$

$$\alpha\frac{\partial\psi_3}{\partial t}=\frac{1}{2}\frac{\partial^3\psi_3}{\partial x^3}-\frac{3}{2}\frac{\partial}{\partial x}\left(\psi_1\frac{\partial\psi_2}{\partial x}\right)+\frac{3}{2}\frac{\partial}{\partial x}(\psi_2^2\psi_3)+\frac{\partial}{\partial x}(\psi_1^3). \qquad (30)$$

If we apply case a) of eq. (26) we get the same set of MKdV equations with ψ_1,ψ_2 and ψ_3 purely real functions.

In the case b) we put $\psi_1=-\psi_3^*=u$ and $\psi_2=-\psi_2^*=iv$ and get:

$$\begin{aligned}\alpha\frac{\partial v}{\partial t}&=-\frac{1}{4}\frac{\partial^3 v}{\partial x^3}+\frac{3}{4i}\frac{\partial^2}{\partial x^2}(u^2-u^{*,2})-3\frac{\partial}{\partial x}(|u|^2v)+\frac{1}{2}\frac{\partial}{\partial x}(v^3),\\ \alpha\frac{\partial u}{\partial t}&=\frac{1}{2}\frac{\partial^3 u}{\partial x^3}-i\frac{3}{2}\frac{\partial}{\partial x}\left(u^*\frac{\partial v}{\partial x}\right)-\frac{3}{2}\frac{\partial}{\partial x}(uv^2)-\frac{\partial}{\partial x}((u^*)^3),\end{aligned} \qquad (31)$$

where u is a complex function, but v is a purely real function.

In the case c):

$$\alpha\frac{\partial u}{\partial t}=\frac{1}{2}\frac{\partial^3 u}{\partial x^3}-\frac{\partial}{\partial x}(u^3), \qquad (32)$$

where u is a complex function, we recover the well known MKdV equation. And finally in the case of $\mathfrak{sl}(6,\mathbb{C})$ algebra with $\mathbb{D}_6$-reduction in the case c) we find

$$\begin{aligned}\alpha\frac{\partial u}{\partial t}&=2\frac{\partial^3 u}{\partial x^3}-2\sqrt{3}\frac{\partial}{\partial x}\left(u\frac{\partial v}{\partial x}\right)-6\frac{\partial}{\partial x}(uv^2),\\ \alpha\frac{\partial v}{\partial t}&=\sqrt{3}\frac{\partial^2}{\partial x^2}(u^2)-6\frac{\partial}{\partial x}(u^2v),\end{aligned} \qquad (33)$$

where u and v are complex functions.

7 Discussion and conclusions

In the present paper we have derived the systems of MKdV equations related to the classical series of $\mathfrak{sl}(N,\mathbb{C})$ Lie algebras. These equations belong to the hierarchy which contains the two-dimensional Toda field theories related to $\mathfrak{sl}(N,\mathbb{C})$ discovered by Mikhailov [8]. The corresponding Lax operator L is endowed with a $\mathbb{Z}_N$-reduction [8]. We have also demonstrated several examples that are obtained from the generic MKdV by imposing additional $\mathbb{Z}_2$-reductions.

These results can be extended also to the other classical series of simple Lie algebras.

Acknowledgements: The work is supported in part by the ICTP-SEENET-MTP project PRJ-09.

Bibliography

[1] M. J. Ablowitz, D. J. Kaup, A. C. Newell, H. Segur. *The inverse scattering transform -- Fourier analysis for nonlinear problems.* Studies in Appl. Math. **53**, n. 4, 249-315, 1974.

[2] F. Calogero, A. Degasperis. *Spectral transform and solitons. Vol. I.* North Holland, Amsterdam, 1982.

[3] L. D. Faddeev, L. A. Takhtadjan. *Hamiltonian methods in the theory of solitons.* (Springer Verlag, Berlin, 1987).

[4] V. S. Gerdjikov, P. P. Kulish. *The generating operator for the $n \times n$ linear system.* Physica D, **3D**, n. 3, 549-564, 1981.

[5] V. Drinfel'd, V. V. Sokolov. *Lie Algebras and equations of Korteweg - de Vries type.* Sov. J. Math. **30**, 1975-2036 (1985).

[6] V. S. Gerdjikov. Generalised Fourier transforms for the soliton equations. Gauge covariant formulation. Inverse Problems **2,** n. 1, 51-74, (1986).

[7] V. E. Zakharov, S. V. Manakov, S. P. Novikov, L. I. Pitaevskii. *Theory of solitons: the inverse scattering method.* (Plenum, N.Y.: Consultants Bureau, 1984).

[8] A. V. Mikhailov *The reduction problem and the inverse scattering problem.* Physica D, **3D**, n. 1/2, 73-117, 1981.

[9] V. S. Gerdjikov. *Z_N--reductions and new integrable versions of derivative nonlinear Schrödinger equations.* In: Nonlinear evolution equations: integrability and spectral methods, Ed. A. P. Fordy, A. Degasperis, M. Lakshmanan, Manchester University Press, (1981), p. 367-372.

[10] V. S. Gerdjikov. Derivative Nonlinear Schrödinger Equations with $\mathbb{Z}_N$ and $\mathbb{D}_N$--Reductions. Romanian Journal of Physics, **58**, Nos. 5-6, 573-582 (2013).

[11] D. J. Kaup, A. C. Newell. Soliton equations, singular dispersion relations and moving eigenvalues. Adv. Math. **31**, 67-100, 1979.

[12] V. S. Gerdjikov, A. B. Yanovski On soliton equations with $\mathbb{Z}_h$ and $\mathbb{D}_h$ reductions: conservation laws and generating operators. J. Geom. Symmetry Phys. **31**, 57-92 (2013).

ON A ONE-PARAMETER FAMILY OF MKDV EQUATIONS RELATED TO THE $\mathfrak{so}(8)$ LIE ALGEBRA

V. S. GERDJIKOV, D. M. MLADENOV, A. A. STEFANOV AND S. K. VARBEV

1 Introduction

The inverse scattering method [1, 2, 3], combined with Mikhailov's group of reductions [4], has led to the discovery of classes of important integrable nonlinear evolution equations (NLEE). Some of the most interesting examples of such equations are the 2-dimensional Toda field theories [4] and higher representatives from their hierarchies [5, 6, 7, 8, 9, 10].

Our aim is to derive a one-parameter family of MKdV equations related to the simple Lie algebra $\mathfrak{so}(8)$ using the procedure introduced by Mikhailov [4]. They admit a Lax pair

$$L\psi \equiv i\frac{\partial\psi}{\partial x} + U(x,t,\lambda)\psi = 0, \quad M\psi \equiv i\frac{\partial\psi}{\partial t} + V(x,t,\lambda)\psi = \psi C(\lambda), \tag{1}$$

satisfying the reduction condition

$$C(U(x,t,\lambda)) = U(x,t,\omega\lambda), \qquad C(V(x,t,\lambda)) = V(x,t,\omega\lambda). \tag{2}$$

A key motivation for choosing $\mathfrak{so}(8)$ is the unique symmetry of its Dynkin diagram [11,12], see Fig.1. This is combined with the fact that $\mathfrak{so}(8)$ is the only simple Lie algebra of rank 4 that has 3 as a double-valued exponent. For more details about the root system and the Cartan-Weyl basis of $\mathfrak{so}(8)$ see the Appendix. Of course these special properties of the algebra $\mathfrak{so}(8)$ will be reflected in the properties of the resulting MKdV equations.

The paper is organized as follows. Section 2 contains some preliminaries needed to derive the equations. We start with the Lax representation which is subject to a $\mathbb{Z}_h$ – reduction group [4], where h=6 is the Coxeter number of $\mathfrak{so}(8)$. In Section 3 we derive the one-parameter family of MKdV equations. In the next Section we derive the time-dependence of the scattering matrix of the Lax operator L. We end with a discussion on the possibilities of imposing additional $\mathbb{Z}_2$-reductions on the equations. The Appendix contains the relevant information about the root system of $\mathfrak{so}(8)$ and its Cartan-Weyl basis.

2 Preliminaries

We assume that the reader is familiar with the theory of semisimple Lie algebras [11, 12]; see also the Appendix. By H_i we will denote elements of the Cartan subalgebra, by E_β the Weyl generator corresponding to the root β, and by α_i the simple roots. The Coxeter number for $\mathfrak{so}(8)$ is 6, and its rank is 4. We denote the Killing form of X and Y by <X,Y>.

The Coxeter automorphism is given by

$$C(X) = cXc^{-1} \tag{3}$$

for every generator X, where c is

$$c = \begin{pmatrix} 0 & -1 & 0 & 0 & 0 & 0 & 0 & 0 \\ 0 & 0 & 0 & 0 & 0 & -1 & 0 & 0 \\ 1 & 0 & 0 & 0 & 0 & 0 & 0 & 0 \\ 0 & 0 & 0 & 0 & 1 & 0 & 0 & 0 \\ 0 & 0 & 0 & 1 & 0 & 0 & 0 & 0 \\ 0 & 0 & 0 & 0 & 0 & 0 & 0 & 1 \\ 0 & 0 & 1 & 0 & 0 & 0 & 0 & 0 \\ 0 & 0 & 0 & 0 & 0 & 0 & 1 & 0 \end{pmatrix}. \tag{4}$$

The Coxeter automorphism introduces a grading in $\mathfrak{so}(8)$ by

$$\mathfrak{so}(8) = \bigoplus_{k=0}^{5} \mathfrak{g}^{(k)}, \tag{5}$$

such that

$$C(X) = \omega^k X, \quad \omega = \exp\left(\frac{2\pi i k}{6}\right), \quad \forall X \in \mathfrak{g}^{(k)}. \tag{6}$$

The grading condition also holds

$$[\mathfrak{g}^{(k)}, \mathfrak{g}^{(l)}] \subset \mathfrak{g}^{(k+l)}, \tag{7}$$

where $k + l$ is taken as modulo h. The Cartan-Weyl basis is introduced in the Appendix. Here we introduce a basis compatible with the grading

$$\mathfrak{so}(8) = l.c.\{\mathcal{E}_i^{(k)}, \mathcal{H}_j^{(k)}\}, \tag{8}$$

where

$$\mathcal{E}_i^{(k)} = \frac{1}{6}\sum_{s=0}^{5} \omega^{-sk} C^s(E_{\alpha_i}), \qquad \mathcal{H}_j^{(k)} = \frac{1}{6}\sum_{s=0}^{5} \omega^{-sk} C^s(H_j). \tag{9}$$

Note that $\mathcal{H}_j^{(k)}$ is non – vanishing only if k is an exponent. The exponents of $\mathfrak{so}(8)$ are: 1,3,5,3. Since 3 is a double-valued exponent, $\mathfrak{g}^{(3)}$ will contain two Cartan elements. In particular:

$$\begin{aligned} \mathcal{H}_1^{(1)} &= H_{e_1} + \omega^{-1} H_{e_2} + \omega H_{e_3}, & \mathcal{H}_1^{(3)} &= H_{e_1} + H_{e_2} + H_{e_3}, \\ \mathcal{H}_2^{(3)} &= H_{e_4}, & \mathcal{H}_1^{(5)} &= H_{e_1} + \omega H_{e_2} + \omega^{-1} H_{e_3}. \end{aligned} \tag{10}$$

3 Derivation of the equations

We start with a Lax pair of the form

$$\begin{aligned} L &= i\,\partial_x + Q(x,t) - \lambda J, \\ M &= i\,\partial_t + V^{(0)}(x,t) + \lambda V^{(1)}(x,t) + \lambda^2 V^{(2)}(x,t) - \lambda^3 K, \end{aligned} \tag{11}$$

where

$$Q(x,t) \in g^{(0)}, \quad V^{(k)}(x,t) \in g^{(k)}, \quad K \in g^{(3)}, \quad J = 3\mathcal{H}_1^{(1)}. \qquad (12)$$

We will assume that $Q(x,t)$ and $V^{(k)}(x,t)$ have the following form:

$$Q(x,t) = \sum_{i=1}^{4} q_i(x,t) 6\mathcal{E}_i^{(0)}, \qquad V^{(0)}(x,t) = \sum_{i=1}^{4} v_i^{(0)}(x,t) 6\mathcal{E}_i^{(0)}, \qquad (13)$$

$$\begin{aligned} V^{(1)}(x,t) &= 6\sum_{i=1}^{4} v_i^{(1)}(x,t)\mathcal{E}_i^{(1)} + 6v_5^{(1)}\mathcal{H}_1^{(1)}, \\ V^{(2)}(x,t) &= 6\sum_{i=1}^{4} v_i^{(2)}(x,t)\mathcal{E}_i^{(2)}. \end{aligned} \qquad (14)$$

To simplify the notation we will omit writing any explicit dependence on x and t.

We require that $[L,M] = 0$ for any λ. The first consequence of this is $[J,K] = 0$. Since $K \in \mathrm{g}^{(3)}$ we have:

$$K = a3\mathcal{H}_1^{(3)} + b\mathcal{H}_2^{(3)}. \qquad (15)$$

We can always absorb one of the parameters by redefining t, so we will have effectively a one − parameter set of matrices K, which, as we shall see below, determine the dispersion law of the relevant NLEE. Thus we will get a one-parameter family of equations of MKdV type.

The condition $[L,M] = 0$ leads to a set of recurrent relations (see [2,9,10]) which allow us to determine $V^{(k)}(x,t)$ in terms of the potential $Q(x,t)$ and its x-derivatives. Skipping the details we give the result:

$$\begin{aligned} v_1^{(2)} &= 2\omega a q_1, & v_2^{(2)} &= 0, \\ v_3^{(2)} &= -\omega(a+b)q_3, & v_4^{(2)} &= -\omega(a-b)q_4. \end{aligned} \qquad (16)$$

In calculating $V^{(1)}$ we have to take into account that $g^{(1)}$ has a nontrivial intersection with the Cartan subalgebra $\mathfrak{h}$: $g^{(1)} \times \mathfrak{h} \neq \emptyset$. Thus along with the off-diagonal elements of $V^{(1)}$

$$\begin{aligned} v_1^{(1)} &= -\tfrac{2}{3}\sqrt{3}a(\omega+1)(\partial_x q_1 - \sqrt{3}q_4 q_3 + \sqrt{3}q_2 q_1) \\ v_2^{(1)} &= 2aq_1^2 + (a+b)q_3^2 + (a-b)q_4^2, \\ v_3^{(1)} &= (\omega+1)(a+b)\left(\tfrac{\sqrt{3}}{3}\partial_x q_3 + q_1 q_4 - q_2 q_3\right), \\ v_4^{(1)} &= (\omega+1)(a-b)\left(\tfrac{\sqrt{3}}{3}\partial_x q_4 + q_1 q_3 - q_2 q_5\right). \end{aligned} \quad (17)$$

We have to calculate also $v_5^{(1)}$. Using a well-known technique from the theory of recursion operators [2, 7, 10] we solve a simple differential equation with the result:

$$v_5^{(1)} = aq_1^2 + \tfrac{1}{2}(a+b)q_3^2 + \tfrac{1}{2}(a-b)q_4^2. \quad (18)$$

Thus for $v_i^{(0)}$ we obtain:

$$\begin{aligned} v_1^{(0)} &= 2a(\partial_x^2 q_1 - \sqrt{3}q_1\,\partial_x q_2) - \sqrt{3}((3a+b)q_4\,\partial_x q_3 + (3a-b)q_3\,\partial_x q_4) \\ &\quad -3q_1(2aq_2^2 + (a-b)q_3^2 + (a+b)q_4^2), \\ v_2^{(0)} &= \sqrt{3}a\,\partial_x q_1^2 + \frac{\sqrt{3}}{2}(a+b)\,\partial_x q_3^2 + \frac{\sqrt{3}}{2}(a-b)\,\partial_x q_4^2 \\ &\quad -3q_2(2aq_1^2 + (a+b)q_3^2 + (a-b)q_4^2), \\ v_3^{(0)} &= -(a+b)(\partial_x^2 q_3 - \sqrt{3}q_3\,\partial_x q_2) - \sqrt{3}((3a+b)q_4\,\partial_x q_1 + 2bq_1\,\partial_x q_4) \\ &\quad +3q_3(2aq_4^2 + (a-b)q_1^2 + (a+b)q_2^2), \\ v_4^{(0)} &= -(a-b)(\partial_x^2 q_4 - \sqrt{3}q_4\,\partial_x q_2) - \sqrt{3}((3a-b)q_3\,\partial_x q_1 - 2bq_1\,\partial_x q_3) \\ &\quad +3q_4(2aq_3^2 + (a-b)q_2^2 + (a+b)q_1^2). \end{aligned} \quad (19)$$

And finally, the λ-independent terms in the Lax representation provide the equations

$$\partial_t q_1 = 2a\,\partial_x^3 q_1 - 3\,\partial_x[q_1(2aq_2^2 + (a-b)q_3^2 + (a+b)q_4^2)]$$

$$-\sqrt{3}[(3a+b)\,\partial_x(q_4\,\partial_x q_3) + (3a-b)\,\partial_x(q_3\,\partial_x q_4) - 2a\,\partial_x(q_1\,\partial_x q_2)], \quad (20)$$

$$\partial_t q_2 = \sqrt{3}a\,\partial_x^2 q_1^2 + \tfrac{\sqrt{3}}{2}(a+b)\,\partial_x^2 q_3^2 + \tfrac{\sqrt{3}}{2}(a-b)\,\partial_x^2 q_4^2$$

$$-3\,\partial_x[q_2(2aq_1^2+(a+b)q_3^2+(a-b)q_4^2)], \tag{21}$$

$$\partial_t q_3 = -(a+b)\,\partial_x^3 q_3 - \sqrt{3}[(3a+b)\,\partial_x(q_4\,\partial_x q_1) + 2b\,\partial_x(q_1\,\partial_x q_4)]$$

$$+3\,\partial_x[q_3(2aq_4^2+(a-b)q_1^2+(a+b)q_2^2)] + \sqrt{3}(a+b)\,\partial_x(q_3\,\partial_x q_2), \tag{22}$$

$$\partial_t q_4 = -(a-b)\,\partial_x^3 q_4 - \sqrt{3}[(3a-b)\,\partial_x(q_3\,\partial_x q_1) - 2b\,\partial_x(q_1\,\partial_x q_3)]$$

$$+3\,\partial_x[q_4(2aq_3^2+(a-b)q_2^2+(a+b)q_1^2)] + \sqrt{3}(a-b)\,\partial_x(q_4\,\partial_x q_2). \tag{23}$$

As we mentioned above, we can rescale $t \to \tau = t/a$, which will result in replacing in the above equations $a = 1$. However, the second parameter $b \to b/a$ will remain.

We end this section by a particular representative of this family obtained by putting $a = b = 1$:

$$\partial_t q_1 = 2\big[\partial_x^3 q_1 - \sqrt{3}\,\partial_x(q_1\,\partial_x q_2) - \sqrt{3}(2\,\partial_x(q_4\,\partial_x q_3)$$

$$+\,\partial_x(q_3\,\partial_x q_4)) - 3\,\partial_x(q_1(q_2^2+q_4^2))] \tag{24}$$

$$\partial_t q_2 = \sqrt{3}\,\partial_x^2(q_1^2+q_3^2) - 6\,\partial_x(q_2(q_1^2+q_3^2)), \tag{25}$$

$$\partial_t q_3 = -2\big[\partial_x^3 q_3 - \sqrt{3}\,\partial_x(q_3\,\partial_x q_2) - \sqrt{3}(2\,\partial_x(q_4\,\partial_x q_1)$$

$$+\,\partial_x(q_1\,\partial_x q_4)) + 3\,\partial_x(q_3(q_2^2+q_4^2))], \tag{26}$$

$$\partial_t q_4 = -2\left[\sqrt{3}\,\partial_x(q_3\,\partial_x q_1) - \sqrt{3}(\partial_x(q_1\,\partial_x q_3)) + 3\,\partial_x(q_4(q_3^2 + q_1^2))\right]. \tag{27}$$

4 The evolution of the scattering matrix

Let us introduce the Jost solutions $\phi_{\pm}(x,t,\lambda)$ of the Lax pair by:

$$\lim_{x\to-\infty} \phi_-(x,t)e^{i\lambda Jx} = \mathbb{1}, \qquad \lim_{x\to\infty} \phi_+(x,t)e^{i\lambda Jx} = \mathbb{1}. \tag{28}$$

The Jost solutions are related by:

$$\phi_-(x,t,\lambda) = \phi_+(x,t,\lambda)T(\lambda,t), \tag{29}$$

where $T(\lambda,t)$ is known as the scattering matrix. Both $\phi_{\pm}(x,t,\lambda)$ satisfy equations (1). Let us now calculate the limit

$$\lim_{x\to\infty} M\phi_+(x,t) \quad = (i\frac{\partial}{\partial t} - \lambda^3 K)e^{-i\lambda J_0^{(1)}x} = e^{-i\lambda J_0^{(1)}x}C(\lambda). \tag{30}$$

Assuming that the definitions of the Jost solutions are t-independent we get

$$C(\lambda) = -\lambda^3 K. \tag{31}$$

Now we calculate

$$\lim_{x\to\infty} M\phi_-(x,t) = (i\frac{\partial}{\partial t} - C(\lambda))e^{-i\lambda J_0^{(1)}x}T(\lambda,t)$$

$$= e^{-i\lambda J_0^{(1)}x}\left(i\frac{\partial T}{\partial t} - C(\lambda)T(\lambda,t)\right) = e^{-i\lambda J_0^{(1)}x}T(\lambda,t)C(\lambda). \tag{32}$$

Thus, if $Q(x,t)$ satisfies the MKdV equations (20)-(23) the scattering matrix $T(\lambda,t)$ must satisfy the linear evolution equation:

$$i\frac{\partial T}{\partial t} - \lambda^3[K,T(\lambda,t)] = 0, \tag{33}$$

whose solution is

$$T(\lambda, t) = e^{-i\lambda^3 K t} T(\lambda, 0) e^{i\lambda^3 K t}. \qquad (34)$$

Thus $T(\lambda, 0)$ can be viewed as the Cauchy data for initial conditions of the scattering matrix. In other words, solving the MKdV eqs.(20)-(23) is reduced to solving the direct and the inverse scattering problem for the Lax operator L - see [4, 9, 10].

5 Discussions and conclusions

The derived equations reflect the unique symmetry of $\mathfrak{so}(8)$. They are integrable and possess soliton solutions. The next steps are to build their soliton solutions and to analyze their Hamiltonian structure.

Along with the $\mathbb{Z}_6$ −reduction (2), we can introduce one of the following involutions ($\mathbb{Z}_2$-reductions):

$$\begin{array}{lll} a) & K_0^{-1} U^\dagger(x, t, \kappa_1(\lambda)) K_0 & = U(x, t, \lambda), \\ b) & K_0^{-1} U^*(x, t, \kappa_2(\lambda)) K_0 & = -U(x, t, \lambda), \\ c) & U^T(x, t, -\lambda) & = -U(x, t, \lambda), \end{array} \qquad (35)$$

where K_0 is an involutive automorphism $K_0^2 = \mathbb{1}$ and $\kappa_1(\lambda)$ and $\kappa_2(\lambda)$ are appropriate conformal mappings. As possible choices for K_0 we may consider: i) an inner automorphism $\mathfrak{so}(8)$ related to a Weyl reflection, or ii) an outer automorphism of $\mathfrak{so}(8)$.

Other MKdV-type equations can be derived using other inequivalent gradings of $\mathfrak{so}(8)$. They will be published elsewhere.

Acknowledgements: The work is supported in part by the ICTP-SEENET-MTP project PRJ-09.

Appendix

Here we will describe the well-known results about the root system and Cartan-Weyl basis of the algebra $\mathfrak{so}(8)-$ see [11, 12]. The root system is given by $\Delta \equiv \{\pm e_j \pm e_k, 1 \leq j \neq k \leq 4\}$. The simple roots are $\alpha_1 = e_1 - e_2, \alpha_2 = e_2 - e_3, \alpha_3 = e_3 - e_4$ and $\alpha_4 = e_3 + e_4$ and the Dynkin diagram is given in Figure 1.

The typical representation of $\mathfrak{so}(8)$ is 8-dimensional; the Cartan-Weyl basis is given by:

$$\begin{aligned} H_k &= E_{k,k} - E_{9-k,9-k}, \qquad E_{e_k - e_j} = E_{kj} - (-1)^{k+j} E_{9-j,9-k}, \\ E_{e_k + e_j} &= E_{k,9-j} - (-1)^{k+j} E_{9-k,j}, \qquad E_{-\alpha} = E_\alpha^T, \end{aligned} \tag{36}$$

where $1 \leq k \neq j \leq 4$.

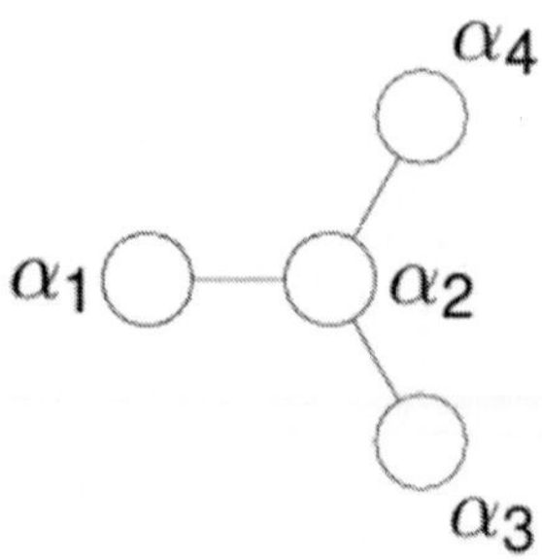

Figure 1: The Dynkin diagram of $\mathfrak{so}(8)$.

We will also need the Coxeter automorphism which may be represented as a composition of two Weyl reflections:

$$C = w_1 w_2, \qquad w_1 = S_{\alpha_1} S_{\alpha_3} S_{\alpha_4}, \qquad w_2 = S_{\alpha_2}. \tag{37}$$

It is easy to check that

$$Ce_1 = e_2, \quad Ce_2 = -e_3, \quad Ce_3 = e_1, \quad Ce_4 = -e_4, \tag{38}$$

i.e. in the 4-dimensional root space C is represented by the matrix

$$C^T = \begin{pmatrix} 0 & 1 & 0 & 0 \\ 0 & 0 & -1 & 0 \\ 1 & 0 & 0 & 0 \\ 0 & 0 & 0 & -1 \end{pmatrix}. \tag{39}$$

It is easy to check that $C^6 = \mathbb{1}$ and $C^3 = -\mathbb{1}$. The exponents of the algebra are determined by the eigenvalues $\zeta_k, k = 1, \dots, 4$ of C which in this case can be calculated with the result

$$\zeta_1 = \omega, \quad \zeta_2 = \omega^3, \quad \zeta_3 = \omega^5, \quad \zeta_1 = \omega^3, \tag{40}$$

i.e. the exponents are equal to 1,3,5,3.

We also remind readers that the Coxeter automorphism can be viewed as an inner automorphism of the algebra $\mathfrak{so}(8)$. In other words,

$$C(E_\alpha) \quad = E_{C(\alpha)} = cE_\alpha c^{-1}, \quad C(H_{e_k}) = H_{C(e_k)} = cH_{e_k}c^{-1}, \tag{41}$$

where the 8×8 matrix c is given in eq. (4).

Bibliography

[1] V. E. Zakharov, S. V. Manakov, S. P. Novikov, L. I. Pitaevskii. *Theory of solitons: the inverse scattering method.* (Plenum, N.Y.: Consultants Bureau, 1984).

[2] M. J. Ablowitz, D. J. Kaup, A. C. Newell, H. Segur. *The inverse scattering transform -- Fourier analysis for nonlinear problems.* Studies in Appl. Math. **53**, n. 4, 249-315, 1974.

[3] V. S. Gerdjikov, P. P. Kulish. *The generating operator for the $n \times n$ linear system.* Physica D, **3D**, n. 3, 549-564, 1981.

[4] A. V. Mikhailov *The reduction problem and the inverse scattering problem.* Physica D, **3D**, n. 1/2, 73-117, 1981.

[5] D. J. Kaup, A. C. Newell, *An Exact Solution for a Derivative Nonlinear Schrödinger Equation*, J. Math. Phys. **19**, 798-801 (1978).

[6] V. Drinfel'd, V. V. Sokolov. *Lie Algebras and equations of Korteweg - de Vries type.* Sov. J. Math. **30**, 1975-2036 (1985).

[7] V. S. Gerdjikov. *Generalised Fourier transforms for the soliton*

equations. Gauge covariant formulation. Inverse Problems **2,** n. 1, 51-74, (1986).

[8] V. S. Gerdjikov. *Z_N--reductions and new integrable versions of derivative nonlinear Schrödinger equations.* In: Nonlinear evolution equations: integrability and spectral methods, Ed. A. P. Fordy, A. Degasperis, M. Lakshmanan, Manchester University Press, (1981), p. 367-372.

[9] V. S. Gerdjikov. Derivative Nonlinear Schrödinger Equations with $\mathbb{Z}_N$ and $\mathbb{D}_N$--Reductions. Romanian Journal of Physics, **58**, Nos. 5-6, 573-582 (2013).

[10] V. S. Gerdjikov, A. B. Yanovski *On soliton equations with $\mathbb{Z}_h$ and $\mathbb{D}_h$ reductions: conservation laws and generating operators.* J. Geom. Symmetry Phys. **31**, 57-92 (2013).

[11] N. Bourbaki. Groupes et Algebres de Lie. Elements de Mathematique. Hermann, Paris (1968).

[12] S. Helgasson, Differential geometry, Lie groups and symmetric spaces (Academic Press, New York, 1978).

ASYMPTOTIC BEHAVIOR OF MANAKOV SOLITONS: EFFECTS OF SHALLOW AND WIDE POTENTIAL WELLS AND HUMPS

A. V. KYULDJIEV, V. S. GERDJIKOV AND M. D. TODOROV

Introduction

The Gross-Pitaevski (GP) equation and its multicomponent generalizations are important tools for analyzing and studying the dynamics of the Bose-Einstein condensates (BEC): see the monographs [17, 20, 26] and the numerous references therein, among which we mention [3, 18, 21, 22, 27, 28]. In the 3Dimensional case these equations can be analyzed solely by numerical methods. If we assume that BEC is quasi-one-dimensional then the GP equations mentioned above may be reduced to the nonlinear Schrödinger equation (NLSE) perturbed by the external potential $V(x)$

$$iu_t + \frac{1}{2}u_{xx} + |u|^2u(x,t) = V(x)u(x,t), \qquad (1)$$

or to its vector generalizations (VNLSE)

$$i\vec{u}_t + \frac{1}{2}\vec{u}_{xx} + (\vec{u}^\dagger, \vec{u})\vec{u}(x,t) = V(x)\vec{u}(x,t). \qquad (2)$$

The Manakov model [24] (MM) is a two-component VNLSE (2) with $V(x) = 0$; for more details see [15].

The analytical approach to the N-soliton interactions was proposed by Zakharov and Shabat [35, 25] for the scalar NLSE. They treated the case of the exact N-soliton solution where all solitons had different velocities. They calculated the asymptotics of the N-soliton solution for $t \to \pm\infty$ and proved that both asymptotics are sums of N one-soliton solutions with the same sets of amplitudes and velocities. The effects of the interaction were shifts in the relative center of masses and phases of the solitons. The same approach, however, is not applicable to the MM, because the asymptotics of the soliton solution for $t \to \pm\infty$ do not commute.

The present paper is an extension of [7, 12, 13] where the main result is that the N-soliton interactions in the adiabatic approximation for the Manakov model ($V(x) = 0$) can also be modeled by the complex Toda chain (CTC) [10, 14, 8]. For $V(x) \neq 0$ we derived a perturbed CTC (PCTC) [7, 12, 3, 13, 16]. Below we concentrate on wide but shallow *sech*-like potentials, *i.e.*:

$$V(x) = \sum_{s=1}^{A} c_s V_s(x, x_s), \qquad V_s(x, x_s) = \frac{1}{\cosh^2(2\nu_0 x - x_s)}, \tag{3}$$

where $x_{s+1} - x_s = 1$ and the quantity A is large, so that initially the whole N-soliton train is in the potential well/hump (see Figure 1).

We also consider soliton trains with varying distances between the initial positions of the solitons. Thus we extend the results in [3, 6, 23, 7, 5, 11]. The corresponding vector N-soliton train is a solution of (2) determined by the initial condition:

$$\vec{u}(x, t = 0) = \sum_{k=1}^{N} u_k(x, t = 0)\vec{n}_k, \; u_k(x, t) = 2\nu_k e^{i\phi_k} \cosh(z_k), \tag{4}$$

where $u_k(x, t)$ is the scalar soliton solution with

$$\begin{aligned} z_k &= 2\nu_k(x - \xi_k(t)), & \xi_k(t) &= 2\mu_k t + \xi_{k,0}, \\ \phi_k &= \mu_k \nu_k z_k + \delta_k(t), & \delta_k(t) &= 2(\mu_k^2 + \nu_k^2)t + \delta_{k,0}. \end{aligned} \tag{5}$$

The s-component polarization vector $\vec{n}_k = \left(n_{k,1}e^{i\beta_{k,1}}, \dots, n_{k,s}e^{i\beta_{k,s}}\right)^T$ is normalized by the conditions

$$\langle \vec{n}_k^\dagger, \vec{n}_k \rangle \equiv \sum_{p=1}^{s} n_{k,p}^2 = 1, \qquad \sum_{p=1}^{s} \beta_{k;p} = 0. \tag{6}$$

The adiabatic approximation holds true if the soliton parameters satisfy [19]:

$$|\nu_k - \nu_0| \ll \nu_0, \quad |\mu_k - \mu_0| \ll \mu_0, \quad |\nu_k - \nu_0||\xi_{k+1,0} - \xi_{k,0}| \gg 1, \quad (7)$$

for all k, where $\nu_0 = 1N\sum_{k=1}^{N} \nu_k$ and $\mu_0 = 1N\sum_{k=1}^{N} \mu_k$ are the average amplitude and velocity, respectively. In fact we have two different scales:

$$|\nu_k - \nu_0| \simeq \varepsilon_0^{1/2}, \qquad |\mu_k - \mu_0| \simeq \varepsilon_0^{1/2}, \qquad |\xi_{k+1,0} - \xi_{k,0}| \simeq \varepsilon_0^{-1/2}.$$

We remind the reader that the basic idea of the adiabatic approximation is to derive a dynamical system for the soliton parameters which would describe their interaction. This idea was initiated by Karpman and Solov'ev [19] and modified by Anderson and Lisak [1]. Later this idea was generalized to N-soliton interactions of scalar NLS solitons [14, 10, 8], and then to the Manakov model, see [3, 5, 7, 13, 11].

In Section 2 we formulate the PCTC model [7, 5] for $sech$-type external potentials. In Section 3 we remind the reader about the asymptotic regimes of the soliton trains predicted by the CTC [10, 8]. Section 4 is dedicated to the comparison between the numeric solutions of the perturbed VNLSE (2) and the predictions of the PCTC model. To this end we solve the VNLSE numerically by using an implicit scheme of Crank-Nicolson type in complex arithmetic. The concept of internal iterations is applied (see [2]) in order to ensure implementation of the conservation laws on the difference level within the round-off error of the calculations [31, 32, 33]. The solutions of the relevant PCTC have been obtained using Maple. Knowing the numeric solution $\vec{u}$ of the perturbed VNLSE we calculate the maxima of $(\vec{u}^{\,\dagger}, \vec{u})$, compare them with the (numeric) solutions for $\xi_k(t)$ of the PCTC and plot the trajectories predicted by both models for each of the solitons.

The effects of the sech-like potentials on CTC

The effects of the external potentials of the form (3) modify the CTC to the following PCTC system:

$$\frac{d\lambda_k}{dt} = -4\nu_0\left(e^{q_{k+1}-q_k}(\vec{n}_{k+1}^{\dagger},\vec{n}_k) - e^{q_k-q_{k-1}}(\vec{n}_k^{\dagger},\vec{n}_{k-1})\right) + M_k + iN_k,$$

$$\frac{dq_k}{dt} = -4\nu_0\lambda_k + 2i(\mu_0 + i\nu_0)\Xi_k - iX_k, \qquad d\vec{n}_k dt = \mathcal{O}(\varepsilon), \tag{8}$$

where $\lambda_k = \mu_k + i\nu_k$, $X_k = 2\mu_k\Xi_k + D_k$ and

$$N_k = -\frac{1}{2}\int_{-\infty}^{\infty}\frac{dz_k}{\cosh z_k}\,\Im(V(y_k)u_k e^{-i\phi_k}),$$

$$M_k = \frac{1}{2}\int_{-\infty}^{\infty}\frac{dz_k \sinh z_k}{cosh^2 z_k}\,\Re(V(y_k)u_k e^{-i\phi_k}),$$

$$\Xi_k = -\frac{1}{4}\nu_k^2\int_{-\infty}^{\infty}\frac{dz_k\, z_k}{\cosh z_k}\,\Im\,(V(y_k)u_k e^{-i\phi_k}),$$

$$D_k = \frac{1}{2\nu_k}\int_{-\infty}^{\infty}\frac{dz_k(1-z_k\tanh z_k)}{\cosh z_k}\,\Re(V(y_k)u_k e^{-i\phi_k}),$$

and $y_k = z_k/(2\nu_0) + \xi_k$. As a result for our specific choice of $V(x)$ we get $N_k = 0$, $\Xi_k = 0$, and:

$$M_k = \textstyle\sum_s 2c_s\nu_k P(\Delta_{k,s}), \qquad D_k = \sum_s c_s R(\Delta_{k,s}). \tag{9}$$

Here $\Delta_{k,s} = 2\nu_0\xi_k - y_s$ and the integrals describing the interaction of the solitons with the potential are given by

$$\begin{aligned} P(\Delta) &= \frac{\Delta+2\Delta\cosh^2(\Delta)-3\sinh(\Delta)\cosh(\Delta)}{\sinh^4(\Delta)}, \\ R(\Delta) &= \frac{6\Delta\sinh(\Delta)\cosh(\Delta)-(2\Delta^2+3)\sinh^2(\Delta)-3\Delta^2}{2\sinh^4(\Delta)}, \end{aligned} \tag{10}$$

see Figure 1. The corrections to N_k and P_k, coming from the linear terms in u depend only on the parameters of the k-th soliton; *i.e.* they are 'local' in k. The details of deriving the integrals can be found in [16].

CTC and the Asymptotic Regimes of *N*-soliton Trains

The dynamics of the N-soliton trains for both the scalar NLSE and the MM are modeled by an integrable model - CTC. It allows one to predict the asymptotic behavior of the solitons. The method of doing so [10] is based on its Lax representation $\dot{L} = [B, L]$ where

$$L = \sum_{k=1}^{N} \left(b_k E_{kk} + a_k (E_{k,k+1} + E_{k-1,k}) \right),$$

$$(11)$$

$$B = \sum_{k=1}^{N} a_k \left(E_{k,k+1} - E_{k-1,k} \right),$$

the matrices $(E_{kn})_{pq} = \delta_{kp}\delta_{nq}$, and $E_{kn} = 0$ whenever one of the indices becomes 0 or $N+1$. The other notations in (11) are as follows:

$$a_k = 12\sqrt{\langle \vec{n}_{k+1}, \vec{n}_k \rangle} e^{(q_{k+1}-q_k)/2}, \qquad b_k = 12(\mu_k + i\nu_k). \qquad (12)$$

The first consequence of the Lax representation is that the CTC has N complex-valued integrals of motion provided by the eigenvalues of L which we denote by $\zeta_k = \kappa_k + i\eta_k$, $k = 1, \dots, N$. Indeed the Lax equation means that the evolution of L is isospectral, *i.e.* $d\zeta_k/dt = 0$. Another important consequence from the results of Moser is that one can write down explicitly the solutions of CTC in terms of the scattering data of L, which consist of $\{\zeta_k, r_k\}_{k=1}^{N}$ where r_k are the first components of the properly normalized eigenvectors of L_0 [30]. Using them one can calculate the asymptotics of these solutions for $t \to \pm\infty$ and show that κ_k determine the asymptotic velocities of the solitons according to:

$$\lim_{t\to\infty} (\xi_k + 2\kappa_k t) = \text{const}. \qquad (13)$$

Thus we are able to classify the regimes of asymptotic behavior as follows:

AFR) The asymptotically free regime takes place if $\kappa_k \neq \kappa_j$ for $k \neq j$, *i.e.*, the asymptotic velocities are all different. Then we have asymptotically separating, free solitons - see also [14, 8];

BSR) The bound state regime takes place for $\kappa_1 = \cdots = \kappa_N = 0$, when all *N* solitons move with the same mean asymptotic velocity;

MAR) a variety of mixed asymptotic regimes happens when one group (or several groups) of particles move with the same mean asymptotic velocity; then they would form one (or several) bound state(s) and the rest of the particles will have free asymptotic motion.

The PCTC taking into account the effects of the sech-like potentials to the best of our knowledge is not integrable and does not allow Lax representation. Therefore we are applying numeric methods to solve it. Our main aim here is to find out potential configurations which result in transition from one asymptotic regime to another.

Comparison between the PCTC Model and Manakov Soliton Interactions

In this Section we will compare how well the PCTC model predicts the soliton interactions of the MM with the external potentials of kind (3).

Let us first describe the types of initial soliton configurations. Below we consider only 3-soliton trains with vanishing initial velocities $\mu_{k,0} = 0$, $k = 1,2,3$. Each of the initial polarization vectors $\vec{n}_{k,0}$ will be parameterized by $\vec{n}_{k,0} = \left(e^{i\gamma_{k,0}}\cos(\theta_{k,0}), e^{-i\gamma_{k,0}}\sin(\theta_{k,0})\right)$ (see Section 1 above), so that the scalar products $(\vec{n}^{\dagger}_{k+1}, \vec{n}_k) = \cos(\theta_{k+1,0} - \theta_{k,0})$ and $\theta_{k,0} = (4-k)\pi/10$. Thus all scalar products just mentioned are equal to $\cos\left(\frac{\pi}{10}\right) \simeq 0.951$; The initial amplitudes are chosen as $\nu_{k,0} = \nu_0 + (2 - k)\Delta\nu$, $\nu_0 = 0.5$.

Finally, we will use two types of initial phase configurations:

$$a)\quad \delta_{1,0} = 0, \quad \delta_{2,0} = \pi, \quad \delta_{3,0} = 0, \quad \Delta\nu = 0.01,$$

$$b)\quad \delta_{1,0} = \delta_{2,0} = \delta_{3,0} = 0, \quad \Delta\nu = 0.025,$$

which will determine the corresponding asymptotic regime. In the case a) we will have AFR provided $\Delta\nu < \nu_{cr}$ - see formula (14) below; for $r_0 = 8$ $\nu_{cr} = 0.02526$. If $\Delta\nu > \nu_{cr}$, then the soliton undergoes a change into BSR. In the case b) we will have BSR if $\Delta\nu > 0$.

One of our aims is to consider also initial soliton configurations which *are not* equidistant for $t = 0$. Such tests, which to the best of our knowledge have not been done until now, will give evidence as to the regions of validity of the CTC.

Remark 4.1 We have given above only examples of 3-soliton configurations which ensure AFR and BSR of Manakov solitons. A complete list of all possible asymptotic regimes for the Manakov case can be done by analogy with [8]. The analytic formula for ν_{cr} (14) has been derived assuming that $r_{01} = r_{02}$, where $r_{0k} = \xi_{0;k+1} - \xi_{0;k}$. For the cases $r_{01} \neq r_{02}$ the corresponding expressions become more complicated. To this end we list the values of $\kappa_k = \Re\zeta_k$ for each of the configurations below.

Let us recall the well-known result (see [10, 14, 8]) that the 3-soliton systems allow for three types of dynamical regimes for large times, namely an asymptotically free regime of 3 solitons takes place if the initial amplitudes are such that $\Delta\nu < \nu_{cr}$:

$$\nu_{cr} = 2\sqrt{2\cos(\theta_1 - \theta_2)}\,\nu_0 \exp(-\nu_0 r_0), \tag{14}$$

AFR) the phases are as in a) above, see [8]. If $r_{01} = r_{02} = 8$ we have $\nu_{cr} = 0.0246$. Such asymptotic regimes for $r_{01} \neq r_{02}$ are shown on the left panels of Figures 1 and 3.

MAR) mixed asymptotic regime: two of the solitons form a bound state and the third soliton goes away from them with a different velocity. Such a regime takes place if the amplitudes are chosen as in (14) and the phases are as in b) above; see the left panel of Figure 4.

BSR) bound state regime when all solitons move asymptotically with the same velocity. Such a regime takes place for amplitudes with $\Delta\nu > \nu_{cr}$ and the phases are as in a). Such an asymptotic regime is shown on the left panel of Figure 5.

It is natural to analyze separately all three regimes and to see what would be the effect of external wells/humps on them. In particular, one can determine for which positions and intensities of the external potentials the solitons will undergo a change from one asymptotic regime to another.

Remark 4.2 The CTC and its perturbative version PCTC use the adiabatic approximation. If we assume that the distance between the solitons is $r_0 = 8$, then the adiabatic parameter $\simeq; 0.01$, so one can expect that the CTC model will hold true up to times of the order of $1/\varepsilon \simeq 100$. Rather surprisingly, quite often we find that the models work well until $t \simeq 1000$ or even longer.

Since the PCTC model is not integrable we will solve it numerically to find the predicted solitons trajectories $\xi_k(t)$. Besides which, we will solve numerically the MM with the initial condition (4) and extract the trajectories of $\max(|u_1|^2 + |u_2|^2)$, where $\vec{u} \equiv (u_1, u_2)$.

On the right panel of Figure 1 we plot samples of a potential well with width 40 composed of 33 wells with depth $c_s = -0.1$ distributed uniformly between abscissas -16 and 16 and with distance between them $h = 1$.

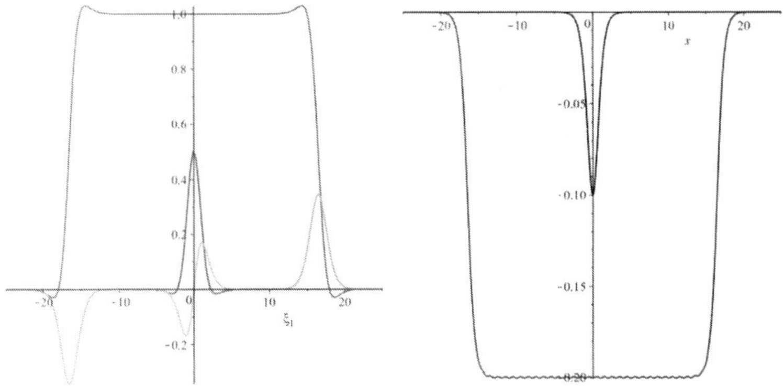

Figure 1: Graphs of P and R functions: for a single sech-potential centered at the origin -- in *cyan* and *red* colors; and for the superposed potential at the neighboring panel -- in *green* and *brown* colors. (left); Single sech-potential in *black* color *vs.* superposed external potential $V(x) = -\sum_{s=0}^{32} 0.1\, sech^2(x - x_s)$, $x_s = -16 + sh$, -- in *blue* color. The superposed potential forms a wide well (right).

Evidently each Manakov soliton solution is parameterized by 6 parameters and four of them are the usual velocity, position, amplitude and phase. Two more parameters fix up the polarization vector. Having in mind the big parametric phenomenology of the solutions we fix the velocities, positions and polarization vectors and vary the initial amplitudes and phases in order to ensure one or another asymptotic regime [8]. Even with only three solitons configuration and wide potential wells/humps with $A = 13$ to $A = 33$ we have a large variety of combinations.

Potential wells, especially when broad enough, attract the solitons and may be used to stabilize them in a bound state. Potential humps repel the solitons; choosing their positions appropriately one can split a soliton in a bound state into free solitons.

In what follows we compare the PCTC models with the numeric solutions of the corresponding (perturbed) MM. We mark the PCTC solutions by dashed lines, and the numeric solutions of the MM and the perturbed MM by solid lines.

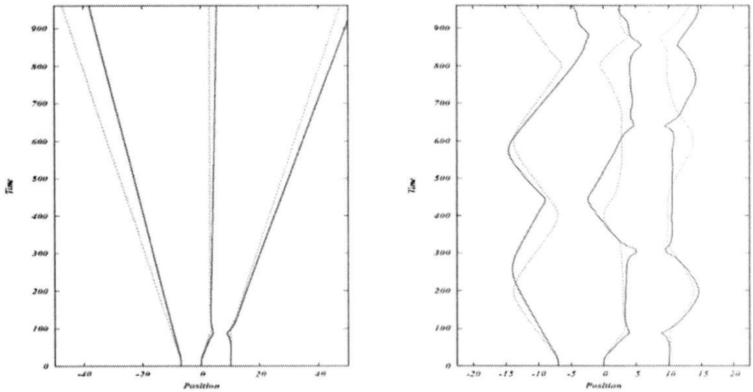

Figure 2: AFR: Free potential behavior with $\Delta\nu = 0$ corresponding to $\kappa_1 = 0.01067$, $\kappa_2 = 0$, $\kappa_3 = -0.01067$ (left panel); External potential well $V(x) = -\sum_{s=0}^{32} 0.01 \operatorname{sech}^2(x - x_s)$, $x_s = -16 + s$ (right panel).

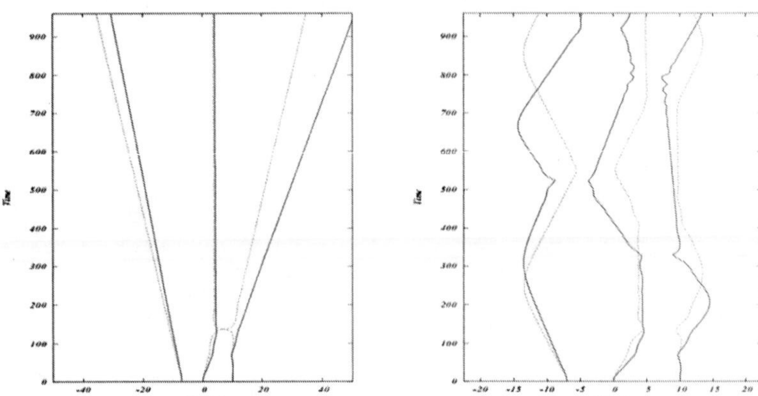

Figure 3: AFR: Free potential behavior with $\Delta\nu = 0.05$ corresponding to $\kappa_1 = 0.00750$, $\kappa_2 = -0.00008$, $\kappa_3 = -0.00742$ (left panel); External potential well $V(x) = -\sum_{s=0}^{32} 0.01 \operatorname{sech}^2(x - x_s)$, $x_s = -16 + s$ (right panel).

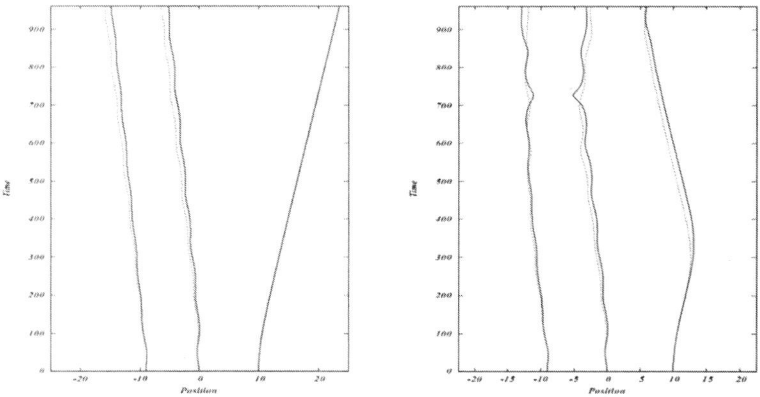

Figure 4: MAR: Free potential behavior with $\Delta\nu = 0.01$ corresponding to $\kappa_1 = 0.003756$, $\kappa_2 = \kappa_3 = -0.001878$, $\delta_1 = \pi/4.064$, $\delta_2 = \pi/2$, $\delta_3 = -\pi/2$ (left panel); External potential well $V(x) = -\sum_{s=0}^{32} 0.01 \operatorname{sech}^2(x - x_s)$, $x_s = -16 + s$ (right panel).

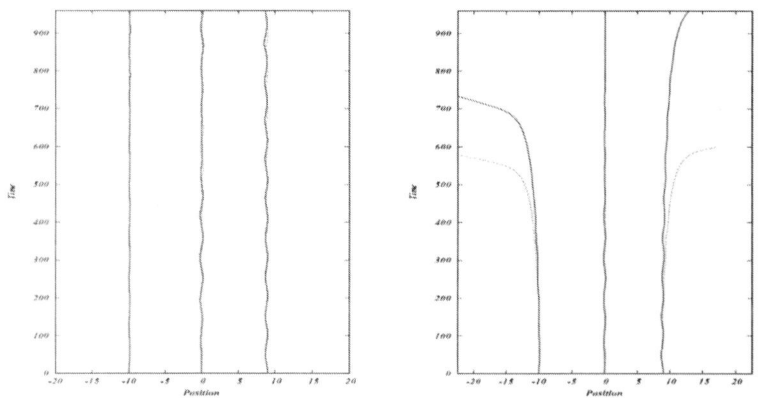

Figure 5: BSR: Free potential behavior with $\Delta\nu = 0.025$ corresponding to $\kappa_1 = \kappa_2 = \kappa_3 = 0$ (left panel); External potential hump $V(x) = \sum_{s=0}^{29} 0.01 \operatorname{sech}^2(x - x_s)$, $x_s = -15 + s$ (right panel).

In the above figures we show some examples of 3-soliton systems. On the figures we plot the trajectories. The first example (Figure 1) clearly demonstrates the role of the external well on the stability of the asymptotically free 3-soliton configuration. The potential (shaded strip) does not allow the solitons to leave the well; they oscillate and form a bound state. The initial positions are -7, 0, and 10 and $\Delta\nu = 0$. Let us note that for non-equal initial distances this case is not singular. The behavior in the next Figure 3 is similar but corresponds to non-trivial deviation $\Delta\nu = 0.05$.

In Figure 4 is demonstrated the influence of external potential on the second possible regime - mixed asymptotic regime. In potential free configuration we have two bound-state solitons and one freely propagating, initially placed at positions -9, 0, 10. The adding of an external potential as a wide well with $A = 33$ and amplitude $c_s = -0.01$ leads to a bound state behavior of all three solitons.

In the next Figure 5 the potential free regime is a bound state. The influence of a potential hump of width 30 and amplitude $c_s = 0.01$ leads to fast violation of this regime and transition to asymptotically free behavior of the lateral solitons. The initial positions are -10, 0, 9. We do not consider the influence of a potential well because it will obviously keep the initial bound state regime.

The comparison of the numerical predictions of both the models in all asymptotic cases is very good.

Conclusion

We have analyzed the effects of external potential wells and humps on the VNLSE soliton interactions using the PCTC model. The comparison with the predictions of the more general VNLSE model [32]

$$i\vec{u}_t + \frac{1}{2}\vec{u}_{xx} + (\vec{u}^{\dagger}, \vec{u})\vec{u} + \alpha\vec{U}(x,t) = 0, \qquad (15)$$

where $\vec{U} = (|u_2|^2 u_1, |u_1|^2 u_2)^T$ and quantity α is the cross-modulation magnitude, is an excellent validation of the consistency and applicability of PCTC.

The superposition of a large number of wells/humps influences strongly the motion of the soliton envelopes and can cause a transition from asymptotically free and mixed asymptotic regimes to a bound state regime and vice versa. Such external potentials are easy to implement in experiments and can be used to control the soliton motion in a given direction and to achieve a predicted motion of the optical pulse. A general feature of the numerical experiments we have conducted is that the predictions of both CTC and PCTC match very well with the MM numeric for long-time evolution, often much longer than expected: see Remark 4.2. This means that PCTC is a reliable *dynamical* model for predicting the evolution of the multisoliton solutions of the Manakov model in adiabatic approximation.

Acknowledgements. The investigation is supported partially by the Bulgarian Science Foundation under grant DDVU02/71.

These results were presented at the 8-th Annual Meeting of the Bulgarian Section of SIAM, 18--19 December 2013.

Bibliography

[1] D. Anderson, and M. Lisak. Nonlinear asymmetric self-phase modulation and self-steepening of pulses in long optical waveguides *Phys. Rev. A*, 27:1393-1398, 1983; D. Anderson, M. Lisak, and T. Reichel. Approximate analytical approaches to nonlinear pulse propagation in optical fibers: A comparison. *Phys. Rev. A*, 38: 1618--1620, 1988.

[2] C. I. Christov, S. Dost, and G. A. Maugin. Inelasticity of soliton collisions in systems of coupled NLS equations. *Physica Scripta*, 50:449-454, 1994.

[3] V. S. Gerdjikov, B. B. Baizakov, and M. Salerno. Modelling adiabatic N-soliton interactions and perturbations. *Theor. Math. Phys.*, 144(2):1138-1146, 2005.

[4] V. S. Gerdjikov. On soliton interactions of vector nonlinear Schrödinger equations., In M. D. Todorov and C. I. Christov (eds.), *AMiTaNS'11, AIP CP1404*, pages 57-67, AIP, Melville, NY, 2011.

[5] V. S. Gerdjikov. Modeling soliton interactions of the perturbed vector nonlinear Schrödinger equation. *Bulgarian J. Phys.*, 38:274-283, 2011.

[6] V. S. Gerdjikov, B. B. Baizakov, M. Salerno, and N. A. Kostov. Adiabatic N-soliton interactions of Bose-Einstein condensates in external potentials. *Phys. Rev. E.*, 73:046606, 2006.

[7] V. S. Gerdjikov, E. V. Doktorov, and N. P. Matsuka. N-soliton Train and Generalized Complex Toda Chain for Manakov System. *Theor. Math. Phys.*, 151(3):762-773, 2007.

[8] V. S. Gerdjikov, E. G. Evstatiev, D. J. Kaup, G. L. Diankov, and I. M. Uzunov. Stability and quasi-equidistant propagation of NLS soliton trains. *Phys. Lett. A*, 241:323-328, 1998.

[9] V. S. Gerdjikov, G. G. Grahovski. On soliton interactions of vector nonlinear Schrödinger equations. In M. D. Todorov and C. I. Christov (eds.) *AMiTaNS'11, AIP CP1404*, pages 57-67, AIP, Melville, NY, 2011.

[10] V. S. Gerdjikov, D. J. Kaup, I. M. Uzunov, and E. G. Evstatiev. Asymptotic behavior of N-soliton trains of the nonlinear Schrödinger equation. *Phys. Rev. Lett.*, 77:3943-3946, 1996.

[11] V. S. Gerdjikov, N. A. Kostov, E. V. Doktorov, and N. P. Matsuka. Generalized Perturbed Complex Toda Chain for Manakov System and exact solutions of the Bose-Einstein mixtures. *Mathematics and Computers in Simulation*, 80:112-119, 2009.

[12] V. S. Gerdjikov and M. D. Todorov. N-soliton interactions for the Manakov system. Effects of external potentials. In R. Carretero-Gonzalez *et al.* (eds.), *Localized Excitations in Nonlinear Complex Systems, Nonlinear Systems and Complexity* 7, pages 147--169, Springer International Publishing Switzerland, 2014, doi 10.1007/978-3-319-02057-0___7.

[13] V. S. Gerdjikov and M. D. Todorov. On the effects of sech-like potentials on Manakov solitons. In M. D. Todorov (ed.) *AMiTaNS'13, AIP CP1561*, pages 75-83, AIP, Melville, NY, 2013, doi 10.1063/1.4827216.

[14] V. S. Gerdjikov, I. M. Uzunov, E. G. Evstatiev, and G. L. Diankov. Nonlinear Schrödinger equation and N-soliton interactions: Generalized Karpman-Soloviev approach and the complex Toda chain. *Phys. Rev. E*, 55(5):6039-6060, 1997.

[15] V. S. Gerdjikov, G. Vilasi, and A. B. Yanovski. Integrable Hamiltonian Hierarchies. Spectral and Geometric Methods. *Lecture Notes in Physics* vol. 748, Springer Verlag, Berlin-Heidelberg-New York, 2008, ISBN: 978-3-540-77054-1.

[16] V. S. Gerdjikov, M. D. Todorov, and A. V. Kyuldjiev, submitted to *Mathematics and Computers in Simulation*, 2013.

[17] A. Griffin, T. Nikuni, and E. Zaremba. *Bose-Condensed Gases at Finite Temperatures*. Cambridge University Press, Cambridge, UK, 2009.

[18] T.-L. Ho. Spinor Bose condensates in optical traps. *Phys. Rev. Lett.*, 81:742, 1998.

[19] V. I. Karpman and V. V. Solov'ev. A Perturbational approach to the two-solition systems. *Physica D*, 3:487-502, 1981.

[20] P. G. Kevrekidis, D. J. Frantzeskakis, and R. Carretero-Gonzalez (eds.). Emergent Nonlinear Phenomena in Bose-Einstein Condensates: Theory and Experiment. Springer, Vol. 45, 2008.

[21] N. A. Kostov, V. Z. Enol'skii, V. S. Gerdjikov, V. V. Konotop, and M. Salerno. On two-component Bose-Einstein condensates in periodic potential. *Phys. Rev. E*, 70:056617, 2004.

[22] N. A. Kostov, V. S. Gerdjikov, and T. I. Valchev. Exact solutions for equations of Bose-Fermi mixtures in one-dimensional optical lattice. *SIGMA*, 3:paper 071, 14 pages, 2007, ArXiv: nlin.SI/0703057, http://www.emis.de/journals/SIGMA/

[23] T. I. Lakoba and D. J. Kaup. Perturbation theory for the Manakov soliton and its applications to pulse propagation in randomly birefringent fibers. *Phys. Rev. E*, 56:6147-6165, 1997.

[24] S. V. Manakov. On the theory of two-dimensional stationary self-focusing of electromagnetic waves. *Zh. Eksp. Teor. Fiz.*, 65: 1392, 1973. English translation: *Sov. Phys. JETP* 38:248, 1974.

[25] S. P. Novikov, S. V. Manakov, L. P. Pitaevski, and V. E. Zakharov. "Theory of Solitons, the Inverse Scattering Method", Consultant Bureau, New York, 1984.

[26] L. P. Pitaevskii and S. Stringari. "Bose-Einstein Condensation". Oxford University Press, Oxford, UK, 2003.

[27] T. Ohmi and K. Machida. Bose-Einstein condensation with internal degrees of freedom in alkali atom gases. *J. Phys. Soc. Jpn.*, 67:1822, 1998.

[28] M. Modugno, F. Dalfovo, C. Fort, P. Maddaloni, and F. Minardi. Dynamics of two colliding Bose-Einstein condensates in an elongated magnetostatic trap. *Phys. Rev. A*, 62: 063607, 2000.

[29] V. M. Perez-Garcia, H. Michinel, J. I. Cirac, M. Lewenstein, and P. Zoller. Dynamics of Bose-Einstein condensates: Variational solutions of the Gross-Pitaevskii equations. *Phys. Rev. A*, 56:1424-1432, 1997.

[30] M. Toda. ``Theory of Nonlinear Lattices.", Springer Verlag, Berlin, 1989.

[31] M. D. Todorov and C. I. Christov. Conservative numerical scheme in complex arithmetic for coupled nonlinear Schrodinger equations. *Discrete and Continuous Dynamical Systems*, Supplement 2007, 982-992.

[32] M. D. Todorov and C. I. Christov. Impact of the large cross-modulation parameter on the collision dynamics of quasi-particles governed by vector NLSE. *Mathematics and Computers in Simulation*, 80:46-55, 2008.

[33] M. D. Todorov and C. I. Christov. Collision dynamics of elliptically polarized solitons in coupled nonlinear Schrödinger equations. *Mathematics and Computers in Simulation*, 82:1221-1232, 2012.

[34] M. Uchiyama, J. Ieda, and M. Wadati. Multicomponent bright solitons in $F = 2$ spinor Bose-Einstein condensates. *J. Phys. Soc. Japan*, 76(7):74005, 2007.

[35] V. E. Zakharov and A. B. Shabat. Exact theory of two-dimensional self-focusing and one-dimensional self-modulation of waves in nonlinear media. English translation: *Soviet Physics-JETP*, 34:62-69, 1972.

CHAPTER NINE

LINEAR ALGEBRA APPLICATIONS

Stability Issues of a Partitioning Algorithm for Tridiagonal and Banded Linear Systems

Velizar Pavlov

Introduction

Tridiagonal or banded linear systems appear in many applications, such as finite elements, difference schemes to differential equations, power distribution systems, etc. Such systems we can solve in parallel by so-called partition methods [3, 4, 6, 8, 9, 11, 12, 13].

A typical member of the partition methods for solving tridiagonal systems is the method of Wang [12]. This method gives an efficient parallel algorithm for solving such systems [4]. Full round-off error analysis of this algorithm can be found in [14].

The generalized partition algorithm of Wang for banded linear systems is considered in [4, 6]. Some aspects of stability analysis of this algorithm are dealt with in [7].

So, this is a review paper where we present the main results of the componentwise stability analysis of Wang's parallel partition method for banded and tridiagonal linear systems.

The Algorithm for Banded Linear Systems

Let the linear system under consideration be denoted by

$$Ax = d,$$

where A is a square matrix of size n ($A \in \mathcal{R}^{n \times n}$) whose bandwidth is $2j + 1$. Here j is the number of superdiagonals, which is equal to the number of subdiagonals. For simplicity we assume also that $n = ks - j$ for some integer k, if s is the number of the parallel processors we want to use. These assumptions are not essential for the consideration.

Now we make the following partitioning of the matrices A, x and d

$$A = \begin{pmatrix} B_1 & \bar{c}_1 & & & & & & \\ a_{1k} & b_{1k} & c_{1k} & & & & & \\ & \bar{a}_2 & B_2 & \bar{c}_2 & & & & \\ & & a_{2k} & b_{2k} & c_{2k} & & & \\ & & & \ddots & \ddots & \ddots & & \\ & & & & \bar{a}_{s-1} & B_{s-1} & \bar{c}_{s-1} & \\ & & & & & a_{(s-1)k} & b_{(s-1)k} & c_{(s-1)k} \\ & & & & & & \bar{a}_s & B_s \end{pmatrix}$$

$$x = (X_1,\ x_{1k},\ X_2,\ x_{2k}, \dots ,\ X_{s-1},\ x_{(s-1)k},\ X_s)^T,$$

$$d = (D_1,\ d_{1k},\ D_2,\ d_{2k}, \dots ,\ D_{s-1},\ d_{(s-1)k},\ D_s)^T,$$

where $B_i \in \mathcal{R}^{(k-j)\times(k-j)}, i = 1,2, \dots, s,$ are band matrices with the same bandwidth as matrix A, $\bar{a}_i, \bar{c}_i$ are matrices of the following kind

$$\bar{a}_i = (a_{(i-1)k+1}, 0, \dots ,0)^T \in \mathcal{R}^{(k-j)\times j}, \quad i = 2, \dots, s,$$

$$\bar{c}_i = (0, \dots ,0, c_{ik-1})^T \in \mathcal{R}^{(k-j)\times j}, \quad i = 1, \dots, s-1,$$

whose elements $a_{(i-1)k+1}, c_{ik-1} \in \mathcal{R}^{j\times j}, \quad a_{ik}, b_{ik}, c_{ik} \in \mathcal{R}^{j\times j} \quad for\ i =$ $1,2, \dots, s-1$, and finally

$$X_i, D_i \in \mathcal{R}^{(k-j)\times 1},\ i = 1,2, \dots, s, \quad x_{ik}, d_{ik} \in \mathcal{R}^{j\times 1},\ i = 1,2, \dots, s-1.$$

Here we present Wang's algorithm in a block form which is more appropriate for the following analysis. For this purpose we define the following permutation

$$[1{:}\, k-j, \dots, (i-1)k+1{:}\, ik-j, \dots, (s-1)k+1{:}\, sk-j,$$

$$k-j+1:k,\dots,ik-j+1:ik,\dots,(s-1)k-j+1:(s-1)k],$$

of the numbers $[1,\dots,sk-j]$, and denote the corresponding permutation matrix by $\mathcal{P}$. By applying this permutation to the rows and columns of matrix A we obtain the system

$$\mathcal{A}\mathcal{P}x=\mathcal{P}d,\qquad \mathcal{A}=\mathcal{P}A\mathcal{P}^T=\begin{pmatrix}A_{11} & A_{12}\\ A_{21} & A_{22}\end{pmatrix},$$

where

$$A_{12}=\begin{pmatrix}\bar{c}_1 & & & \\ \bar{a}_2 & \bar{c}_2 & & \\ & \ddots & \ddots & \\ & & \ddots & \bar{c}_{s-1}\\ & & & \bar{a}_s\end{pmatrix}\in\mathcal{R}^{s(k-j)\times j(s-1)},$$

$$A_{21}=\begin{pmatrix}0 & \cdots & a_k & c_k & \cdots & 0 & & & & & & \\ & & & 0 & \cdots & a_{2k} & c_{2k} & \cdots & & 0 & & \\ & & & & \ddots & & \ddots & \ddots & & & \ddots & \\ & & & & & 0 & \cdots & a_{(s-1)k} & & c_{(s-1)k} & \cdots & 0\end{pmatrix},$$

here $A_{21}\in\mathcal{R}^{j(s-1)\times s(k-j)}$ and

$$A_{11}= \; diag\,\{B_1,B_2,\dots,B_s\}\in\mathcal{R}^{s(k-j)\times s(k-j)},$$

$$A_{22}= \; diag\,(b_k,b_{2k},\dots,b_{(s-1)k})\in\mathcal{R}^{j(s-1)\times j(s-1)}.$$

We will distinguish between the two matrices A (original) and $\mathcal{A}$ (permuted). Evidently, the permutation does not influence the round-off error analysis. The permuted vectors $\mathcal{P}x$ and $\mathcal{P}d$ are not frequently used in the paper. We will stay with the same notation, i.e. x and d, and give explicitly its permuted components when necessary, or write $\mathcal{P}x$ and $\mathcal{P}d$ for the permuted vectors. Otherwise we will need some error bounds on x with respect to the infinity norm, but it is clear that these bounds are not influenced by the permutation, and this will not lead to confusion.

The algorithm can be presented as follows.

Stage 1. Obtain the block LU-factorization

$$A = \begin{pmatrix} A_{11} & A_{12} \\ A_{21} & A_{22} \end{pmatrix} = LU = \begin{pmatrix} A_{11} & 0 \\ A_{21} & I_{j(s-1)} \end{pmatrix} \begin{pmatrix} I_{s(k-j)} & R \\ 0 & S \end{pmatrix}$$

by the following steps:

1. Obtain the LU-factorization of $A_{11} = \mathcal{P}_1 L_1 U_1$ with partial pivoting, if necessary. Here $\mathcal{P}_1$ is a permutation matrix, L_1 is unit lower triangular, and U_1 is upper triangular.
2. Solve $A_{11}R = A_{12}$ using the LU-factorization from the previous item, and compute $S = A_{22} - A_{21}R$, which is the Schur complement of A_{11} in A.

Stage 2. Solve $Ly = d$ by using the LU-factorization of A_{11} (Stage 1).

Stage 3. Solve $Ux = y$ by applying Gaussian elimination (with pivoting, if necessary) to the block S.

Because of the block diagonal structure of A_{11} most of the computations are well parallelized. Let us note that the blocks L_1 and U_1 inherit the block diagonal structure of A_{11}. The block R is quite sparse, and is also structured. If we take into account the structure of A_{12}, then it is clear that

$$R = \begin{pmatrix} p^{(1)} & & & \\ q^{(2)} & p^{(2)} & & \\ & \ddots & \ddots & \\ & & \ddots & p^{(s-1)} \\ & & & q^{(s)} \end{pmatrix} \in \mathcal{R}^{s(k-j)\times j(s-1)},$$

where

$$p^{(i)} = (p_{(i-1)k+1}, p_{(i-1)k+2}, \dots, p_{ik-1})^T \in \mathcal{R}^{(k-j)\times j},$$

$$q^{(i)} = (q_{(i-1)k+1}, q_{(i-1)k+2}, \dots, q_{ik-1})^T \in \mathcal{R}^{(k-j)\times j}.$$

So, the most of the computations at Stage 3 are also well parallelized, because of the block structure of submatrix R.

Let us note that matrix S (the so-called reduced matrix) is block tridiagonal, and banded with bandwidth $4j-1$. We shall need in the following an explicit notation for its entries, which are dense matrices of size $j \times j$. So, we assume that

$$S = \begin{pmatrix} v_1 & w_1 & & & \\ u_2 & v_2 & w_2 & & \\ & \ddots & \ddots & \ddots & \\ & & \ddots & \ddots & w_{s-2} \\ & & & u_{s-1} & v_{s-1} \end{pmatrix} \in \mathcal{R}^{j(s-1)\times j(s-1)},$$

where the entries are computed in the following way

$$u_i = -a_{ik}q_{ik-1}, \quad v_i = b_{ik} - a_{ik}p_{ik-1} - c_{ik}q_{ik+1}, \quad w_i = -c_{ik}p_{ik+1}.$$

Main stability results for banded systems

In the following we denote by a hat the computed quantities. By δT we denote the error of the computation of an arbitrary matrix T, i.e. $\hat{T} = T + \delta T$. By ΔT we denote an equivalent perturbation in matrix T. Finally, by ρ_0 we denote the round-off unit (see [2]).

The general results for banded matrices are given in the following theorem

Theorem 3.1 *For the partitioning algorithm we have that*

$(\mathcal{A} + \Delta\mathcal{A})\mathcal{P}\hat{x} = \mathcal{P}d$, *where*

$|\Delta\mathcal{A}| \le |\mathcal{A}|[(K_1+K_2)f(\rho_0) + h_1(\rho_0)] + |\mathcal{A}||N|[(3K_1+2K_2)f(\rho_0) + h_2(\rho_0)],$

where

$$h_1(\rho_0) = (K_1+K_2)f(\rho_0)g(\rho_0) + K_1K_2f^2(\rho_0) + K_1K_2f^2(\rho_0)g(\rho_0),$$

$$h_2(\rho_0) = (K_1 + K_2)f(\rho_0)g(\rho_0) + 2K_1K_2f^2(\rho_0) + K_1K_2f^2(\rho_0)g(\rho_0),$$

are the terms of higher order in ρ_0, *and*

$$\frac{\|\delta x\|_\infty}{\|\hat{x}\|_\infty} = \frac{\|\hat{x}-x\|_\infty}{\|\hat{x}\|_\infty} \le cond(A,\hat{x})[(K_1+K_2)f(\rho_0) + h_1(\rho_0)]$$

$$+cond^*(A,x^*)r[(3K_1+2K_2)f(\rho_0) + h_2(\rho_0)].$$

In the above theorem $r = \max\{\| \hat{R} \|_\infty, 1\}, K_1 = \max\{k_1, 1\}, K_2 = \max\{k_2, 1\}$, where k_1 bounds growth of the elements when we obtain the LU factorization of A_{11} (Stage 1), k_2 bounds growth of the elements of the Gaussian elimination for the reduced system (Stage 3), $f(\rho_0), g(\rho_0)$ are functions of the following kind

$$f(\rho_0) = \gamma_{j+1} = \gamma_{2j+1}, \quad g(\rho_0) = \gamma_{j+1} + \rho_0,$$

where $\gamma_n = n\rho_0/(1-n\rho_0)$, and $N = \begin{pmatrix} 0 & \hat{R} \\ 0 & I_{j(s-1)} \end{pmatrix}$.

The condition number $cond^*(A,x^*)$ is defined below

$$cond^*(A,x^*) = \frac{\| \, |A^{-1}| \, |A| \, x^* \, \|_\infty}{\|\hat{x}\|_\infty},$$

where the vector x^* is constructed in the following way

$$x^* = (\| \hat{x}_k \|_\infty \, e, |\hat{x}_k^T|, \max\{\| \hat{x}_k \|_\infty, \| \hat{x}_{2k} \|_\infty\}e, \ldots,$$

$$\left|\hat{x}_{(s-1)k}^T\right|, \max\left\{\| \hat{x}_{(s-2)k} \|_\infty, \| \hat{x}_{(s-1)k} \|_\infty\right\}e)^T.$$

Here $e = (1,1,\ldots,1) \in \mathcal{R}^{1\times(k-1)}$. The other condition number is known as a Skeel's conditioning number (see [10])

$$cond(A,\hat{x}) = \frac{\| \, |A^{-1}| \, |A| \, |\hat{x}| \, \|_\infty}{\|\hat{x}\|_\infty}.$$

The condition number $cond^*(A, x^*)$ is introduced to make the obtained bounds more realistic in some cases. As we shall see, in the bounds of the forward error the condition number $cond^*(A, x^*)$ is multiplied by the factor r (which can be large sometimes) while the condition number $cond(A, \hat{x})$ is not. So, when $cond^*(A, x^*)$ is small the influence of r should be negligible. An example of such a case is presented in Section 5, which shows that our bounds are tight.

Now we consider more precisely the case when the matrix A belongs to one of the following types: diagonally dominant, symmetric positive definite, or M-matrix.

It is not difficult to see that the permuted matrix $\mathcal{A}$ is s.p.d., diagonally dominant, and an M-matrix, if the original matrix A is of that type.

For the following bounds of $\| \hat{R} \|_\infty$ and k_2 we need to analyze what is the type of the reduced matrix S if matrix A belongs to one of the above mentioned classes. First we analyze the type of S in exact arithmetic because we need this to bound $\| \hat{R} \|_\infty$. Then at the end of this section we consider the round-off error implementation and comment on the growth of the constant k_2.

The next theorem is true not only for banded but also for general dense matrices (see [1]).

Theorem 3.2 *Let $A \in \mathcal{R}^{n \times n}$. If matrix A is either*

- *symmetric positive definite, or*
- *a nonsingular M-matrix,*

then the reduced matrix S (the Schur complement) preserves the same property.

Hence, it remains to prove that when A is a row diagonally dominant matrix then S preserves this property. Let us note that the case when A is a block row diagonally dominant matrix is considered in [5]. Here we extend the class of diagonally dominant matrices as we consider matrices with standard row diagonally dominance.

Theorem 3.3 Let $A \in \mathcal{R}^{n\times n}$ be a nonsingular row diagonally dominant band matrix. Then the reduced matrix S (the Schur complement) preserves the same property.

As we saw in Theorem 3.1 the error bound depends not only on the growth factors K_1 and K_2, but also on the quantity r, which measures the growth in the matrix $\hat{R}$. Clearly, when some of the blocks B_i are ill conditioned (although the whole matrix A is well conditioned) the factor r can be large. This will lead to large errors even for well conditioned matrices. So, we need some bounds for r, or equivalently $\| \hat{R} \|_\infty$. In the following we show that $\| \hat{R} \|_\infty$ is bounded by not large constants for the above mentioned three classes of matrices.

Theorem 3.4 Let $A \in \mathcal{R}^{n\times n}$ be a nonsingular band M-matrix and

$k_1 cond(A) f(\rho_0) < 1$. Then it is true that

$$\| \hat{R} \|_\infty \leq cond(A)1 - k_1 cond(A_{11}) f(\rho_0) \leq cond(A)1 - k_1 cond(A) f(\rho_0).$$

Theorem 3.5 Let $A \in \mathcal{R}^{n\times n}$ be a nonsingular, row diagonally dominant, band matrix and

$k_1 cond(A) f(\rho_0) < 1$. Then we have

$$\| \hat{R} \|_\infty \leq 11 - k_1 cond(A_{11}) f(\rho_0) \leq 11 - 2k_1 cond(A) f(\rho_0).$$

Theorem 3.6 Let $A \in \mathcal{R}^{n\times n}$ be a symmetric positive definite band matrix and

$k_1(k-1) cond_2(A) f(\rho_0) < 1$, where $cond_2(A) = \| A^{-1} \|_2 \| A \|_2$. Then we have

$$\| \hat{R} \|_\infty \leq \sqrt{j(s-1) cond_2(A)} 1 - k_1 cond(A_{11}) f(\rho_0) \leq \sqrt{j(s-1) cond_2(A)} 1 - k_1(k-1) cond_2(A) f(\rho_0).$$

Theorems 3.4 - 3.6 show that $\| \hat{R} \|_\infty$ is bounded by not large constants for the three classes of matrices, if the whole matrix A is well-conditioned. In order to bound k_2 we can use Theorems 3.2 - 3.3 and the already obtained bounds for the Gaussian elimination in [5]. However, in practice we obtain the computed matrix $\hat{S}$ instead of the exact one. It is important to know what is the distance between S and $\hat{S}$. This question is answered in Theorem 3.7.

Theorem 3.7 For the error $\Omega S = \hat{S} - S$ in the computed reduced matrix $\hat{S}$ it holds that

$$\frac{\|\Omega S\|_\infty}{\|S\|_\infty} \leq K_1 cond(A) r f(\rho_0).$$

The theorems in this section show that $\| \hat{R} \|_\infty$ (and r, respectively) is not large for the three types of matrices when the original matrix A is well conditioned. So, the error in S is also bounded by a not large constant, if matrix A is well conditioned. Consequently, the constant k_2 is close to the theoretical constants (see [5]). For other types of matrices this conclusion may not be true, and the error ΩS may grow.

It is clear that all of these theorems concern the case of banded linear systems. Similar theorems are proved in [14] for tridiagonal linear systems. So, the main conclusions of this section hold in the tridiagonal case.

Main stability results for tridiagonal systems

The general results for tridiagonal matrices are given in the following theorem

Theorem 4.1 *For the partitioning algorithm we have that*

$(\mathcal{A} + \Delta\mathcal{A})\mathcal{P}\hat{x} = \mathcal{P}d,$ *where*

$$|\Delta\mathcal{A}| \leq |\mathcal{A}| h_1(\rho_0) + |\mathcal{A}||N| h_2(\rho_0),$$

where

$$h_1(\rho_0) = K_1 f(\rho_0) + K_2 h(\rho_0) + K_1 K_2 f(\rho_0) h(\rho_0)$$

$$+K_1 f(\rho_0)g(\rho_0) + K_2 h(\rho_0)g(\rho_0) + K_1 K_2 f(\rho_0)h(\rho_0)g(\rho_0),$$

$$h_2(\rho_0) = 3K_1 f(\rho_0) + 2K_2 h(\rho_0) + 2K_1 K_2 f(\rho_0)h(\rho_0)$$

$$+3K_1 f(\rho_0)g(\rho_0) + 3K_2 h(\rho_0)g(\rho_0) + 3K_1 K_2 f(\rho_0)h(\rho_0)g(\rho_0)$$

$$+K_1 f(\rho_0)g^2(\rho_0) + K_2 h(\rho_0)g^2(\rho_0) + K_1 K_2 f(\rho_0)h(\rho_0)g^2(\rho_0),$$

and for the forward error it is true that

$$\frac{\|\delta x\|}{\|\hat{x}\|} = \frac{\|\hat{x}-x\|_\infty}{\|\hat{x}\|_\infty} \leq cond(A,\hat{x})h_1(\rho_0) + cond^*(A,x^*)rh_2(\rho_0).$$

Here the sense of the constants K_1, K_2 and r is the same as in Theorem 3.1. Now the functions $f(\rho_0), g(\rho_0)$ are of the following kind

$$f(\rho_0) = 4\rho_0 + 3\rho_0^2 + \rho_0^3, \quad g(\rho_0) = 3\rho_0 + 3\rho_0^2 + \rho_0^3.$$

The definition of $cond^*(\mathcal{A},x^*)$ is adapting to the new conditions.

Numerical experiments

The numerical experiments in this section are done in MATLAB, where the round-off unit is $\rho_0 \approx 2.22E-16$. We measure two types of errors:

1. The relative forward error

$$FE = \frac{\|\hat{x}-x\|_\infty}{\|\hat{x}\|_\infty},$$

where $\hat{x}$ is the computed solution.

2. The componentwise backward error (see [5])

$$BE = \max_{1\leq i\leq n} \frac{(|A\hat{x}-d|)_i}{(|A||\hat{x}|+|d|)_i}.$$

Let us consider the following examples.

Example 1. $A = tridiag\ (1, b, 1)$, where $b = (\varepsilon, \dots, \varepsilon, 2)$. In this way A becomes very well conditioned. The exact solution is $x = (1,1, \dots, 1)^T$. We can notice how the backward and forward errors grow when $\varepsilon \to 0$, although the matrix A is very well conditioned and we use partial pivoting. This is because $\| \hat{R} \|_\infty$ grows infinitely when $\varepsilon \to 0$, which fact is predicted by our theory. We report the results in Table 1 for different values of ε.

A similar example in the case of banded linear systems can be found in [7].

Example 2. A is the matrix from Example 1 with $\varepsilon =$ 1E-16 (a number less than the round-off unit). A is well conditioned again. The exact solution is

$$x = (1, \dots, 1,0; 1, \dots, 1,0; \dots, 1, \dots, 1,0; 1, \dots, 1)^T,$$

where $x_k = x_{2k} = \dots = x_{(s-1)k} = 0$. We report the results of our example in Table 2 when $\varepsilon =$ 1E-16, $s = 10$ for different values of k. This example shows why we have introduced the condition number $cond^*(A, x^*)$. Here we have a large r factor in Theorem 1 ($\approx 1E$+16) but as can be seen from Table 2 the errors are very small. This is because $cond^*(A, x^*) \approx 0$ for this example, and the influence of r is not essential although the blocks B_i are almost singular. So, a large r does not necessarily mean large errors as could be expected intuitively. The LU-factorization of B_i is done with partial pivoting again to make the constants k_1 and k_2 small. In this way we can see the importance of introducing the second condition number $cond^*(A, x^*)$.

A similar example in the case of banded linear systems can be found in [7].

Table 1: The forward and backward error for the matrix A for Example 1 when $k = 6, s = 10$.

ε	1E--5	1E--10	1E--15
BE	8.73E--11	1.19E--5	0.42
FE	1.74E--10	2.38E--5	2.06

Table 2: The forward and backward error for the matrix A on Example 2 when $s = 10$.

k	6	56	256	556
BE	1.44E--15	1.11E--16	1.66E--16	1.14E--16
FE	3.33E--15	1.99E--15	1.31E--14	1.55E--15

Example 3. Let the matrix A be defined as follows:

$$A = \begin{pmatrix} 4 & -1 & -1 & & & & \\ -1 & 4 & -1 & -1 & & & \\ -1 & -1 & 4 & -1 & -1 & & \\ & \ddots & \ddots & \ddots & \ddots & & \\ & & -1 & -1 & 4 & -1 & \\ & & & -1 & -1 & 4 & \end{pmatrix}.$$

Let us note that A is a nonstrictly row diagonally dominant and symmetric positive matrix. The results which we obtained are given in Table 3.

Table 3: The forward and backward error of Example 3 when $k = 60, s = 8, j = 2$.

x	x_α	e	$rand$	$randn$
BE	3.62E--16	2.58E--16	4.04E--16	1.44E--16
FE	3.54E--11	2.28E--12	5.15E--11	8.08E--11
$cond(A, \hat{x})$	$3.11E + 4$	$6.24E + 4$	$7.56E + 4$	$7.23E + 4$
$cond^*(A, x^*)$	$4.54E + 5$	$1.88E + 4$	$2.32E + 5$	$4.38E + 5$
r	8.98	8.98	8.98	8.98

The exact solutions here are chosen as $x_\alpha = (1, \alpha, \alpha^2, \dots, 10^{-5})^T$, $\alpha = 10^{-5/(n-1)}$, $e = (1,1,\dots,1)^T$, and 'rand' and 'randn' are exact solutions generated by the coresponding MATLAB functions. Again, as predicted by our theoretical results, the BE is small because A is a row diagonally dominant and s.p.d. matrix. The FE is larger because matrix A is not so well conditioned as can be seen from Table 3. The forward error is almost equal to the theoretical bound from Theorem 3.1, which shows that our bounds cannot be significantly improved.

A similar example in the case of tridiagonal linear systems can be found in [14].

Bibliography

[1] Axelsson, O., *Iterative solution methods*, Cambridge University Press, New York, 1994.

[2] Golub, G., C. van Loan, *Matrix Computations*, 3rd ed., The John Hopkins University Press, 1996.

[3] Hajj, I., S. Skelboe, A multilevel parallel solver for block tridiagonal and banded linear systems, *Parallel Computing*, 15 (1990), pp. 21--45.

[4] Heath, M., *Parallel Numerical Algorithms for Banded and Tridiagonal Systems*, University of Illinois at Urbana-Champaign, 2012.

[5] Higham, N., Accuracy and Stability of Numerical Algorithms, SIAM, Philadelphia, 1996.

[6] Meier, U. A parallel partition method for solving banded linear systems. *Parallel Comput.*, 2 (1985), pp. 33--43.

[7] Pavlov, V., Stability of a parallel partitioning algorithm for special classes of banded linear systems, *Lecture Notes in Computer Science*, Spinger (2001), pp. 658--665.

[8] Polizzi, E., A. Sameh, A parallel hybrid banded system solver, *Parallel Computing* 32 (2006), pp. 177--194.

[9] Santos, E., Optimal and efficient parallel tridiagonal solvers using direct methods, *J. Supercomputing* 30 (2004), pp. 97-115.

[10] Skeel, R., Scaling for numerical stability in Gaussian elimination, *J. Assoc. Comput. Mach.*, 26 (1979), pp. 494--526.

[11] Sun, X., W. Zhang, A parallel two-level hybrid method for tridiagonal systems and its application to fast Poisson solvers, *IEEE Trans. Parallel Distrib. Sys.* 15 (2004), pp. 97--106.

[12] Wang, H., A parallel method for tridiagonal linear systems, *ACM Transactions on Mathematical Software*, 7 (1981), pp. 170--183.

[13] Wright, S., Parallel algorithms for banded linear systems, *SIAM J. Sci. Stat. Comput.*, 12 (1991), pp. 824--842.

[14] Yalamov, P., V. Pavlov, On the stabilty of a partitioning algorithm for tridiagonal systems, *SIAM J. Matrix Anal. Appl.*, (2006) online.

On the Stability of a Pentadiagonal Solver

Velizar Pavlov

Introduction

Linear systems with pentadiagonal matrices arise often when solving differential equations numerically. For this reason the developing of specialized algorithms for solving such systems is a particular research interest [2, 3]. In this connection, stability analysis of the algorithms for pentadiagonal linear systems is important.

In this paper we present a round-off error analysis of the LU-decomposition for linear systems with pentadiagonal matrices. In our approach we use the dependence graph of the algorithm and its parallel form [6, 7]. The notion of equivalent perturbation is introduced for every piece of data (input, intermediate and output) in contrast to the generally used backward analysis (see [4]). Then a linear system

$$B\varepsilon = \eta \tag{1}$$

with respect to the vector of equivalent perturbations ε is derived, and the solution of this system gives a first order approximation of the equivalent perturbations. Here matrix B consists of the Frechet-derivatives of all the operations and of elements which are equal to 0 or to -1, and η is the vector of all local absolute round-off errors. Giving values to the equivalent perturbations of the output data we can estimate successively, level by level (see [6, 7]), all the other equivalent perturbations. We are interested in the equivalent perturbations of the input data which are the results of the backward analysis.

The estimates of backward analysis can be written in a simple analytical form, while the estimates of forward analysis depend strongly on intermediate results. Besides, backward analysis needs much fewer operations when the estimates are defined numerically.

Description of the LU-decomposition

The algorithm is described in [5]. Let us consider the following system

$$Ax = f, \tag{2}$$

where

$$A = \begin{bmatrix} c_1 & d_1 & e_1 & & & & & \\ b_2 & c_2 & d_2 & e_2 & & & & \\ a_3 & b_3 & c_3 & d_3 & e_3 & & & \\ & & & \cdots & & & & \\ & & & a_{n-1} & b_{n-1} & c_{n-1} & d_{n-1} \\ & & & & a_n & b_n & c_n \end{bmatrix}, \quad f = \begin{bmatrix} f_1 \\ f_2 \\ \vdots \\ \vdots \\ f_n \end{bmatrix}.$$

We look for a solution of the following kind:

$$\begin{aligned} x_i &= \alpha_i x_{i+1} + \beta_i x_{i+2} + \gamma_i, \quad i = 1, \dots, n-2 \\ x_{n-1} &= \alpha_{n-1} x_n + \gamma_{n-1}, \\ x_n &= \gamma_n. \end{aligned} \tag{3}$$

Let us note that $\alpha_1, \beta_1, \gamma_1$ can be derived from the first equation of system (2), and then using the representation (3) for x_{i-2}, x_{i-1} we get the coefficients $\alpha_i, \beta_i, \gamma_i$ from the $i-th$ equation as follows:

$$\Delta_1 = c_1, \ \alpha_1 = -d_1/\Delta_1, \ \beta_1 = -e_1/\Delta_1, \ \gamma_1 = f_1/\Delta_1,$$

$$\Delta_2 = c_2 + b_2\alpha_1, \ \alpha_2 = -(d_2 + b_2\beta_1)/\Delta_2, \ \beta_2 = -e_2/\Delta_2, \tag{4}$$

$$\gamma_2 = (f_2 - b_2\gamma_1)/\Delta_2,$$

$$\Delta_i = c_i + (a_i\alpha_{i-2} + b_i)\alpha_{i-1} + a_i\beta_{i-2},$$

$$\alpha_i = -[d_i + (a_i\alpha_{i-2} + b_i)\beta_{i-1}]/\Delta_i,$$

$$\beta_i = -e_i/\Delta_i, \tag{5}$$

$$\gamma_i = [f_i - (a_i\alpha_{i-2} + b_i)\gamma_{i-1} - a_i\gamma_{i-2}]/\Delta_i,$$

$$i = 1, \dots, n,$$

where $e_{n-1} = e_n = d_n = 0$. Equalities (4) are called forward elimination, and equalities (3) are called back substitution. Actually, (3) and (4,5) realize the following decomposition $A = LU$, where

$$L = \begin{bmatrix} \Delta_1 & & & & & \\ b_2 & \Delta_2 & & & & 0 \\ a_3 & a_3\alpha_1 + b_3 & \Delta_3 & & & \\ & \ddots & \ddots & \ddots & & \\ & 0 & & & & \\ & & a_n & a_n\alpha_{n-2} + b_n & \Delta_n \end{bmatrix},$$

$$U = \begin{bmatrix} 1 & -\alpha_1 & -\beta_1 & & 0 \\ & 1 & -\alpha_2 & -\beta_2 & \\ & & \ddots & \ddots & \\ & 0 & & 1 & -\alpha_{n-1} \\ & & & & 1 \end{bmatrix}.$$

From (4,5) we obtain the triangular system

$$U_x = \gamma, \quad \gamma = L^{-1}f, \tag{6}$$

where $\gamma = (\gamma_1, \dots, \gamma_n)^T$, and then the recurrence relations (3) produce the solution x.

The round-off error analysis is done under the assumptions that matrix A is diagonally dominant, i.e

$$|c_i| \geq |a_i| + |b_i| + |d_i| + |e_i|, \quad i = 1, \dots, n, \tag{7}$$

for $a_1 = b_1 = a_2 = e_{n-1} = e_n = d_{n-1} = 0$, and that at least for one i the inequality is strict. Under these assumptions it can be shown that the algorithm is correct (see [5]) and that the following estimate is valid

$$|\alpha_i| + |\beta_i| \leq 1, \quad i = 1, \dots, n. \tag{8}$$

Backward analysis of the back substitution

We shall do the backward analysis of the forward elimination and the back substitution separately. Let us consider the back substitution first. The dependence graph of this part of the algorithm is given in Figure 1, where $q_i = (\alpha_i, \beta_i, \gamma_i)$. In each vertex only one term of the recurrence relation (3) is computed. The vectors q_i are inputs for the back substitution. Now using the method described in [7] we see that matrix B from (1) has the following structure

$$\begin{bmatrix} \tilde{x}_n & 1 & & & & 0 & & & & \vdots & \tilde{\alpha}_{n-1} & -1 & & & \\ & & & \tilde{x}_{n-1} & \tilde{x}_n & 1 & & & & \vdots & \tilde{\beta}_{n-2} & \tilde{\alpha}_{n-2} & -1 & & \\ & & \cdots & \cdots & & & & & & \vdots & & \cdots & \cdots & & \\ & 0 & & & & & & & & \vdots & 0 & & & & \\ & & & & & \tilde{x}_2 & \tilde{x}_3 & 1 & & \vdots & & & \tilde{\beta}_1 & \tilde{\alpha}_1 & -1 \end{bmatrix}.$$

The wave denotes that the elements are computed with round-off errors. The size of matrix B is $(n-1) \times (4n-4)$ and it has a full rank. System (1) has a set of solutions and we have a choice.

Using floating-point arithmetic operations we assume that

$$fl(x * y) = (x * y)(1 + \rho), \qquad |\rho| \leq \rho_0$$

for $* \in \{+, -, \times, /\}$, where ρ_0 is the round-off unit (see [1]).

Further on, by the lower indices of ε and η we denote the corresponding equivalent perturbations and absolute round-off errors. Then neglecting terms of second order of ρ_0 simple round-off analysis gives that

$$\eta_{x_{n-1}} = \tilde{\alpha}_{n-1}\tilde{x}_n(\rho_1^{(n-1)} + \rho_2^{(n-1)}) + \tilde{\gamma}_{n-1}\rho_2^{(n-1)},$$

$$\eta_{x_i} = \tilde{\alpha}_i\tilde{x}_{i+1}(\rho_1^{(i)} + \rho_3^{(i)} + \rho_4^{(i)}) + \tilde{\beta}_i\tilde{x}_{i+1}(\rho_2^{(i)} + \rho_3^{(i)} + \rho_4^{(i)}) + \tilde{\gamma}_i\rho_4^{(i)},$$

$$|\rho_j^{(i)}| \leq \rho_0, \quad i = n-2, \dots, 1, \quad j = 1, \dots, 4.$$

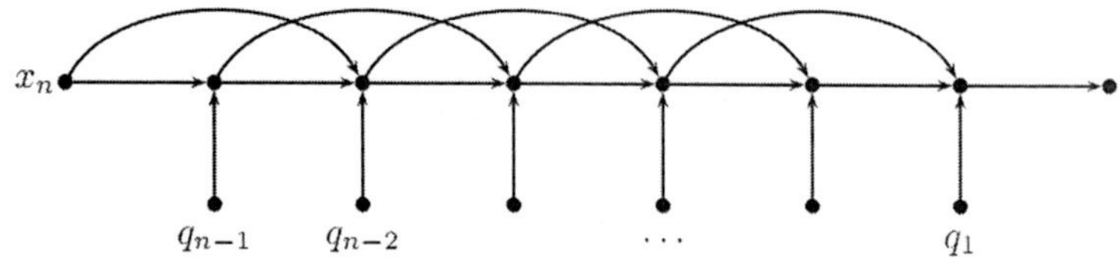

Figure 1: Dependence graph of the back substitution

Now we choose the following solution of system (1)

$$\varepsilon_{x_i} = 0, \quad i = n, \dots, 1,$$

$$\varepsilon_{\alpha_{n-1}} = \tilde{\alpha}_{n-1}(\rho_1^{(n-1)} + \rho_2^{(n-1)}),$$

$$\varepsilon_{\gamma_{n-1}} = \tilde{\gamma}_{n-1}\rho_2^{(n-1)},$$

$$\varepsilon_{\alpha_1} = \tilde{\alpha}_i(\rho_1^{(i)} + \rho_3^{(i)} + \rho_4^{(i)}), \tag{9}$$

$$\varepsilon_{\beta_i} = \tilde{\beta}_i(\rho_2^{(i)} + \rho_3^{(i)} + \rho_4^{(i)}),$$

$$\varepsilon_{\gamma_i} = \tilde{\gamma}_i\rho_4^{(i)},$$

$$|\rho_j^{(i)}| \le \rho_0, \quad i = n-1, \dots, 1, \quad j = 1, \dots, 4.$$

Besides, let us have $\varepsilon_{\gamma_n} = 0$. From (9) for $i = n-2, \dots, 1$ we obtain the estimates

$$|\varepsilon_{\alpha_{n-1}}| \le 2|\tilde{\alpha}_{n-1}|\rho_0,$$

$$|\varepsilon_{\gamma_{n-1}}| \le \rho_0,$$

$$|\varepsilon_{\alpha_i}| \le 3|\tilde{\alpha}_i|\rho_0, \tag{10}$$

$$|\varepsilon_{\beta_i}| \le 3|\tilde{\beta}_i|\rho_0,$$

$$|\varepsilon_{\gamma_i}| \leq |\tilde{\gamma}_i| \rho_0.$$

So, the backward analysis of the back substitution gives very good estimates of equivalent perturbations.

Backward analysis of the forward elimination

The dependence graph of this part of the algorithm is given in Figure 2. There operations (4) are placed in the first two vertices, and the operation (5) is placed in every other vertex in the graph, where the vectors

$r_i = (a_i, b_i, c_i, d_i, e_i, f_i)^T$ are the inputs.

In this case matrix B has the following structure

$$B = \begin{bmatrix} H_1 & -I & & & & & & & \\ & & G_2 & H_2 & -I & & & & \\ & & F_3 & 0 & G_3 & H_3 & -I & & \\ & & & \dots\dots & & & & & \\ & & & & & F_n & 0 & G_n & H_n & -I \end{bmatrix},$$

where F_i, G_i, H_i are Frechet-derivatives of the operations in every vertex of the graph with respect to the vectors $(\tilde{\alpha}_{i-2}, \tilde{\beta}_{i-2}, \tilde{\gamma}_{i-2})^T$, $(\tilde{\alpha}_{i-1}, \tilde{\beta}_{i-1}, \tilde{\gamma}_{i-1})^T$ and r_i. Blocks F_i, G_i, H_i are given below:

$$F_i = \frac{-a_i}{\tilde{\Delta}_i} \begin{bmatrix} \tilde{\beta}_{i-1} - \tilde{\alpha}_{i-1}\tilde{\alpha}_i & -\tilde{\alpha}_i & 0 \\ -\tilde{\alpha}_{i-1}\tilde{\beta}_i & -\tilde{\beta}_i & 0 \\ \tilde{\gamma}_{i-1} + \tilde{\alpha}_{i-1}\tilde{\gamma}_i & -\tilde{\gamma}_i & -1 \end{bmatrix}, \tag{11}$$

$$G_i = \frac{a_i\tilde{\alpha}_{i-2} + b_i}{\tilde{\Delta}_i} \begin{bmatrix} -\tilde{\alpha}_i & 1 & 0 \\ \tilde{\beta}_i & 0 & 0 \\ \tilde{\gamma}_i & 0 & -1 \end{bmatrix}, \tag{12}$$

$$H_i = \begin{bmatrix} \frac{\partial\tilde{\alpha}_i}{\partial a_i} & \frac{\partial\tilde{\alpha}_i}{\partial b_i} & \frac{1}{\tilde{\Delta}_i} & 0 & \frac{-\tilde{\alpha}_i}{\tilde{\Delta}_i} & 0 \\ \frac{\partial\tilde{\beta}_i}{\partial a_i} & \frac{\partial\tilde{\beta}_i}{\partial b_i} & 0 & \frac{-1}{\tilde{\Delta}_i} & \frac{-\tilde{\beta}_i}{\tilde{\Delta}_i} & 0 \\ \frac{\partial\tilde{\gamma}_i}{\partial a_i} & \frac{\partial\tilde{\gamma}_i}{\partial b_i} & 0 & 0 & \frac{-\tilde{\gamma}_i}{\tilde{\Delta}\ i} & \frac{1}{\tilde{\Delta}_i} \end{bmatrix}.$$

Here we assume that $\tilde{\Delta} = 0,\ i = 2, \ldots, n$. The derivatives with respect to a_i and b_i are not necessary in the further investigation, so they are not written explicitly. The equivalent perturbations $\varepsilon_{\alpha_i}, \varepsilon_{\beta_i}, \varepsilon_{\gamma_i}$ are already defined in Section 2. Then from the structure of matrix B in this section it is clear that we have to solve a system with the block diagonal matrix diag $\{H_i\}_{i=1}^{n}$ in order to obtain the equivalent perturbations of the vector r_i. Here we consider only the i-th block equation. It looks as follows

$$H_i\varepsilon_{r_i} = \eta_{q_i} - F_i\varepsilon_{q_{i-2}} - G_i\varepsilon_{q_{i-1}} + \varepsilon_{q_i}. \quad (13)$$

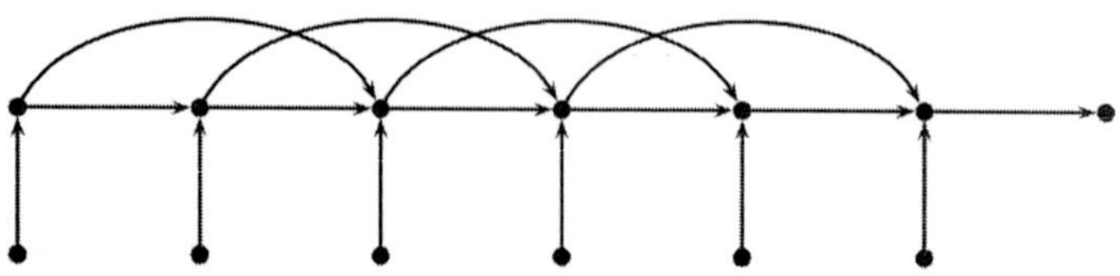

Figure 2: Dependence graph of the forward substitution

Neglecting terms of second order of ρ_0 simple round-off error analysis gives the estimates of $\eta_{q_i} = (\eta_{\alpha_i}, \eta_{\beta_i}, \eta_{\gamma_i})^T$

$$|\eta_{\alpha_i}| \leq (|d_i| + 2.5|a_i\tilde{\alpha}_{i-2}\tilde{\beta}_{i-1}| + 2|b_i\tilde{\beta}_{i-1}|)|\tilde{\Delta}_i^{-1}|\ \rho_0$$

$$+|\tilde{\alpha}_i|\ |\tilde{\Delta}_i^{-1}|\ |\eta_{\Delta_i}|,$$

$$|\eta_{\beta_i}| \leq |e_i|\ |\tilde{\Delta}_i^{-1}|\rho_0 + |\tilde{\beta}_i|\ |\tilde{\Delta}_i^{-1}|\ |\eta_{\Delta_i}|, \quad (14)$$

$$|\eta_{\gamma_i}| \leq (1.5|f_i| + 3|a_i\tilde{\gamma}_{i-1}\tilde{\alpha}_{i-2}| + 2.5|b_i\tilde{\gamma}_{i-1}|$$

$$+1.5|a_i\tilde{\gamma}_{i-2}|)\ |\tilde{\Delta}_i^{-1}|\rho_0 + |\tilde{\gamma}_i|\ |\tilde{\Delta}_i^{-1}||\eta_{\Delta_i}|,$$

where

$$|\eta_{\Delta_i}| \leq (|c_i| + 2.5|a_i\tilde{\alpha}_{i-1}\tilde{\alpha}_{i-2}| + 2|b_i\tilde{\alpha}_{i-1}| + |a_i\tilde{\beta}_{i-2}|)\rho_0.$$

System (13) has a set of solutions. Let us choose $\varepsilon_{a_i} = \varepsilon_{b_i} = \varepsilon_{c_i} = 0$. Then the rest of the unknown $\varepsilon_{ri}^* = (\varepsilon_{d_i}, \varepsilon_{e_i}, \varepsilon_{f_i})^T$ are defined uniquely

$$\varepsilon_{\mathrm{ri}}^* = \tilde{\Delta}_i(\eta_{q_i} - F_i\varepsilon_{q_{i-2}} - G_i\varepsilon_{q_{i-1}} + \varepsilon_{q_i}).$$

In all the following estimates neglecting terms of second order of ρ_0 we can consider that

$$|\tilde{\alpha}_i + \tilde{\beta}_i| \leq 1. \quad (15)$$

Now from (10), (11), (12) and (14) after some computations we can obtain the following estimates

$$\parallel \varepsilon_A \parallel_\infty \leq \max_i(5|c_i| + |d_i| + 14|a_i| + 10|b_i| + 0.5|e_i|)\ \rho_0 = BE_1 \leq$$

$$\leq 9.5 \parallel A \parallel_\infty\ \rho_0, \quad (16)$$

$$\parallel \varepsilon_f \parallel_\infty \leq \max_i[1.5|f_i| + (13|a_i| + 7|b_i| + 1.5|c_i|) \parallel \tilde{\gamma} \parallel_\infty]\ \rho_0 \leq$$

$$\leq (1.5 \parallel f \parallel_\infty + 7.25 \parallel A \parallel_\infty \parallel \tilde{\gamma} \parallel_\infty)\ \rho_0. \quad (17)$$

The last estimate depends on the intermediate data $\tilde{\gamma}$. Two other estimates follow from (17) and (6) depending only on the input or output data

$$\parallel \varepsilon_f \parallel_\infty \leq (2 \parallel f \parallel_\infty + 11.5 \parallel A \parallel_\infty \parallel \tilde{x} \parallel_\infty)\rho_0 = BE_2, \quad (18)$$

$$\parallel \varepsilon_f \parallel_\infty \leq (2 + 11.5 \parallel A \parallel_\infty \parallel A^{-1} \parallel_\infty \parallel f \parallel_\infty)\rho_0.$$

Here we use the fact that $\parallel U \parallel_\infty \leq 2$ and $L^{-1} = UA^{-1}$. The estimates thus obtained depend only on the condition of A and do not depend explicitly on n. This shows that the algorithm is stable and backward analysis depends only on the condition of problem (2).

Let us note that forward analysis can be obtained from system (1) using the representations of the blocks F_i, G_i, H_i , but it depends on the quantities

$$\frac{|a_i|}{|\Delta_i|}, \frac{|a_i \alpha_{i-2}+b_i|}{|\Delta_i|},$$

which cannot be estimated analytically so easily. Besides, backward analysis uses $(7n-6)$ comparisons and 16 multiplications and additions, while forward analysis would use $0(n)$ arithmetic operations.

Table 1: Coefficients of the experimental matrices

	a_i	b_i	c_i	d_i	e_i
$M1(n)$	-1	-1	4	-1	-1
$M2(n)$	$-s$	$-s$	$1+4s$	$-s$	$-s$
$M3(5)$	-1	-1	$2, i=1$ $102, i=2$ $3+10^{2i-2}, i=3,4$ $2, i=5$	-10^{2i-2}	-1
$M4(10)$	-1	-1	$2, i=1$ $12, i=2$ $3+10^{i-1}, i=3,\dots,9$ $2, i=10$	-10^{i-1}	-1

Numerical experiments

The numerical experiments are done in MATLAB where the round-off unit is $\rho_0 \approx$ 2.22E-16. We measure two types of errors:

1. The forward error

$$FE = \| x - \tilde{x} \|_\infty,$$

where $\tilde{x}$ is the computed solution.

2. The backward error

$$BE = BE_1 + BE_2,$$

where BE_1 and BE_2 are defined in (16) and (18).

The algorithm is tested with matrices of order

$$n = 20{,}50{,}100{,}200{,}500{,}1000{,}2000,$$

the coefficients of which are given in Table 1. The right part f is chosen so that the exact solution is $x = (1, \dots, 1)^T$ in all examples. The forward and backward error are compared in all the tests, where $\tilde{x}$ is the solution of (2) with round-off errors and x is the exact solution.

Table 2: Results for the class $M1(n)$ for different n

n	BE	FE
20	$2.7E-16$	$1.1E-15$
50	$3.3E-16$	$2.6E-15$
100	$4.1E-16$	$3.9E-15$
200	$5.5E-16$	$1.8E-14$
500	$7.4E-16$	$6.6E-14$
1000	$1.2E-15$	$7.8E-14$
2000	$4.2E-15$	$2.9E-13$

Table 3: Results for the class $M2(n)$ for different s

s	BE	FE
0.001	$5.7E-16$	$1.1E-16$
0.12	$1.3E-15$	$1.2E-16$
0.25	$2.1E-15$	$1.2E-16$
0.5	$3.5E-15$	$1.4E-16$
1	$6.7E-15$	$1.8E-16$
2	$1.3E-14$	$3.4E-16$
4	$2.4E-14$	$1.3E-15$

Table 4: Results for the class $M2(n)$ when $s = 100$ for different n

n	BE	FE
20	$2.3E-14$	$1.2E-15$
50	$3.2E-14$	$2.6E-15$
100	$4.5E-14$	$2.7E-15$
200	$1.2E-13$	$1.6E-14$
500	$2.7E-13$	$2.3E-14$
1000	$3.6E-13$	$2.6E-14$
2000	$6.1E-13$	$3.2E-14$

For the class $M1(n)$ the results are presented in Table 2. So, the small BE shows that the algorithm is stable. At the same time the FE is growing because the norm $\| A^{-1} \|_\infty$ is growing with n.

For the class $M2(n)$ in the case $n = 2000$ for different s the results are presented in Table 3. The quantities BE and FE rarely change with the growth of s.

The results for $M2(n)$ when $s = 100$ are given separately in Table 4 because the FE changes for different n. Although these matrices are ill-conditioned ($\| A^{-1} \|_\infty \geq 10^6$) Table 4 shows that the equivalent perturbations describe the behavior of the round-off error quite well.

Finally for the matrix $M3(5)$ we have that $BE = 5.4E-1, FE = 3.4E-2$, and for the matrix $M4(10)$ we have that $BE = 73.2$, $FE = 3.1$. The equivalent perturbations describe the real situation quite well again. The last two examples also show that although matrices $M3(5)$ and $M4(10)$ are diagonally dominant and the diagonal dominance is strict for one row, the result is far away from the exact solution x. The explanation is that for these matrices the coefficients α_i are approaching 1, and the coefficients β_i and the elements d_i are growing very fast. For this reason we have $\Delta_n \approx 0$ ($\Delta_5 = 1.9E-15$ for $M3(5)$, $\Delta_{10} = 2.2E-16$ for $M4(10)$), and $\gamma_n = x_n$ is computed with a big round-off error.

Bibliography

[1] Golub, G., C. van Loan, *Matrix Computations*, 3rd ed., The John Hopkins University Press, 1996.

[2] Jia, J., Q. Kong, T. Sogabe, A fast numerical algorithm for solving pentadiagonal linear systems, *Int. Journal of Computer Mathematics*, v. 89, pp. 851-860, 2012.

[3] Karawia, A., A computational algorithm for solving periodic pentadiagonal linear systems, *Appl. Mathematics and computation*, v. 174, pp. 613-618, 2006.

[4] Higham, N., Accuracy and stability of Numerical Algorithms, SIAM, Philadelphia, 1996.

[5] Samarskiy, A.A, E. Nikolaev, *Methods for solving grid equations*, Moscow, 1978.

[6] Voevodin, V., P. Yalamov, A new method of round-off error estimation, *Parallel and Distibuted Proc.*, Elsevier, 1990.

[7] Voevodin, V.V. Mathematical models and methods in parallel processing, Moscow, 1986 (in Russian).